Analytic Geometry

Sixth Edition

Douglas F. Riddle

*St. Joseph's University
and
Huron University/Japan*

PWS Publishing Company

I(T)P An International Thomson Publishing Company

Boston · Albany · Bonn · Cincinnati · Detroit · London · Madrid · Melbourne · Mexico City
New York · Paris · San Francisco · Singapore · Tokyo · Toronto · Washington

PWS PUBLISHING COMPANY

20 Park Plaza, Boston, MA 02116-4324

International Thomson Publishing
The trademark ITP is used under license.

For more information, contact:

PWS Publishing Company
20 Park Plaza
Boston, MA 02116

International Thomson
  Publishing Europe
Berkshire House 168–173
High Holborn
London WC1V 7AA
England

Thomas Nelson Australia
102 Dodds Street
South Melbourne, 3205
Victoria, Australia

Nelson Canada
1120 Birchmount Road
Scarborough, Ontario
Canada M1K 5G4

International Thomson
  Editores
Campos Eliseos 385, Piso 7
Col. Polanco
11560 Mexico D.F., Mexico

International Thomson
  Publishing GmbH
Königswinterer Strasse 418
53227 Bonn, Germany

International Thomson
  Publishing Asia
221 Henderson Road
#05-10 Henderson Building
Singapore 0315

International Thomson
  Publishing Japan
Hirakawacho Kyowa Building,
  31
2-2-1 Hirakawacho
Chiyoda-ku, Tokyo 102
Japan

Sponsoring Editor: David Dietz
Editorial Assistant: Julia Chen
Production Coordinator: Patricia Adams
Production: Greg Hubit Bookworks
Market Development Manager: Marianne C.P. Rutter
Manufacturing Coordinator: Marcia A. Locke
Interior Design: Carol Rose
Interior Illustrator: Lotus Art
Cover Designer: Patricia Adams
Cover Photo: © Steven Hunt/The Image Bank
Compositor: Best-set Typesetter Ltd., Hong Kong
Text Printer and Binder: RR Donnelley

Library of Congress Cataloging-in-Publication Data

Riddle, Douglas F.
  Analytic geometry / Douglas F. Riddle. — 6th ed.
    p. cm.
  Includes index.
  ISBN-13: 978-0-534-94854-2
  ISBN-10: 0-534-94854-5
  1. Geometry, Analytic.   I. Title.
  QA551.R48   1996
  516.3′—dc20   95-40781
                                          CIP

Contents

Preface

Analytic Geometry, Sixth Edition, designed for students with a reasonably sound background in algebra, geometry, and trigonometry, contains more than enough material for a three-semester-hour or five-quarter-hour course in analytic geometry.

As in previous editions, my aim has been to write a text that students find to be interesting, clear, and readable and that instructors find will enhance their presentations of the material. Let us consider the distinctive features of this text and this edition under the following general headings: the presentation of the material, problems and their answers, and the coverage of the subject.

THE PRESENTATION

1. **Historical Notes**. Historical notes are included to portray those who have helped to develop analytic geometry as well as to show the position of analytic geometry in the general framework of mathematical development. These take two forms: brief comments within the text itself, and longer notes presented in a sidebar. These notes have been updated and expanded slightly.

2. **Applications**. Although analytic geometry lays a foundation for the study of calculus, it also has many direct applications in physical situations. Some of the most important applications are considered in the text as well as in the problems. This is especially true in the case of the conic sections with their important reflective properties, orbits of heavenly bodies, and engineering applications.

3. **Graphing Calculators**. The introduction of graphing calculators and computer graphing programs has had an influence upon the way that mathematics is taught, and we can only expect that this influence will increase considerably in the next few years. The treatment of graphing by means of a graphing calculator has been expanded considerably in the present edition with many examples and problems. A graphing calculator introduction has been added to acquaint the student with these calculators generally. However, with one exception, individual keystrokes are not included in this introduction or in the examples; this allows the student to use any such calculator with this text. The exception mentioned above is in the graphing of conic sections. Since they are represented by second-degree equations rather than functions, a program is needed in order to graph them. Such a program has been included for three of the most frequently used graphing calculators: Texas Instruments TI-82, Sharp EL-9300C, and Casio fx-9700GE. These programs graph the conic section and its asymptotes where

appropriate. Although the graphing calculator can be used on any problem requiring a graph, those specified for it are those that take advantage of its special characteristics, allowing us to make comparisons among several graphs. There has been no attempt to designate (nongraphing) calculator problems, because it has been my experience that almost all students have calculators and regard *all* problems as calculator problems.

4. **Functions**. Although analytic geometry, with its emphasis on the conic sections, tends to ignore the important topic of functions, they are considered in some detail in Section 1.7.

5. **Important Results**. As in previous editions, important results have been given in theorems, most of which are proved in the text.

6. **Vectors**. As in the previous editions, a vector approach is used in those places in which the presentation is enhanced by it. Since it is recognized that some prefer a nonvector approach, the vector arguments and solutions are supplemented by nonvector ones. Thus the instructor has the option of taking either approach without having to depart from the text. More examples of vector proofs have been included in this edition.

7. **Examples**. The examples include illustrations of all (or almost all) of the A and B type problems. They are given in some detail, usually with a figure, so that the student will have little difficulty following them.

8. **Other Changes in the Sixth Edition**: The explanation of analytic proofs has been expanded and clarified. The idea that three conditions determine a circle has also been expanded and clarified. A second, alternate approach to equations of rotation has been given. The approach to radicals and their restriction on the domain of a function has been simplified. Simultaneous, rather than sequential, graphing of a pair of polar equations is used to clarify the idea that two curves intersect by crossing a given point at different "times." The section on quadric equations in three dimensions has been expanded to cover more general equations, with a graphing calculator used to find traces in the coordinate planes and in planes parallel to them. In order to prevent the book from growing longer with every revision, the former Section 9.7 on the distance from a point to a line or plane, and the angles between two lines or planes, has been eliminated. The distance portion has been absorbed into the previous section. Finally, the numeric tables have been eliminated from the appendices. It is felt that the widespread use of calculators has made them unnecessary. Their place has been taken by a brief review of trigonometry, which some students may need before considering Chapter 8 on polar coordinates.

PROBLEMS AND ANSWERS

While many students skip the exposition and some skip the examples, they must, of necessity, spend a great deal of time on the problems. Thus the problems, which some authors look upon as unimportant, are in reality the most important part of the text.

1. **Organization of the Problems**. The problem sets have been divided into five sections labeled A, B, C, GRAPHING CALCULATOR, and APPLICATIONS. Section A consists of routine problems that every student can be expected to master. Those of section B are less routine but still not a great challenge. The average student can be expected to master most of them. The C problems present something of a challenge; only the better students can be expected to work them. Of course, such a system of classification is quite subjective; individual instructors can gauge their students' abilities much more precisely. Thus it should be taken as a rough guide only. The remaining two sections are self-explanatory. Although there has been no attempt to classify them with regard to their degree of difficulty, the GRAPHING CALCULATOR problems tend to be at the A and B levels while the APPLICATIONS problems are mostly at the B and C levels.

2. **Answers**. Answers to the odd-numbered problems are provided at the back of the book. There are two exceptions to this odd-number rule. The first of these occurs when there are many parts to consecutive problems (as in the first few problems of Section 8.2). In this case, answers are given for the first half of all such problems, rather than for all parts of one problem and no part of the next. The other exception is the answers to the review problems. Since many students use these problems to prepare for examinations, answers are given for all of them. The answers have been independently checked for accuracy.

3. **Instructor's Answers Manual**. An answers manual, consisting of answers to all problems in the book, is available to instructors upon adoption of the text.

4. **Students' Solutions Manual**. A manual providing complete solutions to odd-numbered exercises is available for student purchase.

COVERAGE

Finally, some attempt has been made to hold in check the tendency for textbooks to grow larger in each successive edition. Those sections and chapters that users have indicated that they do not cover have been deleted from the text. In many such cases, problems, with short explanations, have been taken from the deleted sections and incorporated into other sections so that students can still get a taste of some of this material. In particular, the section on the reflective properties of conics has been deleted; but these properties are discussed in the applications of the various conic sections. Thus, what seemed to be of only theoretical importance in past editions is shown to be quite significant outside of mathematics. In spite of these omissions, there is still more material here than can be covered in a three-semester-hour or five-quarter-hour course. The following is one suggested outline of what may realistically be covered:

Chapter	Sections
1	1–7
2	1–3
3	1–3
4	1–4
5	1–4
6	1, 3, 4
7	1–6
8	1–6
9	1–6, 9

ACKNOWLEDGMENTS

My thanks go to the following reviewers for their helpful comments and suggestions: Roy Dean Alston, Stephen F. Austin State University; Zachary Franco, Texas A&M University; Yizeng Li, University of Texas; Lance L. Littlejohn, Utah State University; Sally Ann Low, Angelo State University; Robert A. Rider, Ohio University; Sam Rodgers, Mountain View College; Brian Siebenaler, Ball State University; Cathryn K. Stark, Collin County Community College.

DOUGLAS F. RIDDLE

Introduction to the Graphing Calculator

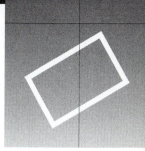

Before using a graphing calculator, you should know how it carries out the graphing. The screen of many of the currently used graphing calculators is made up of 5985 dots, which can be turned on or off individually. These dots are called picture elements, or pixels. They are arranged in a rectangular pattern that is 95 pixels from side to side and 63 from top to bottom. Each pixel is assigned a pair of coordinates, which the user determines by the range set. The leftmost and right-most pixels have x values x_{min} and x_{max}, respectively. The distance from one pixel to the next is determined by the expression

$$\Delta x = \frac{x_{max} - x_{min}}{94}.$$

Note that we divide by 94—not 95. Δx represents the space between the pixels. With 95 pixels, there are only 94 spaces between them. Δx is added to the x value of a given pixel to get the x value of the next one to the right. This continues across the screen.

Suppose, for example, that $x_{min} = -10$ and $x_{max} = 10$. Then

$$\Delta x = \frac{10 - (-10)}{94} = 0.212766 \ldots .$$

Thus the pixels have x coordinates

$$-10.0$$
$$-9.787234$$
$$-9.574468$$
$$\cdot$$
$$\cdot$$
$$\cdot$$
$$-0.425532$$
$$-0.212766$$
$$0.0$$
$$0.212766$$
$$0.425532$$

.

.

.

$$9.574468$$
$$9.787234$$
$$10.0.$$

A similar computation is done to determine the y values of the pixels.

Each of the 95 x values above is substituted into the functional expression the user gives to get the corresponding y value. If the resulting y value is within the given range of y values, the pixel with the x value used and the y value nearest the computed y value is turned on. This gives the "dot" graph of the function. An example of such a graph is shown in Figure I.1. By using the "trace" feature on this graph, we can easily verify that there are approximately 5 dots between any two integer values of x. This graph was generated using the settings

Mode: Dot
$$Y_1 = 6 - x^2$$

Range:
$$x_{min} = -10$$
$$x_{max} = 10$$
$$x_{scl} = 1$$
$$y_{min} = -10$$
$$y_{max} = 10$$
$$y_{scl} = 1$$

The range setting here is the "standard" one.

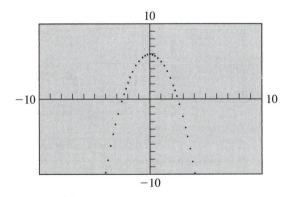

Figure I.1

Because the interval from -10 to 10 is not evenly divisible by 94, the x values as we move from left to right tend to be long decimal expressions, as indicated above. A second problem with these standard settings is that we have the same range of values in both the horizontal and vertical directions. Since the calculator screen is not square, this results in a distortion of the graph. While that does not cause any great difficulty here, it compresses the graph vertically, making circles look like ellipses.

Let us consider how we might avoid such problems. In order to avoid the long decimal fractions assigned to most of the pixels, we simply need to choose x_{min} and x_{max} so that their difference is evenly divisible by 94. (By "evenly divisible" we mean merely that the quotient is a terminating decimal—not necessarily an integer.) This can be done by using $x_{min} = -9.4$ and $x_{max} = 9.4$ instead of -10 and 10. Similarly, we choose $y_{min} = -6.3$ and $y_{max} = 6.3$. These settings not only give more convenient values for the pixels, but they also eliminate the distortion. We shall refer to these range settings as the "normal range."

Be aware that, while these settings are a convenient starting point, we cannot always stay with them. The nature of the graph may require that we use different scales on the x and y axes, resulting in some distortion. Likewise, zooming in or out will change the scales in unpredictable ways. Nevertheless, multiples of 94 and 63 are the most convenient choices, when we are able to make them.

The screen of another popular calculator is made up of 6144 pixels, 96 across and 64 from top to bottom. Δx is then determined by

$$\Delta x = \frac{x_{max} - x_{min}}{95}.$$

The recommended alternate standard settings for such calculators are

$$x_{min} = -9.6$$
$$x_{max} = 9.4$$
$$y_{min} = -6.4$$
$$y_{max} = 6.2.$$

The reason for the unbalanced choice is to make sure that $x = 0$ is one of the pixel values. If x_{min} and x_{max} were taken to be -9.5 and 9.5, respectively, the pixel values near the origin would be $x = -0.1$ and $x = 0.1$.

Since the size of the screen is likely to be increased in the future, let us consider a simple way to get horizontal and vertical pixel counts for any given screen size. Simply count the number of characters that will fit on a line as well as the number of lines that fit on the screen. Presently, a common size is 8 lines with 16 characters per line. A character is 5 pixels wide and 7 pixels high, and there is a 1-pixel space between characters on a line as well as between two lines. Thus 16 characters with 15 spaces between them results in a 95-pixel width $(16 \cdot 5 + 15)$. Similarly, 8 lines with 7 spaces between them gives a 63-pixel height $(8 \cdot 7 + 7)$.

Finally, we see that the dot graph of Figure I.1 is disconnected. For example, there is a gap between $(3, -3)$ and $(3.2, -4.24)$. Although there is no pixel with x value between 3 and 3.2, there is a 5-pixel gap between the y values -3 and -4.2 (the closest pixel value to -4.24). To connect the points, additional pixels in the $x = 3$ and $x = 3.2$ columns are turned on. For example, to bridge the gap between $(3, -3)$ and $(3.2, -4.2)$, the 5 additional pixels—$(3, -3.2)$, $(3, -3.4)$, $(3.2, -3.6)$,

(3.2, −3.8), and (3.2, −4)—are turned on. If we select the "connected" mode, this is done for the entire graph as shown in Figure I.2.

As we can see from this figure, the graph appears to be a series of dots and short line segments. In the future, we can expect the screen to be larger and the pixels to be smaller, giving smoother curves. This can already be seen on computers, with their larger screen and greater resolution.

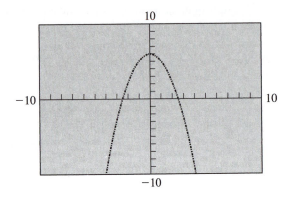

Figure I.2

One problem most graphing calculators have (indeed, most computer programs have, too) is that they graph only functions, not equations. Thus, to graph a conic section whose equation has the form

$$Ax^2 + Bxy + Cy^2 + Dx + Ey + F = 0,$$

we must first solve for y as a function (or pair of functions) of x. This can be done using the quadratic formula if $C \neq 0$ or by solving the linear equation in y if $C = 0$. Then we can graph the resulting function(s).

This process is tedious and time-consuming. However, many graphing calculators have modest programming capabilities. Thus they can be programmed to graph an equation of the above form. The program is based on the following results. If $C \neq 0$, the quadratic formula applied to

$$Cy^2 + (Bx + E)y + (Ax^2 + Dx + F) = 0$$

gives

$$y = \frac{-(Bx + E) \pm \sqrt{(Bx + E)^2 - 4C(Ax^2 + Dx + F)}}{2C}.$$

If $C = 0$, then we have

$$(Bx + E)y + (Ax^2 + Dx + F) = 0,$$

giving

$$y = -\frac{Ax^2 + Dx + F}{Bx + E}.$$

The following program, written for several popular graphing calculators, carries out the graphing of a general second-degree equation. In addition, it also graphs asymptotes if we have a hyperbola. This program can easily be adapted to graphing calculators other than those listed in this introduction.

Before executing, select a convenient window. If the initial window you set is not suitable for your particular equation, it can be changed and the function regraphed without repeating the program. However, if you do so, a vertical asymptote (which is not the graph of a function and must be drawn on the final result) will not appear within the new window.

In addition, the function menu should be cleared. The program will graph all functions in the function menu. This will include not only those that have been created by the program but also those left in the function menu that have not been overwritten.

Although neither indentation nor comments are allowed in the calculator program, they have been included in the following to aid in reading and understanding this program. Do not attempt to put them into the program.

```
PROGRAM:GPHCONIC (* For Texas Instruments TI-82 *)
:Disp "FOR AX² + BXY + CY²"
:Disp "+DX + EY + F = 0"
:Prompt A,B,C,D,E,F

(* Check to see that the given equation is really 2nd degree *)
:If A² + B² + C² = 0 (* A, B, C all 0 *)
:Then
:      Disp "NOT 2ND DEGREE EQUATION"
:      Goto Z
:End

(* Separate into cases: Quadratic in y and Linear in y *)
:If C ≠ 0
:Then              (* use the quadratic formula to solve for y *)
:      "(−(B*X + E) + √((B*X + E)² − 4C(A*X² + D*X + F)))/(2C)" → Y₁
:      "(−(B*X + E) − √((B*X + E)² − 4C(A*X² + D*X + F)))/(2C)" → Y₂
:      If B² − 4A*C > 0
:      Then       (* Hyperbola—graph asymptotes *)
:            PrgmASYMP
:      End
:Else             (* C = 0; solve linear equation for y *)
:      "−(A*X² + D*X + F)/(B*X + E)" → Y₁
:      If B² − 4A*C > 0
:      Then       (* Hyperbola—graph asymptotes *)
:            "−A*X/B + A*E/B² − D/B" → Y₂
:            Vertical −E/B
:      End
:End
```

```
:DispGraph
:Lbl Z
:
```

PROGRAM:ASYMP
(* Approximates the asymptotes by taking lines through points with x coordinates
20 and −20 *)
(* Using x = 20 *)
:20B + E → G
:400A + 20D + F → H
:(−G + √(G² − 4C*H))/(2C) → P
:(−G − √(G² − 4C*H))/(2C) → Q

(* Using x = −20 *)
:−20B + E → I
:400A − 20D + F → J
:(−I + √(I² − 4C*J))/(2C) → R
:(−I − √(I² − 4C*J))/(2C) → S

(* Graph line through (20, P) and (−20, S) *)
:(P − S)/40 → M
:"P+ M(X − 20)" → Y₃

(* Graph line through (20, Q) and (−20, R) *)
:(Q − R)/40 → N
:"Q + N(X −20)" → Y₄
:

gphconic   (* For Sharp EL-9300C *)
—REAL
Print "For Ax² + Bxy + Cy²
Print "+ Dx + Ey + F = 0
Input A
Input B
Input C
Input D
Input E
Input F

(* Check to see that the given equation is really 2nd degree *)
If A² + B² + C² ≠ 0 Goto isconic
        Print "Not 2nd degree equation
        Goto endprog
Label isconic

(* Separate into cases: Quadratic in y and Linear in y *)

If C = 0 Goto lineareq
    (* Use quadratic formula to solve for y *)
    Graph $(-(B*X + E) + \sqrt{((B*X + E)^2 - 4C(A*X^2 + D*X + F)))/(2C)}$
    Graph $(-(B*X + E) - \sqrt{((B*X + E)^2 - 4C(A*X^2 + D*X + F)))/(2C)}$
    If $B^2 - 4A*C < = 0$ Goto nonhyper
        Gosub asymp
    Label nonhyper
End

Label lineareq
    (* C = 0; solve linear equation for y *)
    Graph $-(A*X^2 + D*X + F)/(B*X + E)$
    If $B^2 - 4A*C < = 0$ Goto hypernot
    (* Hyperbola—graph asymptotes *)
    Graph $-A*X/B + A*E/B^2 - D/B$
    Line $-E/B$, 10, $-E/B$, -10
    Label hypernot
Label endprog
End

Label asymp (* Subroutine to graph asymptotes *)
(* Approximates the asymptotes by taking lines through points with x coordinates
20 and -20 *)
(* Using x = 20 *)
g = 20B + E
h = 400A + 20D + F
p = $(-g + \sqrt{(g^2 - 4C*h)})/(2C)$
q = $(-g - \sqrt{(g^2 - 4C*h)})/(2C)$

(* Using x = -20 *)
i = $-20B + E$
j = 400A $- 20D + F$
r = $(-i + \sqrt{(i^2 - 4C*j)})/(2C)$
s = $(-i - \sqrt{(i^2 - 4C*j)})/(2C)$
(* Graph lines using the point-slope form *)
m = (p − s)/40
Graph p + m(X − 20)
n = (q − r)/40
Graph q + n(X − 20)
Return

Prog 0 (* For Casio fx-9700GE *)
"FOR AX^2 + BXY + CY^2"↵
"+DX + EY + F = 0"↵
"A ="? → A ↵
"B ="? → B ↵

"C ="? → C ↵
"D ="? → D ↵
"E ="? → E ↵
"F ="? → F ↵

'CHECK TO SEE THAT THE GIVEN EQUATION IS REALLY 2ND DEGREE ↵
$A^2 + B^2 + C^2 = 0 \Rightarrow$ Goto 1 'A, B, C ALL 0; NOT 2ND DEGREE↵
Goto 2: '≠ 0; 2ND DEGREE↵
      Lbl 1: "NOT 2ND DEGREE EQUATION": Goto 9↵
'SEPARATE INTO CASES: QUADRATIC IN Y AND LINEAR IN Y↵
Lbl 2: $C \neq 0 \Rightarrow$ Goto 3↵
'$C \neq 0$; USE QUADRATIC FORMULA TO SOLVE FOR Y↵
Goto 4:'C = 0; SOLVE LINEAR EQUATION FOR Y↵
     'USE QUADRATIC FORMULA TO SOLVE FOR Y↵
     Lbl 3↵
     Graph $Y = (-(B \times X + E) + \sqrt{(B \times X + E)^2 - 4C(A \times X^2 + D \times X + F)}) \div (2C)$↵
     Graph $Y = (-(B \times X + E) - \sqrt{(B \times X + E)^2 - 4C(A \times X^2 + D \times X + F)}) \div (2C)$↵
     $B^2 - 4A \times C > 0 \Rightarrow$ Prog 1↵
     Goto 9↵

     'C = 0; SOLVE LINEAR EQUATION FOR Y↵
     Lbl 4↵
     Graph $Y = -(A \times X^2 + D \times X + F) \div (B \times X + E)$↵
     $B^2 - 4A \times C > 0 \Rightarrow$ Goto 5↵
     Goto 9↵
       'HYPERBOLA; GRAPH ASYMPTOTES↵
       Lbl 5↵
       Graph $Y = -A \times X \div B + A \times E \div B^2 - D \div B$↵

Lbl 9↵
Prog 1:
'APPROXIMATES ASYMPTOTES BY TAKING LINES THROUGH POINTS WITH X COORDINATES 20 AND −20↵
'USING X = 20↵
$20B + E \to G$↵
$400A + 20D + F \to H$↵
$(-G + \sqrt{G^2 - 4C \times H}) \div (2C) \to P$↵
$(-G - \sqrt{G^2 - 4C \times H}) \div (2C) \to Q$↵

'USING X = −20↵
$-20B + E \to I$↵
$400A - 20D + F \to J$↵
$(-I + \sqrt{I^2 - 4C \times J}) \div (2C) \to R$↵
$(-I - \sqrt{I^2 - 4C \times J}) \div (2C) \to S$↵

'GRAPH LINE THROUGH (20, P) AND (−20, S)↵
(P − S) ÷ 40 → M↵
Graph Y = P + M(X − 20)↵

'GRAPH LINE THROUGH (20, Q) AND (−20, R)↵
(Q − R) ÷ 40 → N↵
Graph Y = Q + N(X − 20)↵

1.1

THE CARTESIAN PLANE

Analytic geometry provides a bridge between algebra and geometry that makes it possible for geometric problems to be solved algebraically (or analytically). It also allows us to solve algebraic problems geometrically, but the former is far more important, especially when numbers are assigned to essentially geometric concepts. Consider, for instance, the length of a line segment or the angle between two lines. Even if the lines and points in question are accurately known, the number representing the length of a segment or the angle between two lines can be determined only approximately by measurement. Algebraic methods provide an exact determination of the number.

The association between the algebra and geometry is made by assigning numbers to points. Suppose we look at this assignment of numbers to the points on a line. First of all, we select a pair of points, O and P, on the line, as shown in Figure 1.1. The point O, which we call the origin, is assigned the number zero, and the point P is assigned the number one. Using $\overline{OP}$ as our unit of length,* we assign numbers to all other points on the line in the following way: Q on the P side of the origin is assigned the positive number x if and only if its distance from the origin is x. A point Q on the opposite side of the origin is assigned the negative number $-x$ if and only if its distance from the origin is x units. In this way every point on the line is assigned a real number, and for each real number there corresponds a point on the line.

Thus, a **scale** is established on the line, which we now call a **coordinate line**. The number representing a given point is called the **coordinate** of that point, and the point is called the **graph** of the number.

Just as points on a line (a one-dimensional space) are represented by single numbers, points in a plane (a two-dimensional space) can be represented by pairs of numbers. Later we shall see that points in a three-dimensional space can be represented by triples of numbers.

---

* We shall use the notation AB for the line segment joining the points A and B, and $\overline{AB}$ for its length.

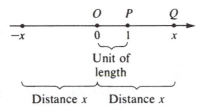

Figure 1.1

In order to represent points in a plane by pairs of numbers, we select two intersecting lines and establish a scale on each line, as shown in Figure 1.2. The point of intersection is the origin. These two lines, called the axes, are distinguished by identifying symbols (usually by the letters x and y). For a given point P in the plane, there corresponds a point P_x on the x axis. It is the point of intersection of the x axis and the line containing P and parallel to the y axis. (If P is on the y axis, this line coincides with the y axis.) Similarly, there exists a point P_y on the y axis which is the point of intersection of the y axis and the line through P that parallels (or is) the x axis. The coordinates of these two points on the axes are the **coordinates** of P. If a is the coordinate of P_x and b is the coordinate of P_y, then the point P is represented by (a, b). In this example, a is called the x **coordinate**, or **abscissa**, of P and b is the y **coordinate**, or **ordinate**, of P.

In a coordinate plane, the following conventions normally apply:

1. The axes are taken to be perpendicular to each other.

2. The x axis is a horizontal line with the positive coordinates to the right of the origin, and the y axis is a vertical line with the positive coordinates above the origin.

3. The same scale is used on both axes.

These conventions, of course, need not be followed when others are more convenient. We shall violate the third rather frequently when considering figures that would be very difficult to sketch if we insisted upon using the same scale on

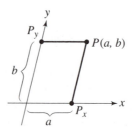

Figure 1.2

both axes. In such cases, we shall feel free to use different scales, remembering that we have distorted the figure in the process. Unless a departure from convention is specifically stated or is obvious from the context, we shall always follow the first two conventions.

We can now identify the coordinates of the points in Figure 1.3. Note that all points on the x axis have the y coordinate zero, while those on the y axis have the x coordinate zero. The origin has both coordinates zero, since it is on both axes.

The axes separate the plane into four regions, called **quadrants**. It is convenient to identify them by the numbers shown in Figure 1.4. The points on the axes are not in any quadrant.

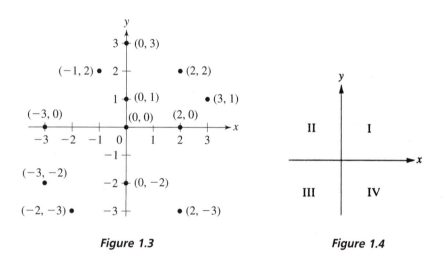

Figure 1.3   Figure 1.4

The coordinates of a point determined in this way are sometimes called cartesian coordinates, after the French mathematician and philosopher René Descartes. In the appendix of a book published in 1637, Descartes gave the first description of analytic geometry. From it came the developments that eventually led to the invention of the calculus. See the historical note on page 97.

1.2

DISTANCE FORMULA

Suppose we consider the distance between two points on a coordinate line. Let P_1 and P_2 be two points on a line, and let them have coordinates x_1 and x_2, respectively. If P_1 and P_2 are both to the right of the origin, with P_2 farther right than P_1 (as in Figure 1.5a on page 4), then

$$\overline{P_1P_2} = \overline{OP_2} - \overline{OP_1} = x_2 - x_1.$$

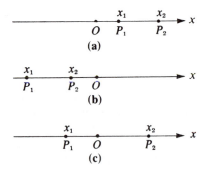

Figure 1.5

Expressing the distance between two points is only slightly more complicated if the origin is to the right of one or both of the points. In Figure 1.5b,

$$\overline{P_1P_2} = \overline{P_1O} - \overline{P_2O} = -x_1 - (-x_2) = x_2 - x_1,$$

and in Figure 1.5c,

$$\overline{P_1P_2} = \overline{P_1O} + \overline{OP_2} = -x_1 + x_2 = x_2 - x_1.$$

Thus, we see that $\overline{P_1P_2} = x_2 - x_1$ in all three of these cases in which P_2 is to the right of P_1. If P_2 were to the left of P_1, then

$$\overline{P_1P_2} = x_1 - x_2,$$

as can be easily verified. Thus, $\overline{P_1P_2}$ can always be represented as the larger coordinate minus the smaller. Since $x_2 - x_1$ and $x_1 - x_2$ differ only in that one is the negative of the other and since distance is always nonnegative, we see that $\overline{P_1P_2}$ is the difference that is nonnegative. Thus,

$$\overline{P_1P_2} = |x_2 - x_1|.$$

This form is especially convenient when the relative positions of P_1 and P_2 are unknown. However, since absolute values are sometimes rather bothersome, we will avoid them whenever the relative positions of P_1 and P_2 are known.

Let us now turn our attention to the more difficult problem of finding the distance between two points in the plane. Suppose we are interested in the distance between $P_1 = (x_1, y_1)$ and $P_2 = (x_2, y_2)$ (see Figure 1.6). A vertical line is drawn through P_1 and a horizontal line through P_2 intersecting at a point $Q = (x_1, y_2)$. Assuming P_1 and P_2 are not on the same horizontal or vertical line, P_1P_2Q forms a right triangle with the right angle at Q. Now we can use the Theorem of Pythagoras to determine the length of P_1P_2. By the previous discussion,

$$\overline{QP_2} = |x_2 - x_1| \qquad \text{and} \qquad \overline{P_1Q} = |y_2 - y_1|$$

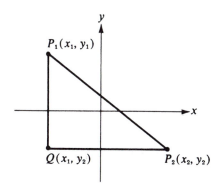

Figure 1.6

(the absolute values are retained here, since we want the resulting formula to hold for *any* choice of P_1 and P_2, not merely for the one shown in Figure 1.6). Now by the Pythagorean Theorem,

$$\overline{P_1P_2} = \sqrt{|x_2 - x_1|^2 + |y_2 - y_1|^2}.$$

But, since $|x_2 - x_1|^2 = (x_2 - x_1)^2 = (x_1 - x_2)^2$, the absolute values may be dropped and we have

$$\overline{P_1P_2} = \sqrt{(x_2 - x_1)^2 + (y_2 - y_1)^2}.$$

Thus we have proved the following theorem.

THEOREM 1.1   *The distance between two points $P_1 = (x_1, y_1)$ and $P_2 = (x_2, y_2)$ is*

$$\overline{P_1P_2} = \sqrt{(x_2 - x_1)^2 + (y_2 - y_1)^2}.$$

In deriving this formula, we assumed that P_1 and P_2 are not on the same horizontal or vertical line; however, the formula holds even in these cases. For example, if P_1 and P_2 are on the same horizontal line, then $y_1 = y_2$ and $y_2 - y_1 = 0$. Thus,

$$\overline{P_1P_2} = \sqrt{(x_2 - x_1)^2} = |x_2 - x_1|.$$

Note that $\sqrt{(x_2 - x_1)^2}$ is *not always* $x_2 - x_1$. Since the symbol $\sqrt{}$ indicates the nonnegative square root, we see that if $x_2 - x_1$ is negative, then $\sqrt{(x_2 - x_1)^2}$ is not equal to $x_2 - x_1$ but, rather, equals $|x_2 - x_1|$. Suppose, for example, that $x_2 - x_1 = -5$. Then $\sqrt{(x_2 - x_1)^2} = \sqrt{(-5)^2} = \sqrt{25} = 5 = |x_2 - x_1|$.

Theorem 1.1 depends upon the convention that the axes are perpendicular. If this convention is not followed, Theorem 1.1 cannot be used, but another more general formula based on the law of cosines can be derived. However, we shall not derive it here, since the convention of using perpendicular axes is so widely observed.

EXAMPLE 1 Find the distance between $P_1 = (1, 4)$ and $P_2 = (-3, 2)$ (see Figure 1.7).

SOLUTION
$$\overline{P_1P_2} = \sqrt{(-3 - 1)^2 + (2 - 4)^2}$$
$$= \sqrt{16 + 4} = \sqrt{4 \cdot 5} = 2\sqrt{5}$$

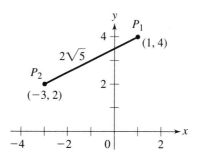

Figure 1.7

EXAMPLE 2 If $P_1 = (x, 0)$, $P_2 = (2, 5)$, and $\overline{P_1P_2} = 5\sqrt{2}$, find x (see Figure 1.8).

SOLUTION
$$\overline{P_1P_2} = \sqrt{(x - 2)^2 + (0 - 5)^2} = 5\sqrt{2}$$
$$\sqrt{x^2 - 4x + 4 + 25} = 5\sqrt{2}$$
$$x^2 - 4x + 29 = 50$$
$$x^2 - 4x - 21 = 0$$
$$(x + 3)(x - 7) = 0$$
$$x = -3 \quad \text{or} \quad x = 7$$

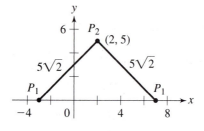

Figure 1.8

EXAMPLE 3

Determine whether $A = (1, 7)$, $B = (0, 3)$, and $C = (-2, -5)$ are collinear.

SOLUTION

$$\overline{AB} = \sqrt{(0 - 1)^2 + (3 - 7)^2} = \sqrt{17}$$
$$\overline{BC} = \sqrt{(-2 - 0)^2 + (-5 - 3)^2} = \sqrt{68} = \sqrt{4 \cdot 17} = 2\sqrt{17}$$
$$\overline{AC} = \sqrt{(-2 - 1)^2 + (-5 - 7)^2} = \sqrt{153} = \sqrt{9 \cdot 17} = 3\sqrt{17}$$

Since $\overline{AC} = \overline{AB} + \overline{BC}$, the three points must be collinear (Figure 1.9a). If they were not, they would form a triangle and any one side would be less than the sum of the other two (Figure 1.9b).

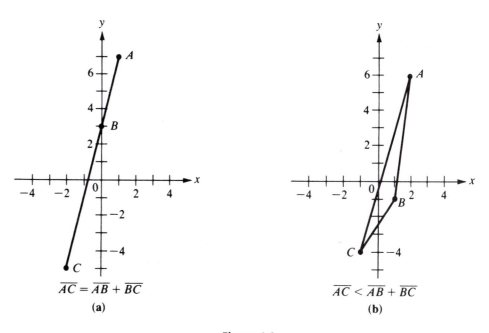

$$\overline{AC} = \overline{AB} + \overline{BC}$$

(a)

$$\overline{AC} < \overline{AB} + \overline{BC}$$

(b)

Figure 1.9

EXAMPLE 4

Show that $(1, 2)$, $(4, 7)$, $(-6, 13)$, and $(-9, 8)$ are the vertices of a rectangle.

SOLUTION

The points are plotted in Figure 1.10. Let us check lengths.

$$\overline{P_1P_2} = \sqrt{(4 - 1)^2 + (7 - 2)^2} = \sqrt{34}$$
$$\overline{P_3P_4} = \sqrt{(-9 + 6)^2 + (8 - 13)^2} = \sqrt{34}$$
$$\overline{P_2P_3} = \sqrt{(-6 - 4)^2 + (13 - 7)^2} = \sqrt{136}$$
$$\overline{P_4P_1} = \sqrt{(1 + 9)^2 + (2 - 8)^2} = \sqrt{136}$$

Although $\overline{P_1P_2} = \overline{P_3P_4}$ and $\overline{P_2P_3} = \overline{P_4P_1}$, we are not justified in saying that we have a rectangle; we can merely conclude that we have a parallelogram. But if the diagonals of a parallelogram are equal, then the parallelogram is a rectangle (see Problem 30). Let us then consider the lengths of the diagonals.

$$\overline{P_1P_3} = \sqrt{(-6 - 1)^2 + (13 - 2)^2} = \sqrt{170}$$

$$\overline{P_2P_4} = \sqrt{(-9 - 4)^2 + (8 - 7)^2} = \sqrt{170}$$

Since the parallelogram has equal diagonals, we may conclude that it is a rectangle.

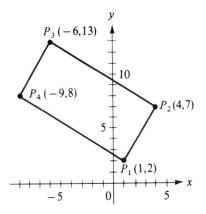

Figure 1.10

A second method of verifying that the parallelogram is a rectangle is to establish that one of the interior angles is a right angle. We can do this by the converse of the Pythagorean Theorem, which states that if $\overline{AB}^2 + \overline{BC}^2 = \overline{AC}^2$, then ABC is a right triangle with right angle at B. Thus, instead of finding the lengths of both diagonals, we find only one, say $\overline{P_1P_3} = \sqrt{170}$. It is then a simple matter to verify that $\overline{P_1P_3}^2 = \overline{P_1P_2}^2 + \overline{P_2P_3}^2$, showing that the angle at P_2 is $90°$.

When we use the methods of analytic geometry to prove geometric theorems, such proofs are called *analytic proofs*. When carrying out analytic proofs, we should recall that a plane does not come fully equipped with coordinate axes—they are imposed upon the plane to make the transition from geometry to algebra. Thus, we are free to place the axes in any position we choose in relation to the given figure. We place them in a way that makes the algebra as simple as possible. However, we must be careful not to make additional assumptions about the figure.

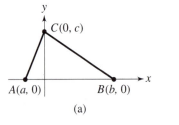

(a)

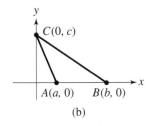

(b)

Figure 1.11

For example, if we are dealing with a triangle, we may put the *x* axis on one side, *AB*, of the triangle and the *y* axis through the third vertex, *C*, as shown in Figure 1.11a. When the picture is drawn in this way, there is a temptation to assume that *a* is negative. However, *a* can be zero or positive as shown in Figure 1.11b. The assumption that it is negative may invalidate your proof. Be aware of such dangers.

EXAMPLE 5

Prove analytically that the diagonals of a rectangle are equal.

SOLUTION

First we place the axes in a convenient position. Let us put the *x* axis on one side of the rectangle and the *y* axis on another, as illustrated in Figure 1.12. Since we have a rectangle, the coordinates of *B* and *D* determine those of *C*.

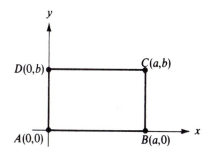

$$\overline{AC} = \sqrt{(a - 0)^2 + (b - 0)^2} = \sqrt{a^2 + b^2}$$
$$\overline{BD} = \sqrt{(0 - a)^2 + (b - 0)^2} = \sqrt{a^2 + b^2}$$

Since $\overline{AC} = \overline{BD}$, the theorem is proved.

Figure 1.12

Note that we have made no assumption here except that the sides of the rectangle meet at a right angle. Be careful that you do not make any additional assumption. For example, we have considered points *B* and *D* with coordinates $(a, 0)$ and $(0, b)$, respectively. We assumed neither that $a = b$ nor that $a \neq b$.

PROBLEMS

A   *In Problems 1–8, find the distance between the given points.*

1. $(1, -3), (2, 5)$ **2.** $(4, 13), (-1, 5)$ **3.** $(3, -2), (3, -4)$

4. $(-5, 1), (0, -10)$ **5.** $(1/2, 3/2), (-5/2, 2)$ **6.** $(2/3, 1/3), (-4/3, 4/3)$

7. $(\sqrt{2}, 1), (2\sqrt{2}, 3)$ **8.** $(\sqrt{3}, -\sqrt{2}), (-3\sqrt{3}, \sqrt{2})$

In Problems 9–14, determine whether the three given points are collinear.

9. $(2, 1), (4, 3), (-1, -2)$ **10.** $(3, 2), (4, 6), (0, -8)$

11. $(-2, 3), (7, -2), (2, 5)$ **12.** $(2, -1), (-1, 4), (5, -6)$

13. $(1, -1), (3, 3), (0, -3)$ **14.** $(1, \sqrt{2}), (4, 3\sqrt{2}), (10, 6\sqrt{2})$

In Problems 15–18, determine whether the three given points are the vertices of a right triangle.

15. $(0, 2), (-2, 4), (1, 3)$ **16.** $(-1, 3), (4, 6), (-3, 1)$

17. $(9, 6), (-5, 4), (7, 10)$ **18.** $(9, -2), (8, 0), (-6, -7)$

B *In Problems 19–22, find the unknown quantity.*

19. $P_1 = (1, 5), P_2 (x, 2), \overline{P_1P_2} = 5$

20. $P_1 = (-3, y), P_2 = (9, 2), \overline{P_1P_2} = 13$

21. $P_1 = (x, x), P_2 = (1, 4), \overline{P_1P_2} = \sqrt{5}$

22. $P_1 = (x, 2x), P_2 = (2x, 1), \overline{P_1P_2} = \sqrt{2}$

23. Show that $(5, 2)$ is on the perpendicular bisector of the segment AB where $A = (1, 3)$ and $B = (4, -2)$.

24. Show that $(-2, 4), (2, 0), (2, 8)$, and $(6, 4)$ are the vertices of a square.

25. Show that $(1, 1), (4, 1), (3, -2)$, and $(0, -2)$ are the vertices of a parallelogram.

26. Find all possible values for y so that $(5, 8), (-4, 11)$, and $(2, y)$ are the vertices of a right triangle.

27. Determine whether each of the following points is inside, on, or outside the circle with center $(-2, 3)$ and radius 5: $(1, 7), (-3, 8), (2, 0), (-5, 7), (0, -1), (-5, -1), (-6, 6), (4, 2)$.

C **28.** Find the center and radius of the circle circumscribed about the triangle with vertices $(5, 1), (6, 0)$, and $(-1, -7)$.

29. Show that a triangle with vertices $(x_1, y_1), (x_2, y_2)$, and (x_3, y_3) has area

$$\frac{1}{2}|x_1y_2 + x_2y_3 + x_3y_1 - x_1y_3 - x_2y_1 - x_3y_2|$$

$$= \left| \frac{1}{2} \begin{vmatrix} x_1 & y_1 & 1 \\ x_2 & y_2 & 1 \\ x_3 & y_3 & 1 \end{vmatrix} \right|.$$

[*Hint:* Consider the rectangle with sides parallel to the coordinate axes and containing the vertices of the triangle.]

30. Prove analytically that if the diagonals of a parallelogram are equal, then the parallelogram is a rectangle. [*Hint:* Place the axes as shown in Figure 1.13 and show that $\overline{AC} = \overline{BD}$ implies that A is the origin.]

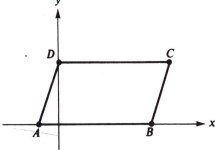

Figure 1.13

31. Prove analytically that the sum of the lengths of two sides of a triangle is greater than the length of the third side.

32. Locations on maps are often determined by grid lines, as shown in Figure 1.14. Suppose that there is a network of streets running along the grid lines together with a diagonal street as shown. If each grid square is 1 mile on a side, find the distance from X to Y.

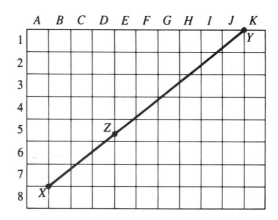

Figure 1.14

33. Suppose that, in Figure 1.14, the portion of the road from X to Z is closed for repairs. What is the distance from X to Y along existing roads?

34. Suppose that, in Figure 1.14, the road from Y to Z is closed for repairs. What is the shortest distance from X to Y along existing roads?

1.3

POINT-OF-DIVISION FORMULAS

Suppose we want to find the point which is some fraction of the way from A to B. Is it possible to express the coordinates of the point we want in terms of the coordinates of A and B? Let $A = (x_1, y_1)$ and $B = (x_2, y_2)$ be given and let $P = (x, y)$ be the point we are seeking. If we let

$$r = \frac{\overline{AP}}{\overline{AB}}$$

(see Figure 1.15), then P is 1/3 of the way from A to B when $r = 1/3$, P is 4/5 of the way from A to B when $r = 4/5$, and so on. Thus we generalize the problem to one in which x and y are to be expressed in terms of x_1, y_1, x_2, y_2, and r. The problem can be simplified considerably by working with the x's and y's separately.

If A, B, and P are projected onto the x axis (see Figure 1.15) to give the points A_x, B_x, and P_x, respectively, we have, from elementary geometry,

$$r = \frac{\overline{AP}}{\overline{AB}} = \frac{\overline{A_x P_x}}{\overline{A_x B_x}} = \frac{x - x_1}{x_2 - x_1}.$$

Solving for x gives

$$x = x_1 + r(x_2 - x_1).$$

By projecting onto the y axis, we have

$$y = y_1 + r(y_2 - y_1).$$

These two results, known as point-of-division formulas, are stated in the following theorem.

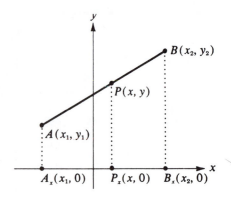

Figure 1.15

THEOREM 1.2   *If $A = (x_1, y_1)$, $B = (x_2, y_2)$, and P is a point such that $r = \overline{AP}/\overline{AB}$, then the coordinates of P are*

$$x = x_1 + r(x_2 - x_1) \qquad and \qquad y = y_1 + r(y_2 - y_1).$$

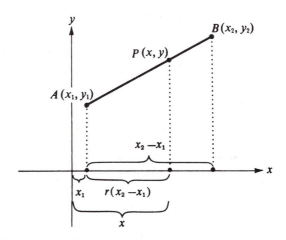

Figure 1.16

Figure 1.16 gives a geometric interpretation of the terms in the point-of-division formula for x. A careful examination of this figure will help you to understand and remember the formulas. Note that if B is to the left of A, $x_2 - x_1$ is negative, but it has the same absolute value as the distance between projections of A and B on the x axis. You might sketch this and compare your sketch with Figure 1.16. Of course, a similar figure can be used to interpret the y terms of the point-of-division formula.

EXAMPLE 1   Find the point one-third of the way from $A = (2, 5)$ to $B = (8, -1)$.

SOLUTION

$$r = \frac{\overline{AP}}{\overline{AB}} = \frac{1}{3}$$

$$x = x_1 + r(x_2 - x_1)$$

$$= 2 + \frac{1}{3}(8 - 2)$$

$$= 4$$

$$y = y_1 + r(y_2 - y_1)$$

$$= 5 + \frac{1}{3}(-1 - 5)$$

$$= 3$$

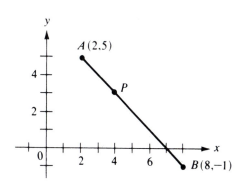

Figure 1.17

Thus the desired point is $(4, 3)$ (see Figure 1.17).

So far we have tacitly assumed that r is between 0 and 1. If r is either 0 or 1, the point-of-division formulas would give us $P = A$ or $P = B$, respectively, a result that $r = \overline{AP}/\overline{AB}$ would lead us to expect. Similarly, if $r > 1$, then $r = \overline{AP}/\overline{AB}$ indicates $\overline{AP} > \overline{AB}$, which is exactly what the point-of-division formulas give. Thus if we wanted to extend the segment AB beyond B to a point P which is r times as far from A as B is, we could still use the point-of-division formulas.

EXAMPLE 2   If the segment AB, where $A = (-3, 1)$ and $B = (2, 5)$, is extended beyond B to a point P twice as far from A as B is (see Figure 1.18), find P.

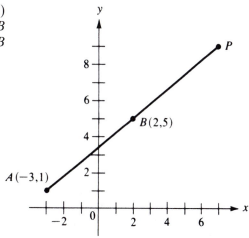

Figure 1.18

SOLUTION

$$r = \frac{\overline{AP}}{\overline{AB}} = 2$$

$$x = x_1 + r(x_2 - x_1) \qquad y = y_1 + r(y_2 - y_1)$$
$$\quad = -3 + 2[2 - (-3)] \qquad \quad = 1 + 2(5 - 1)$$
$$\quad = 7 \qquad\qquad\qquad\qquad = 9$$

Thus $P = (7, 9)$.

EXAMPLE 3   If $A = (2, -4)$, $P = (8, -1)$, and $\overline{AP}/\overline{AB} = 3/5$, find B.

SOLUTION   In this case, we use A and P to find B, rather than using A and B to find P. We still use the same point-of-division formulas; we simply solve for (x_2, y_2) rather than for (x, y).

$$x = x_1 + r(x_2 - x_1) \qquad y = y_1 + r(y_2 - y_1)$$

$$8 = 2 + \frac{3}{5}(x_2 - 2) \qquad -1 = -4 + \frac{3}{5}(y_2 + 4)$$

$$6 = \frac{3}{5}(x_2 - 2) \qquad\qquad 3 = \frac{3}{5}(y_2 + 4)$$

$$30 = 3x_2 - 6 \qquad\qquad 15 = 3y_2 + 12$$
$$3x_2 = 36 \qquad\qquad\qquad 3y_2 = 3$$
$$x_2 = 12 \qquad\qquad\qquad y_2 = 1$$

Thus $B = (12, 1)$ (see Figure 1.19).

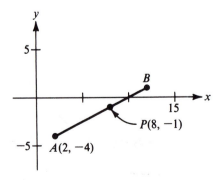

Figure 1.19

EXAMPLE 4   If $A = (-1, 5)$, $B = (7, 1)$, $\overline{AP}/\overline{PB} = 3$, find P.

SOLUTION   Note that we are *not* given the ratio $\overline{AP}/\overline{AB} = r$; we are given $\overline{AP}/\overline{PB}$. Nevertheless, it is a simple matter to find r from the ratio that we are given. Let us assume that $\overline{AP} = 3$ and $\overline{PB} = 1$ to give $\overline{AP}/\overline{PB} = 3$. Then $\overline{AB} = \overline{AP} + \overline{PB} = 3 + 1 = 4$ and $\overline{AP}/\overline{AB} = 3/4$. Of course, we do not know that $\overline{AP} = 3$ and $\overline{PB} = 1$; but we do know that $\overline{AP} = 3\overline{PB}$. Thus if $\overline{PB} = k$, then $\overline{AP} = 3k$. This gives $\overline{AB} = 4k$; and $\overline{AP}/\overline{AB} = 3/4$, as before. We now proceed as we did in the previous examples.

$$x = x_1 + r(x_2 - x_1) \qquad y = y_1 + r(y_2 - y_1)$$
$$= -1 + \frac{3}{4}(7 + 1) \qquad = 5 + \frac{3}{4}(1 - 5)$$
$$= 5 \qquad\qquad = 2$$

Thus $P = (5, 2)$.

While negative values of r do not make sense in $r = \overline{AP}/\overline{AB}$, we find that their use in the point-of-division formulas has the effect of extending the segment AB in the reverse direction—that is, from B through A to P. Suppose, for example, that $r = -2$. Then $\overline{AP}$ is twice $\overline{AB}$, and P and B are on opposite sides of A. However, we can get the same result by reversing the roles of A and B and using a positive value of r.

EXAMPLE 5   Given the segment AB, where $A = (-3, 1)$ and $B = (2, 5)$, is extended beyond A to a point P twice as far from B as A is (see Figure 1.20); find P.

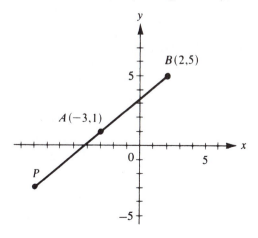

Figure 1.20

SOLUTION   Suppose we use a negative value of r. Since $\overline{AP} = \overline{AB}$ with A between P and B,

$$r = -\frac{\overline{AP}}{\overline{AB}} = -1.$$

Therefore

$$x = x_1 + r(x_2 - x_1) \qquad y = y_1 + r(y_2 - y_1)$$
$$= -3 - 1[2 - (-3)] \qquad = 1 - 1(5 - 1)$$
$$= -8 \qquad\qquad = -3$$

Thus $P = (-8, -3)$.

ALTERNATE SOLUTION   Reversing the roles of A and B, we have

$$r = \frac{\overline{BP}}{\overline{BA}} = 2, B = (x_1, y_1) = (2, 5), \quad \text{and} \quad A = (x_2, y_2) = (-3, 1).$$

$$x = x_1 + r(x_2 - x_1) \qquad y = y_1 + r(y_2 - y_1)$$
$$= 2 + 2(-3 - 2) \qquad = 5 + 2(1 - 5)$$
$$= -8 \qquad\qquad = -3$$

and $P = (-8, -3)$, as before.

One very important special case of the point-of-division formulas arises when $r = 1/2$, which gives the midpoint of the segment AB. Using the point-of-division formulas, we have the following theorem.

THEOREM 1.3   *If P is the midpoint of AB, then the coordinates of P are*

$$x = \frac{x_1 + x_2}{2} \quad and \quad y = \frac{y_1 + y_2}{2}.$$

Thus, to find the midpoint of a segment AB, we merely average both the x and y coordinates of the given points. A moment of thought will reveal the reasonableness of this; the average of two grades is halfway between them, the average of two temperatures is halfway between them, and so forth.

EXAMPLE 6   Find the midpoint of the segment AB, where $A = (1, 5)$ and $B = (-3, -1)$.

SOLUTION

$$x = \frac{x_1 + x_2}{2} \qquad y = \frac{y_1 + y_2}{2}$$
$$= \frac{1 - 3}{2} \qquad = \frac{5 - 1}{2}$$
$$= -1 \qquad\qquad = 2$$

Thus $P = (-1, 2)$.

EXAMPLE 7   Prove analytically that the segment joining the midpoints of two sides of a triangle is parallel to the third side and one-half its length.

SOLUTION   Let us place the axes as indicated in Figure 1.21 and let D and E be the midpoints of AC and BC, respectively. By the midpoint formula, $D = (a/2, c/2)$ and $E = (b/2, c/2)$. Since D and E have identical y coordinates, DE is horizontal and therefore parallel to AB. Finally $\overline{DE} = b/2 - a/2 = (b - a)/2$ and $\overline{AB} = b - a$; thus $\overline{DE} = \overline{AB}/2$.

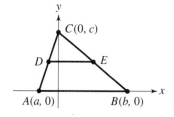

Figure 1.21

EXAMPLE 8   Using the normal range for your calculator (see page xi) and the DRAW or PLOT facility, graph the line segment from $(-4, -2)$ to $(3, 1)$. Estimate the midpoint. Compare your estimate with the computed value.

SOLUTION   For the TI-82, the normal range is $-9.4 \le x \le 9.4$, $-6.2 \le y \le 6.2$. (For the TI-81 it is $-9.6 \le x \le 9.4$, $-6.4 \le y \le 6.2$.) Next we clear all functions from the function menu and graph. The result is a screen that is clear except for the axes. Now we select Line from the DRAW or PLOT menu. Using the cursor keys and ENTER or EXE, we select the two endpoints of the segment. The result is shown in Figure 1.22.

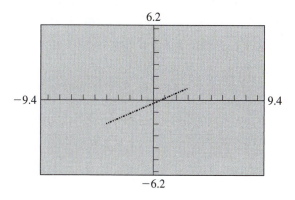

Figure 1.22

From this graph we might estimate the midpoint to be $(-0.2, -0.4)$. The computed value is $(-0.5, -1)$. This difference between the estimated and computed value illustrates the need to solve this problem algebraically rather than geometrically.

PROBLEMS

A   *In Problems 1–6, find the point P such that $\overline{AP}/\overline{AB} = r$.*

1. $A = (3, 4)$, $B = (7, 0)$, $r = 1/4$ **2.** $A = (4, -2)$, $B = (-2, -5)$, $r = 2/3$

3. $A = (5, -1)$, $B = (-4, -5)$, $r = 1/5$ **4.** $A = (2, 4)$, $B = (-5, 2)$, $r = 2/5$

5. $A = (-4, 1)$, $B = (3, 8)$, $r = 3$ **6.** $A = (-6, 2)$, $B = (4, 4)$, $r = 5/2$

In Problems 7–10, find the midpoint of the segment AB.

7. $A = (5, -2)$, $B = (-1, 4)$ **8.** $A = (-3, 3)$, $B = (1, 5)$

9. $A = (4, -1)$, $B = (3, 3)$ **10.** $A = (-1, 4)$, $B = (0, 2)$

B   **11.** If $A = (3, 5)$, $P = (6, 2)$, and $\overline{AP}/\overline{AB} = 1/3$, find B.

12. If $P = (4, 7)$, $B = (2, -1)$, and $\overline{AP}/\overline{AB} = 2/5$, find A.

13. If $P = (2, -5)$, $B = (4, -3)$, and $\overline{AP}/\overline{AB} = 1/2$, find A.

14. If $A = (3, 3)$, $P = (5, 2)$, and $\overline{AP}/\overline{AB} = 3/5$, find B.

In Problems 15–18, find the point P between A and B such that AB is divided in the given ratio.

15. $A = (5, -3)$, $B = (-1, 6)$, $\overline{AP}/\overline{PB} = 1/2$

16. $A = (-1, -3)$, $B = (-8, 11)$, $\overline{AP}/\overline{PB} = 3/4$

17. $A = (2, -1)$, $B = (4, 5)$, $\overline{AP}/\overline{PB} = 2/3$

18. $A = (5, 8)$, $B = (2, -1)$, $\overline{AP}/\overline{PB} = 5/1$

19. If $P = (4, -1)$ is the midpoint of the segment AB, where $A = (2, 5)$, find B.

20. Find the center and radius of the circle circumscribed about the right triangle with vertices $(1, 1)$, $(1, 4)$, and $(7, 4)$.

21. Find the point of intersection of the medians of the triangle with vertices $(5, 2)$, $(0, 4)$, and $(-1, -1)$. (See Problem 26.)

22. Prove analytically that the diagonals of a parallelogram bisect each other.

23. Find the point of intersection of the diagonals of the parallelogram with vertices $(1, 1)$, $(4, 1)$, $(3, -2)$, and $(0, -2)$.

24. Prove analytically that the midpoint of the hypotenuse of a right triangle is equidistant from the three vertices.

25. Prove analytically that the vertex and the midpoints of the three sides of an isosceles triangle are the vertices of a rhombus.

26. Prove analytically that the medians of a triangle are concurrent at a point two-thirds of the way from each vertex to the midpoint of the opposite side.

27. The point $(1, 4)$ is at a distance 5 from the midpoint of the segment joining $(3, -2)$ and $(x, 4)$. Find x.

C 28. The midpoints of the sides of a triangle are $(-1, 3)$, $(1, -2)$, and $(5, -3)$. Find the vertices.

29. Three vertices of a parallelogram are $(2, 5)$, $(-7, 1)$, and $(4, -6)$. Find the fourth vertex. [*Hint:* There is more than one solution. Sketch all possible parallelograms using the three given vertices.]

30. Show that if a triangle has vertices (x_1, y_1), (x_2, y_2), and (x_3, y_3), then the point of intersection of its medians is

$$\left(\frac{x_1 + x_2 + x_3}{3}, \frac{y_1 + y_2 + y_3}{3} \right)$$

31. Prove analytically that the sum of the squares of the four sides of a parallelogram is equal to the sum of the squares of the two diagonals.

GRAPHING CALCULATOR

In Problems 32–35, use the DRAW or PLOT facility to graph the line segments joining the given points. Estimate the midpoint and compare it with the computed value.

32. $(3, 5)$, $(-1, 2)$

33. $(-7, 4)$, $(2, -5)$

34. $(6, 3)$, $(-4, 4)$

35. $(-7, -2)$, $(-3, 3)$

APPLICATIONS

36. Suppose that, in Figure 1.14 on page 11, two cars drive toward each other at the same speed from X and Y. On what grid square will they meet? Suppose we use 2.3 to mean three-tenths of the way down grid band 2 and B.4 to mean four-tenths of the way across grid band B. At what point will the two cars meet?

37. Suppose that, in Problem 36, the car from X is traveling twice as fast as the one from Y. At what point will they meet?

38. Suppose that, in Problem 36, the car from X is traveling at 30 miles per hour and the one from Y is traveling at 40 miles per hour. Where will the two cars meet? At what time?

39. Suppose that, in Problem 38, the car from X starts 15 minutes later than the one from Y. Where will they meet? At what time?

40. Suppose that, in Problem 38, the car from X starts 5 minutes earlier than the one from Y. Where will they meet? At what time?

1.4

INCLINATION AND SLOPE

An important concept in the description of line and one that is used quite extensively throughout calculus has to do with the inclination of a line. First let us recall the convention from trigonometry which states that angles measured in the counterclockwise direction are positive, while those measured in the clockwise direction are negative. Thus we have the following definition.

DEFINITION *The **inclination** of a line that intersects the x axis is the measure of the smallest nonnegative angle which the line makes with the positive end of the x axis. The inclination of line parallel to the x axis is 0.*

We shall use the symbol θ to represent an inclination. The inclination of a line is always less than 180°, or π radians, and every line has an inclination. Thus, for any line,

$$0° \leq \theta < 180° \qquad \text{or} \qquad 0 \leq \theta < \pi.$$

Figure 1.23 shows several lines with their inclinations. Note that the angular measure is given in both degrees and radians. Although there is no reason to show preference for one over the other at this time, radian measure is the preferred way of representing an angle in more advanced courses.

While the inclination of a line may seem like a simple representation, we cannot, in general, find a simple relationship between the inclination of a line and the coordinates of points on it without resorting to tables of trigonometric functions. Thus, we consider another expression related to the inclination—namely, the slope of a line.

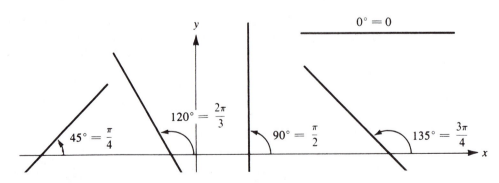

Figure 1.23

DEFINITION *The **slope m** of a line is the tangent of the inclination; thus,*

$$m = \tan\theta.$$

While it is possible for two different angles to have the same tangent, it is not possible for lines having two different inclinations to have the same slope. The reason for this is the restriction on the inclination, $0° \le \theta < 180°$. Nevertheless, one minor problem does arise from the use of slope since the tangent of 90° is not defined. Thus vertical lines have inclination 90° but no slope. *Do not confuse "no slope" with "zero slope."* A horizontal line definitely has a slope and that slope is the number 0, but there is no number at all (not even 0) which is the slope of a vertical line. Some might object to this nonexistence of tan 90° by saying that it is "infinity," or "∞." However, infinity is not a number. Also, while the symbol ∞ is quite useful in calculus when dealing with limits, its use in algebra or in an algebraic development of trigonometry leads to trouble.

While the nonexistence of the slope of certain lines is somewhat bothersome, it is more than counterbalanced by the simple relationship between the slope and the coordinates of a pair of points on the line. Recall that if θ is as shown in either of the two positions in Figure 1.24, then

$$\tan \theta = \frac{y}{x}.$$

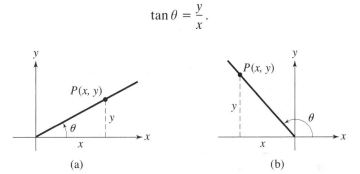

(a)   (b)

Figure 1.24

Unfortunately, the lines with which we are dealing are not always so conveniently placed. Suppose we have a line with a pair of points, $P_1 = (x_1, y_1)$ and $P_2 = (x_2, y_2)$, on it (see Figure 1.25). If we place a pair of axes parallel to the old axes,

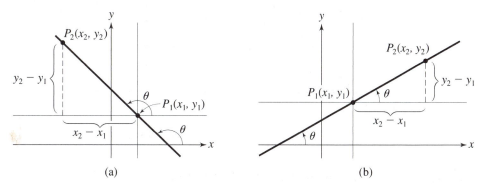

(a)   (b)

Figure 1.25

with P_1 as the new origin, then the coordinates of P_2 with respect to this new coordinate system are $x = x_2 - x_1$ and $y = y_2 - y_1$. Now θ is situated in a position that allows us to use the definition of $\tan \theta$ and state the following theorem.

THEOREM 1.4   *A line through $P_1 = (x_1, y_1)$ and $P_2 = (x_2, y_2)$, where $x_1 \neq x_2$, has slope*

$$m = \frac{y_2 - y_1}{x_2 - x_1} = \frac{y_1 - y_2}{x_1 - x_2}.$$

Note the agreement of the subscripts in the numerator and denominator. The slope of the line joining two points is the difference of the y coordinates divided by the difference of the x coordinates *taken in the same order*. The numerator and denominator are signed (or directed) vertical and horizontal distances.

One description of the slope of a line is that it is the vertical rise of the line divided by the horizontal run, or simply, rise over run. When the "run" is to the left, it is negative, giving a negative slope. This description of the slope as rise over run is one that is often used to describe the pitch of a roof or the grade of a road. They are often given as percentages, where a 1% grade corresponds to a slope of 0.01.

EXAMPLE 1   Find the slope of the line containing $P_1 = (1, 5)$ and $P_2 = (7, -7)$.

SOLUTION

$$m = \frac{y_2 - y_1}{x_2 - x_1} = \frac{-7 - 5}{7 - 1} = \frac{-12}{6} = -2$$

Since

$$m = \tan \theta = -2,$$
$$\theta = \arctan(-2) = 117°$$

The line through P_1 and P_2 is shown in Figure 1.26. A slope of -2 means that, as we move a unit distance to the right on the line, we move down (because of the minus) a distance 2.

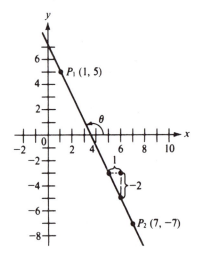

Figure 1.26

hISTORICAL NOTES

MODERN MATHEMATICS AND ANCIENT GREEK THOUGHT

The ancient Greeks have had a profound influence upon Western thought for more than two millennia. While this influence has generally been overwhelmingly positive, there have been negative aspects as well. Examples of both positive and negative influences can be found in philosophy, astronomy, biology, physics, and medicine. Our concern here is with their influence upon mathematics.

Perhaps the greatest contribution that the ancient Greeks made to mathematics was that of Thales and Pythagoras around 600 B.C. They were the first to look upon mathematics as a deductive system in which mathematical statements follow as logical consequences from other statements. Furthermore, they set about mathematizing natural observations. Their thinking was extended and refined for several centuries, resulting in indirect proofs, the geometry of conic sections, trigonometry, and the axiomatic method of Euclid.

However, much of Greek mathematics was clothed in a mysticism that saw magic, and even theology, in numbers. Thus, the discovery of incommensurable segments (irrational numbers) was looked upon by the Pythagoreans as an assault on the gods and was suppressed for a time. The idea that mathematics is a human creation for human use was foreign to them. A second unfortunate turn was the Greek insistence on the exclusive use of synthetic, rather than analytic, methods— of a static, rather than a dynamic, mathematics. To be sure, there was one Greek mathematician, Archimedes, who was not fettered by these ideas; however, he was centuries before his time. Thus, when he died, his ideas died with him.

It was many centuries later that mathematics threw off the limitations that had been imposed by the Greeks and entered the modern era. The advances in elementary algebra in the sixteenth and seventeenth centuries and the analytic methods introduced in the seventeenth century marked the beginning of modern mathematics. By allowing geometric problems to be solved algebraically (or analytically), analytic geometry paved the way for the development of calculus and the subsequent explosion of mathematical thought. Thus, with the study of analytic geometry and calculus, you are entering the world of modern mathematics.

EXAMPLE 2

Graph the line through (2, 1), with slope 3/2.

SOLUTION

Remember that the numerator and denominator of the slope represent vertical and horizontal distances. Starting at the point (2, 1), we proceed horizontally a distance 2 in the positive direction (to the right) and vertically a distance 3, again in the positive direction (upward). This takes us to the point (4, 4). The desired line is the line joining the given point (2, 1) and the point (4, 4) as shown in Figure 1.27.

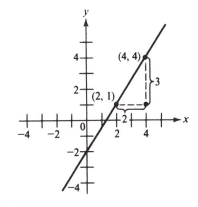

Figure 1.27

Note that if the slope had been $-3/2$, then, putting the minus with the numerator, the only difference would be that the vertical distance would be in the negative direction or downward.

Since a vertical line has no slope, Theorem 1.4 does not hold in that case; however, $x_1 = x_2$ for any pair of points on a vertical line, and the right-hand side of the slope formula is also nonexistent. Thus there is no slope when the right-hand side of the slope formula does not exist.

EXAMPLE 3   Use DRAW or PLOT to draw the line through $(5, 1)$ and $(-2, 5)$.

SOLUTION   Since DRAW or PLOT adds to an existing graph, we must clear the function screen and graph. This gives us a screen that is blank except for the axes. DRAW or PLOT easily allows us to draw a segment from $(5, 1)$ to $(-2, 5)$. This segment can be extended across the entire screen by the following method: To extend downward, again use DRAW or PLOT Line and anchor one end at the *higher* of the two given points, that is, at $(-2, 5)$. Now move the cursor downward and to the right to the bottom or right edge of the screen. In this case, move it to the right edge as shown in Figure 1.28. Now move the cursor along the edge until the new line segment completely hides the first one. When this occurs, anchor the end on the right edge. This method can be repeated to extend the segment to the top or left edge.

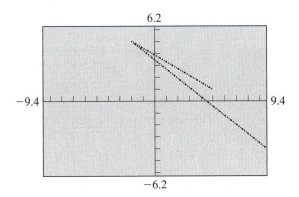

Figure 1.28

Because we have no equation we cannot use ZOOM or TRACE or change the Window settings. Any attempt to do so will result in a blank screen (except for the axes). Nevertheless, the drawn line will be remembered if the calculator is turned off or if we do other things that do not involve graphing.

EXAMPLE 4   Use DRAW or PLOT to draw the line through $(-4, 2)$ with slope -3.

SOLUTION   Given a point and a slope, it is important that we have the same scale on both axes, such as we have using the normal range (but *not* the standard range). Once again we clear the function menu and graph. Then, using DRAW or PLOT Line, we anchor the line at the given point $(-4, 2)$. Now we move the cursor 1 pixel to the right and 3 pixels down (down because the slope is negative). This may be repeated as often as necessary to reach the edge or to come close to the edge. Figure 1.29 shows the result after this much has been done.

This technique may be repeated at the other end by anchoring at $(-4, 2)$ and reversing both directions in moving the cursor—in this case, by repeatedly moving 1 space to the left and 3 spaces up.

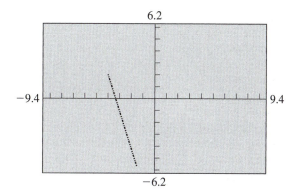

Figure 1.29

1.5

PARALLEL AND PERPENDICULAR LINES

If two nonvertical lines are parallel, they must have the same inclination and, thus, the same slope (see Figure 1.30). If two parallel lines are vertical, then neither one has slope. Similarly, if $m_1 = m_2$ or if neither line has slope, then the two lines are parallel. Thus, two lines are parallel if and only if $m_1 = m_2$ or neither line has slope.

If two nonvertical lines l_1 and l_2 with the respective inclinations θ_1 and θ_2 are perpendicular (see Figure 1.31), then (assuming l_1 to be the line with the larger inclination)

$$\theta_1 - \theta_2 = 90°,$$

and

$$\theta_1 = \theta_2 + 90°.$$

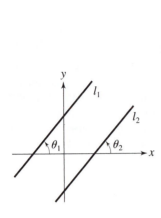

Figure 1.30

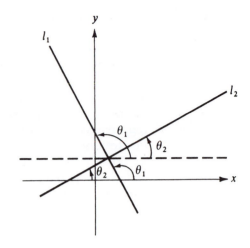

Figure 1.31

Thus

$$\tan \theta_1 = \tan (\theta_2 + 90°) = -\cot \theta_2 = -\frac{1}{\tan \theta_2}$$

or

$$m_1 = -\frac{1}{m_2}.$$

On the other hand, if $m_1 = -1/m_2$, the argument can be traced backward to show that the difference of the inclinations is 90° and the lines are perpendicular. Therefore we have the following theorem.

THEOREM 1.5   *The lines l_1 and l_2 with slopes m_1 and m_2, respectively, are*

(a) parallel or coincident if and only if $m_1 = m_2$,

(b) perpendicular if and only if $m_1 m_2 = -1$.

EXAMPLE 1   Find the slopes of l_1 containing $(1, 5)$ and $(3, 8)$ and l_2 containing $(-4, 1)$ and $(0, 7)$; determine whether l_1 and l_2 are parallel, coincident, perpendicular, or none of these.

SOLUTION

$$m_1 = \frac{8 - 5}{3 - 1} = \frac{3}{2}$$

and

$$m_2 = \frac{7 - 1}{0 + 4} = \frac{6}{4} = \frac{3}{2}$$

We now know that l_1 and l_2 are either parallel or coincident. While it is clear from Figure 1.32 that they are parallel rather than coincident, this would not be so obvious if the lines were closer together. Let us show it analytically. We begin by finding the slope of the line l_3 joining $(1, 5)$ on l_1 and $(-4, 1)$ on l_2.

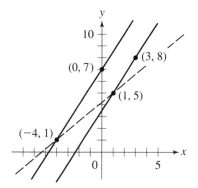

$$m_3 = \frac{5 - 1}{1 + 4} = \frac{4}{5}$$

If l_1 and l_2 were coincident, then l_3 would be coincident to both of them; therefore it would have the same slope as l_1 and l_2. Since its slope is $4/5$ rather than $3/2$, l_1 and l_2 are not coincident; they are parallel.

Figure 1.32

EXAMPLE 2 If the line through $(x, -3)$ and $(3, 1)$ is perpendicular to the line through $(x, -3)$ and $(-1, -2)$, find x (see Figure 1.33).

SOLUTION The slope m_1 of the line through $(x, -3)$ and $(3, 1)$ is

$$m_1 = \frac{-3 - 1}{x - 3} = \frac{-4}{x - 3}.$$

The slope m_2 of the line through $(x, -3)$ and $(-1, -2)$ is

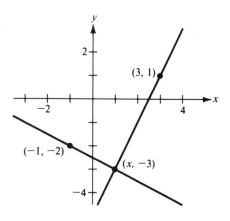

Figure 1.33

$$m_2 = \frac{-3 + 2}{x + 1} = \frac{-1}{x + 1}.$$

Since the lines are perpendicular, $m_1 m_2 = -1$.

$$\frac{-4}{x - 3} \frac{-1}{x + 1} = -1$$

$$\frac{4}{x^2 - 2x - 3} = -1$$

$$-x^2 + 2x + 3 = 4$$

$$x^2 - 2x + 1 = 0$$

$$(x - 1)^2 = 0$$

$$x = 1$$

PROBLEMS

A *In problems 1–8, find the slope (if any) and the inclination of the line through the given points.*

1. $(2, 3), (5, 8)$ **2.** $(-1, 4), (4, 2)$ **3.** $(-2, -2), (4, 2)$

4. $(3, -5), (1, -1)$ **5.** $(-4, 2), (-4, 5)$ **6.** $(2, 3), (-4, 3)$

7. $(a, a), (b, b)$ **8.** $(a, a), (-a, 2a)$

In Problems 9–14, graph the line through the given point and having the given slope.

9. $(5, -2), m = 2$ **10.** $(-2, 4), m = 3/4$

11. $(3, 1), m = -1/3$ **12.** $(0, 4), m = -3$

13. $(4, 2), m = 0$ **14.** $(-3, 1)$, no slope

In Problems 15–24, find the slopes of the lines through the two pairs of points; then determine whether the lines are parallel, coincident, perpendicular, or none of these.

15. $(1, -2), (-2, -11); \quad (2, 8), (0, 2)$ **16.** $(1, 5), (-2, -7); \quad (7, -1), (3, 0)$

17. $(1, 5), (-1, -1); \quad (0, 3), (2, 7)$ **18.** $(1, 3), (-1, -1); \quad (0, 2), (4, -2)$

19. $(1, 1), (4, -1); \quad (-2, 3), (7, -3)$ **20.** $(1, -4), (6, 1); \quad (2, 3), (-1, 6)$

21. $(1, 2), (3, 2); \quad (4, 1), (4, -2)$ **22.** $(1, 5), (1, 1); \quad (-2, 2), (-2, 4)$

23. $(2, 1), (5, -1); \quad (3, 3), (12, -3)$ **24.** $(1, -1), (5, 2); \quad (9, 5), (-3, -4)$

B **25.** If the line through $(x, 5)$ and $(4, 3)$ is parallel to a line with slope 3, find x.

26. If the line through $(x, 5)$ and $(4, 3)$ is perpendicular to a line with slope 3, find x.

27. If the line through $(x, 1)$ and $(0, y)$ is coincident with the line through $(1, 4)$ and $(2, -3)$, find x and y.

28. If the line through $(-2, 4)$ and $(1, y)$ is perpendicular to one through $(-2, 4)$ and $(x, 2)$, find a relationship between x and y.

29. If the line through $(x, 4)$ and $(3, 7)$ is parallel to one through $(x, -1)$ and $(5, 1)$, find x.

30. Show by means of slopes that $(1, 1)$, $(4, 1)$, $(3, -2)$, and $(0, -2)$, are the vertices of a parallelogram.

31. Show by means of slopes that $(-2, 4)$, $(2, 0)$, $(6, 4)$, and $(2, 8)$ are the vertices of a square.

32. A certain section of railroad roadbed rises 100 feet per mile of track. What is the percentage grade for this section of track?

33. Prove analytically that the diagonals of a square intersect at right angles.

34. Prove analytically that one median of an isosceles triangle is an altitude.

35. Prove analytically that the diagonals of a rhombus intersect at right angles.

C   **36.** Prove analytically that the medians of an equilateral triangle are altitudes.

GRAPHING
CALCULATOR

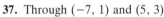

In Problems 37–44, use DRAW or PLOT to draw the line described.

37. Through $(-7, 1)$ and $(5, 3)$

38. Through $(4, 2)$ and $(-3, 4)$

39. Through $(3.2, 5.6)$ and $(0, -2.2)$

40. Through $(-5.3, 4.9)$ and $(2.2, -0.3)$

41. Through $(5, 4)$ with slope 5

42. Through $(-2, -3)$ with slope -2

43. Through $(2, 3.4)$ with slope -1.5

44. Through $(-1.3, 4.2)$ with slope 2.7

APPLICATIONS

45. Find the pitch (slope) of the roof shown in Figure 1.34.

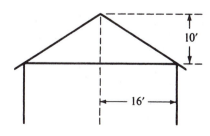

10'

16'

Figure 1.34

46. Find the percentage grade of the section of road shown in Figure 1.35.

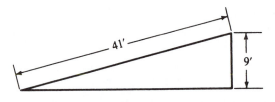

Figure 1.35

47. The direction, or bearing, of an aircraft or sailing vessel is usually expressed in the form N 30°E, which means 30° east of north. If a ship is steaming toward a lighthouse that is 10 miles north and 4 miles east of the ship, what is the ship's bearing?

48. Suppose that, in Problem 47, the ship changes its course when it has steamed halfway to the lighthouse. At that point, it steams toward a waiting ship that is 4 miles due east of the lighthouse. What is the bearing for this second leg?

1.6

ANGLE FROM ONE LINE TO ANOTHER

If l_1 and l_2 are two intersecting lines, then an angle from l_1 to l_2 is any angle measured from l_1 to l_2. If the measurement is in the counterclockwise direction, then the angle is positive; if it is in the clockwise direction, the angle is negative. While there are many angles from l_1 to l_2, all are related (see Figure 1.36) in that, if α is one of them, all can be expressed in the form

$$\alpha + n \cdot 180°,$$

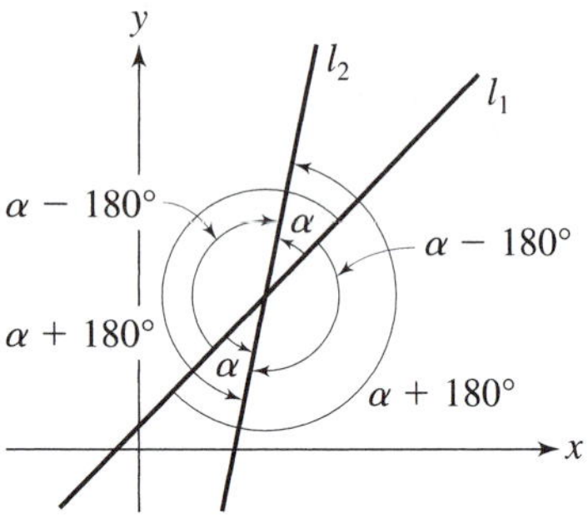

Figure 1.36

where n is an integer (positive, negative, or zero). Since any two of these angles differ from each other by a multiple of 180°, they all have the same tangent.

THEOREM 1.6   *If l_1 and l_2 are nonperpendicular lines with slopes m_1 and m_2, respectively, and α is any angle from l_1 to l_2, then*

$$\tan \alpha = \frac{m_2 - m_1}{1 + m_1 m_2}.$$

PROOF   Let θ_1 and θ_2 be the inclinations of the lines l_1 and l_2, respectively. Figure 1.37 shows that

$$\alpha = \theta_2 - \theta_1$$

for one of the angles α from l_1 to l_2. Thus

$$\tan \alpha = \frac{\tan \theta_2 - \tan \theta_1}{1 + \tan \theta_1 \tan \theta_2}.$$

But, since $m_1 = \tan \theta_1$ and $m_2 = \tan \theta_2$, we have

$$\tan \alpha = \frac{m_2 - m_1}{1 + m_1 m_2}. \quad \blacksquare$$

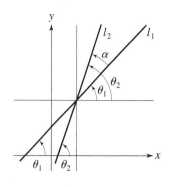

Figure 1.37

We have assumed in this argument that l_1 and l_2 intersect. If they do not, then $m_1 = m_2$. Using $m_1 = m_2$ in Theorem 1.6, we find that $\tan \alpha = 0$ and $\alpha = 0°$. Thus we shall use the convention that $\alpha = 0°$ if l_1 and l_2 are parallel. This is in agreement with the convention that $m = 0$ for horizontal lines.

The trigonometric identity used in this proof is, of course, true only when $\tan \alpha$ and $(m_2 - m_1)/(1 + m_1 m_2)$ both exist. Tan α does not exist if $\alpha = 90°$, but then $m_2 = -1/m_1$ and $1 + m_1 m_2 = 0$, which gives the one case in which $(m_2 - m_1)/(1 + m_1 m_2)$ does not exist. Thus Theorem 1.6 holds for all values of α except $\alpha = 90°$, for which case neither side of the equation exists.

DEFINITION

*The **angle** from l_1 to l_2 is the smallest nonnegative angle from l_1 to l_2.*

EXAMPLE 1

If l_1 and l_2 have slopes $m_1 = 3$ and $m_2 = -2$, respectively, find the angle from l_1 to l_2.

SOLUTION

$$\tan \alpha = \frac{m_2 - m_1}{1 + m_1 m_2} = \frac{-2 - 3}{1 + 3(-2)} = 1$$

Thus $\alpha = \arctan 1 = 45°$. See Figure 1.38.

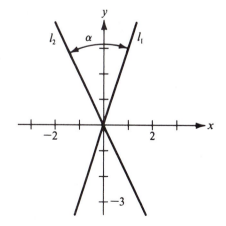

Figure 1.38

Theorem 1.6 assumes that both l_1 and l_2 have slopes. If one of the two lines is vertical, it has no slope; but it does have an inclination of 90°. If the inclination of the other line is θ, then we have one of the four situations shown in Figure 1.39. From (a) and (b) of this figure we see that when l_1 is vertical, the angle α from l_1 to l_2 is either $\theta + 90°$ or $\theta - 90°$. Since these two values differ by 180°, which is the period of the tangent, it follows that

$$\tan \alpha = \tan (\theta + 90°) = \tan (\theta - 90°)$$

$$= -\cot \theta = -\frac{1}{\tan \theta} = -\frac{1}{m},$$

where m is the slope of l_2.

Similarly, when l_2 is vertical, we see from (c) and (d) of Figure 1.39 that $\alpha = 90° - \theta$ or $\alpha = 270° - \theta$. Again they differ by 180°; again their tangents are equal. Thus when l_2 is vertical,

$$\tan \alpha = \tan (90° - \theta) = \cot \theta = \frac{1}{\tan \theta} = \frac{1}{m}.$$

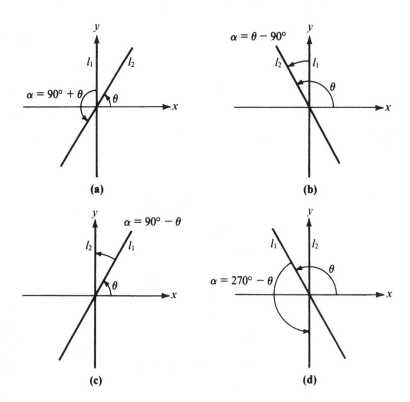

Figure 1.39

EXAMPLE 2   Find the angle from l_1 to l_2, where l_1 is a vertical line and l_2 has slope 1/2.

SOLUTION   The inclination of l_1 is 90°; the inclination of l_2 is θ, where $\tan\theta = 1/2$ (see Figure 1.40). Since we want the angle α from l_1 to l_2, we see that one angle from l_1 to l_2 is

$$\alpha' = \theta + 90°$$

and

$$\tan\alpha = \tan\alpha' = \tan(\theta + 90°)$$
$$= -\cot\theta$$
$$= -\frac{1}{\tan\theta} = -\frac{1}{1/2} = -2.$$

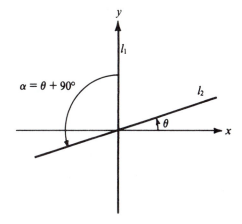

Figure 1.40

Thus the angle we want is the smallest positive angle whose tangent is -2. Using tables or a pocket calculator, we see that α is approximately 117°.

EXAMPLE 3   Find the slope of the line bisecting the angle from l_1, with slope 7, to l_2, with slope 1.

SOLUTION   Let m be the slope of the desired line. Since $\alpha_1 = \alpha_2$ (see Figure 1.41), we have

$$\tan\alpha_1 = \tan\alpha_2$$

and

$$\frac{m - m_1}{1 + m_1 m} = \frac{m_2 - m}{1 + m_2 m}$$
$$\frac{m - 7}{1 + 7m} = \frac{1 - m}{1 + m}$$
$$(m - 7)(1 + m) = (1 + 7m)(1 - m)$$
$$m^2 - 6m - 7 = -7m^2 + 6m + 1$$
$$8m^2 - 12m - 8 = 0$$
$$4(2m + 1)(m - 2) = 0$$
$$m = -1/2 \quad \text{or} \quad m = 2.$$

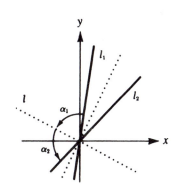

Figure 1.41

We have two answers, but obviously we want only one. Which one? Since one of them is the negative reciprocal of the other, they represent slopes of perpendicular lines, one of which is the bisector of the angle from l_1 to l_2, while the other bisects the angle from l_2 to l_1. An inspection of Figure 1.41 shows that the answer we want is $m = -1/2$.

EXAMPLE 4

Find the slope of the line bisecting the angle from l_1, with slope 2, to l_2, with no slope.

SOLUTION

Let m and θ be the slope and inclination, respectively, of the desired line. Since l_2 has no slope, it is a vertical line (see Figure 1.42). Thus $\alpha_2 = 90° - \theta$ and

$$\tan \alpha_2 = \cot \theta = \frac{1}{\tan \theta} = \frac{1}{m}.$$

Since l and l_1 have slopes m and $m_1 = 2$,

$$\tan \alpha_1 = \frac{m - m_1}{1 + mm_1} = \frac{m - 2}{1 + 2m}.$$

Finally, the equality of α_1 and α_2 gives

$$\tan \alpha_1 = \tan \alpha_2$$

$$\frac{m - 2}{1 + 2m} = \frac{1}{m}$$

$$m^2 - 2m = 1 + 2m$$

$$m^2 - 4m - 1 = 0$$

$$m = \frac{4 \pm \sqrt{16 + 4}}{2} = \frac{4 \pm 2\sqrt{5}}{2}$$

$$= 2 \pm \sqrt{5}.$$

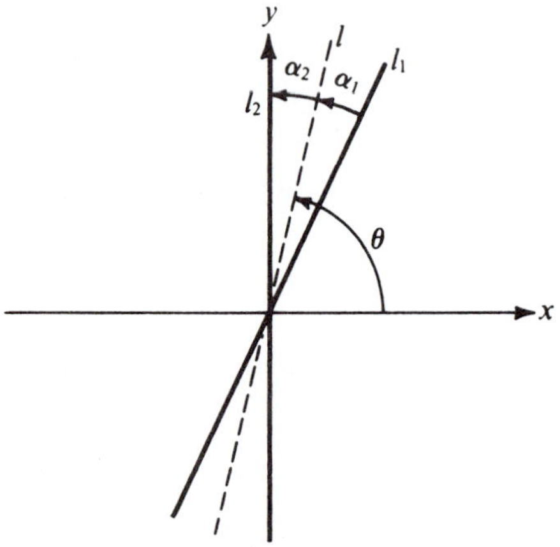

Figure 1.42

Again we have two answers, representing the slopes of the bisectors of the angle from l_2 to l_1 as well as that from l_1 to l_2. It is easily seen from Figure 1.42 that the one we want is $m = 2 + \sqrt{5}$.

PROBLEMS

A *In Problems 1–8, find the angle from l_1 to l_2 with slopes m_1 and m_2, respectively.*

1. $m_1 = -2, m_2 = 3$ **2.** $m_1 = 1, m_2 = 4$ **3.** $m_1 = -3, m_2 = 2$

4. $m_1 = 5, m_2 = -1$ **5.** $m_1 = 10, m_2$ does not exist **6.** $m_1 = 0, m_2 = -1$

7. $m_1 = 2/3, m_2$ does not exist **8.** m_1 does not exist, $m_2 = -2$

In Problem 9–16, find the angle from l_1 to l_2, where l_1 and l_2 contain the points indicated.

9. l_1: $(1, 4), (3, -1)$; l_2: $(3, 2), (5\ -1)$ **10.** l_1: $(2, 5), (-3, 10)$; l_2: $(-1, -3), (3, 3)$

11. l_1: $(4, 5), (1, 1)$; l_2: $(3, -3), (0, 4)$ **12.** l_1: $(1, 1), (0, 5)$; l_2: $(4, 3), (-1, 2)$

13. l_1: $(3, 4), (3, -1)$; l_2: $(2, 5), (-1, 2)$ **14.** l_1: $(-1, 2), (-1, -1)$; l_2: $(-3, 4), (1, 0)$

15. l_1: $(5, 1), (3, -3)$; l_2: $(5, 1), (5, -3)$ **16.** l_1: $(3, -4), (2, 3)$; l_2: $(2, -3), (2, 4)$

B *In Problems 17–24, find the slope of the line bisecting the angle from l_1 to l_2 with slopes m_1 and m_2, respectively.*

17. $m_1 = 3, m_2 = -2$ **18.** $m_1 = 1, m_2 = -7$ **19.** $m_1 = 2, m_2 = 3$

20. $m_1 = -1, m_2 = 2$ **21.** $m_1 = -3, m_2 = 5$ **22.** $m_1 = 2, m_2 = 0$

23. $m_1 = 3/4, m_2$ does not exist **24.** m_1 does not exist, $m_2 = 1$

25. Find the interior angles of the triangle with vertices $A = (1, 5)$, $B = (3, -1)$, and $C = (-1, -1)$.

26. Find the interior angles of the triangle with vertices $A = (3, 2)$, $B = (4, 5)$, and $C = (-1, -1)$.

27. Find the slope of the line l_1 such that the angle from l_1 to l_2 is Arctan 2/3, where l_2 contains $(2, 1)$ and $(-4, -5)$.

28. Find the slope of the line l_1 such that the angle from l_1 to l_2 is 45°, where the slope of l_2 is -2.

29. Find the slope of the line l_2 such that the tangent of the angle from l_1 to l_2 is $-1/2$, where l_1 is a vertical line.

30. A line with slope 1 bisects the angle from l_1 to l_2, where l_2 has slope 2. What is the slope of l_1?

31. A line with slope $3 + \sqrt{10}$ bisects the angle from l_1 to l_2, where l_1 has slope 1. What is the slope of l_2?

32. Show by means of angles that $A = (-1, 0)$, $B = (4, 6)$, and $C = (10, 1)$ are the vertices of an isosceles triangle.

C **33.** Lines l_1 and l_2 have slopes m and $1/m$, respectively. What is the slope of the line bisecting the angle from l_1 to l_2?

34. If lines l_1 and l_2 have slopes 1 and m, respectively, what is the slope of the line bisecting the angle from l_1 to l_2?

1.7

GRAPHS AND POINTS OF INTERSECTION

The graph of an equation in two variables x and y is simply the set of all points (x, y) in the plane whose coordinates satisfy the given equation. The determination of the graph of an equation is one of the principal problems of analytic geometry. Although we shall consider other methods in Chapter 7, we consider only point-by-point plotting here. To do this, we assign a value to either x or y, substitute the assigned value into the given equation, and solve for the other.

EXAMPLE 1

Graph $2x + 3y = 6$.

SOLUTION

x	y
-6	6
-3	4
0	2
3	0
6	-2

Plotting these points and joining them, we have the line of Figure 1.43.

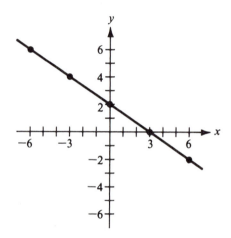

Figure 1.43

EXAMPLE 2

Graph $x^2 + y^2 = 25$.

SOLUTION

x	y
0	± 5
± 1	$\pm 2\sqrt{6}$
± 2	$\pm\sqrt{21}$
± 3	± 4
± 4	± 3
± 5	0

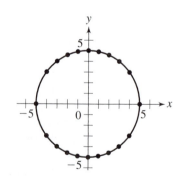

Figure 1.44

Plotting these points and joining them, we have the circle of Figure 1.44.

One obvious question that arises is how many points must be plotted before drawing the graph. There is no specific answer—plot as many as are needed to get a reasonable idea of the appearance of the graph.

Since each point of a graph satisfies the given equation, a point of intersection of two graphs is simply a point that satisfies both equations. Thus, any such point can be found by solving the two equations simultaneously.

EXAMPLE 3   Find all points of intersection of $x^2 + y^2 = 25$ and $x + y = 2$.

SOLUTION   Solving the second equation for y and substituting into the first, we have

$$x^2 + (2 - x)^2 = 25$$
$$2x^2 - 4x - 21 = 0$$
$$x = \frac{4 \pm \sqrt{16 + 168}}{4} = \frac{2 \pm \sqrt{46}}{2} = 1 \mp \frac{1}{2}\sqrt{46}$$
$$y = 2 - x = 1 \mp \frac{1}{2}\sqrt{46}.$$

Thus, the two points of intersection are

$$\left(1 + \frac{1}{2}\sqrt{46},\ 1 - \frac{1}{2}\sqrt{46}\right)$$

and

$$\left(1 - \frac{1}{2}\sqrt{46},\ 1 + \frac{1}{2}\sqrt{46}\right).$$

These are shown in Figure 1.45.

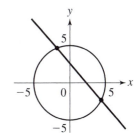

Figure 1.45

Closely related to equations is the idea of a function; in fact, equations are often used to represent functions. A **function** is an association, or pairing, of the values of one set, called the **domain** of the function, with those of another (not necessarily different) set, called the **range**, such that no value in the domain is associated with more than one value in the range.

We have used the word "value" of a set to emphasize the fact that, although the domain and range are often sets of real numbers, they need not be so. For example, we might take the domain to be the set of all circles in the plane and associate its area with each circle. Nevertheless, the functions that we shall consider here all have domains and ranges consisting entirely of real numbers.

As noted above, equations are often used to represent functions. For example, the equation $y = x^2$ describes a pairing of x values with the corresponding y

values. Furthermore, a single value of x gives only one y—there is no way to get two values of y from a single x. Thus, this equation describes y as a function of x. When, as in this case, the domain and range are not specified, the domain is taken to be the set of all real numbers x for which there is a corresponding y; the range is the set of all possible y values. In this case, the domain is the set of all real numbers and the range is the set of all nonnegative real numbers.

In order to emphasize the fact that y is function of x, the above equation is often written $f(x) = x^2$. This says that, for each number x, the corresponding functional value is the square of x. This notation was invented by the Swiss mathematician Leonhard Euler (pronounced OY-ler). Other benefits come with this notation, but they do not concern us and will not be considered here.

The restriction that "no value in the domain is associated with more than one value in the range" implies that a function *must* be single-valued. While $y = x^2$ defines y as a function of x, $y^2 = x$ does not. For the equation $y^2 = x$, $x = 4$ corresponds to both $y = 2$ and $y = -2$. Thus, $y^2 = x$ does not define y as a function of x; it does, however, define x as a function of y. When we consider equations involving x and y, we shall consider only the possibility of y as a function of x— not that of x as a function of y. This implies that functions, which must be single-valued, are easily recognized from their graphs; no vertical line can contain more than one point of the graph. Figure 1.46 shows the difference between $y = x^2$ and $y^2 = x$. In Figure 1.46a, *every* vertical line contains only one point of the graph. On the other hand, Figure 1.46b shows that some vertical lines contain two points of the graph of $y^2 = x$, implying that y is not a function of x. It might be noted that there are some lines (those to the left of the y axis) that contain no point of the graph and one line (the y axis) that contains exactly one point. These facts are immaterial in deciding whether we have a function; the only thing that matters is that there is at least one vertical line containing two or more points of the graph.

Why do we have this preoccupation with single-valuedness? The answer is that the ideas of limits and continuity, which are central to calculus, are much more difficult if we allow multiple-valued functions. Calculus is where the idea of a function is most useful. In fact, functions were invented by Gottfried Wilhelm

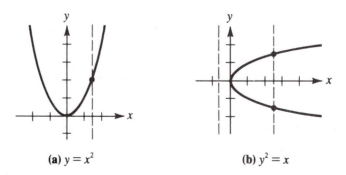

(a) $y = x^2$ **(b)** $y^2 = x$

Figure 1.46

Leibniz (pronounced LIBE-nits), who, along with Sir Isaac Newton, gave the initial development of calculus.

One problem that we have with point-by-point graphing is the amount of tedious arithmetic that is required. Fortunately, we are relieved of this work by graphing programs on computers and by the recent introduction of graphing calculators. Nevertheless, computers and graphing calculators have their drawbacks.

The principal problem is that they do not graph equations—they graph functions. Furthermore, these functions must be given explicitly. What does that mean? Equations such as $2x - 3y = 5$ and $x^2 + y^2 = 25$ imply that y is a function (or a combination of functions) of x. Such a function (or functions) is called an **implicit function**. If we solve these equations for y, giving

$$y = \frac{2x - 5}{3}$$

and

$$y = \pm\sqrt{25 - x^2},$$

we have y given explicitly as a function (or, in the second case, as a pair of functions) of x. Thus, in order to have $2x - 3y = 5$ graphed for us, we must give it as

$$y = (2x - 5)/3.$$

In order to have $x^2 + y^2 = 25$ graphed, we must give the two functions

$$y_1 = \sqrt{(25 - x^2)}$$
$$y_2 = -\sqrt{(25 - x^2)}.$$

Therefore, our first step in using a computer or graphing calculator is to solve our equation for y. Some of the most powerful computer programs can do that for us, but many do not.

A second problem is that calculators or computers have troubles with portions of graphs that are nearly vertical, especially if the function is undefined for values of x near the steep portion of the graph. Thus, in graphing the circle $x^2 + y^2 = 25$, the two halves of the circle might not meet because the nearly vertical portions are not graphed.

Still another disadvantage is their inability to graph discontinuous functions. This is a relatively minor problem, since most of the commonly used functions are continuous. It can be overcome (at least partially) by using Dot rather than Connected MODE.

Other problems are associated with computers and graphing calculators. We shall consider them when we take up curve sketching in more detail in Chapter 7.

We have already mentioned the greatest advantage of calculators and computers—namely, relief from arithmetic computation. Another advantage is the ability to plot several graphs on the same set of axes. This is what allows us to graph an entire circle rather than having to be satisfied with either the top half or the bottom half. Furthermore, it provides a convenient way of comparing certain graphs.

A second advantage is their ability to zoom in on certain portions of the graph. In this way, the graphical determination of points of intersection, which is quite impractical when working by hand except for very rough approximations, can now be used as a practical method.

EXAMPLE 4   Use a graphing calculator to graph $x^2 + y^2 = 25$ and $x^2 + y^2 = 20$ using the normal range. Change the window so that $-10 \le x \le 10$ and regraph.

SOLUTION   As noted above, we must solve the given equations for y. Thus the two equations are entered as the four functions

$$Y_1 = \sqrt{(25 - X^2)} \qquad\qquad Y_3 = \sqrt{(20 - X^2)}$$
$$Y_2 = -\sqrt{(25 - X^2)} \qquad\qquad Y_4 = -\sqrt{(20 - X^2)}$$

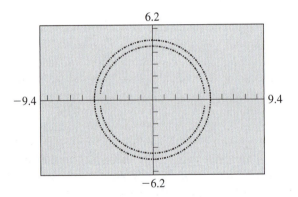

Figure 1.47

Using the normal range, we have the graphs shown in Figure 1.47. Notice that the two halves of the outer circle, representing $x^2 + y^2 = 25$, meet; the two halves of the inner circle, representing $x^2 + y^2 = 20$, do not. What is the reason for this? If we do a TRACE, we see that the leftmost point of Y_1 is $(-5, 0)$, which is a pixel value. Other pixels are turned on to connect this one to the next one on the right, which is $(-4.8, 1.4)$. However, a TRACE of Y_3 shows that the leftmost pixel of that graph is $(-4.4, 0.8)$; the next pixel value to the left of this has $x = -4.6$ for which Y_3 is undefined. Y_3 is defined only for pixel values $-4.4 \le x \le 4.4$ with $y \ge 0.8$. Similarly, Y_4 is defined only for $-4.4 \le x \le 4.4$ with $y \le -0.8$. Thus the two halves do not meet.

Now, when we change the window so that $-10 \le x \le 10$, we see from Figure 1.48 that the situation is reversed; that is, the inner circle is connected while the

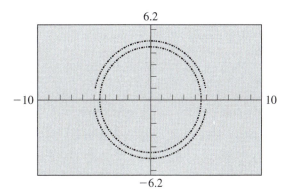

Figure 1.48

outer circle is not. Again, a trace of Y_1 shows that its leftmost point is $(-4.893617, 1.0259203)$ with pixel value $(-4.893617, 1.0)$, which is well above the x axis. On the other hand, the leftmost point of Y_3 is $(-4.468085, 0.19030367)$ corresponding to the pixel $(-4.680851, 0.2)$, which is the first pixel value above the x axis.

Thus we see that there is no one window setting that will make all circles connected; the result depends on the equations graphed as well as the choice of the window.

EXAMPLE 5

Use a graphing calculator to graph $x^2 + y^2 = 25$ and $x + y = 2$. Using TRACE and ZOOM, estimate the leftmost point of intersection. If your calculator has CALC, JUMP, or G SOLVE, use it to find the above point of intersection. Compare its coordinates with the coordinates found in Example 3.

SOLUTION

From Example 3, the leftmost point of intersection is

$$\left(1 - \frac{1}{2}\sqrt{46},\ 1 + \frac{1}{2}\sqrt{46}\right) = (-2.391165, 4.391165)$$

Graphing with the normal range and using TRACE, we have the point $(-2.4, 4.3863424)$ on the circle and $(-2.4, 4.4)$ on the line. If we ZOOM In on this point and use TRACE, we find that we do not improve upon this. Averaging the y coordinates, our best estimate is $(-2.4, 4.393)$, which is correct to two decimal places and only slightly off in the third.

On the other hand, using CALC, JUMP, or G SOLVE, we get $(-2.391165, 4.391165)$, which is accurate to six places.

PROBLEMS

All of the problems given below can be solved either with or without a graphing calculator.

A   *Plot the graphs of the equations in Problems 1–12. Indicate which give y as a function of x.*

1. $y = 2x - 1$ **2.** $x + 2y = 3$ **3.** $3x - 5y = 2$

4. $4x + 2y = 3$ **5.** $y = x^2 + 1$ **6.** $y = 2x - x^2$

7. $y = x^2 - x - 2$ **8.** $y = x^2 - 4x + 3$ **9.** $x^2 + y^2 = 1$

10. $x^2 + y^2 = 16$ **11.** $y = x^3$ **12.** $y = x^3 - 2x^2$

In Problems 13–16, find the points of intersection and sketch the graphs of the equations.

13. $3x - 5y = 2$ **14.** $2x + y = -1$
     $4x + 2y = 1$ $3x + 2y = 0$

15. $3x + 2y = -1$ **16.** $3x - 2y = 10$
     $x - 4y = -12$ $6x + 6y = -1$

B   *Plot the graphs of the equations in Problems 17–24. Indicate which give y as a function of x.*

17. $x^2 - y^2 = 4$ **18.** $x^2 - y^2 = -4$ **19.** $4x^2 + y^2 = 4$

20. $x^2 - 4y^2 = 4$ **21.** $y = \dfrac{x}{x + 1}$ **22.** $y = \dfrac{x + 1}{x}$

23. $y^2 = x^3$ **24.** $\sqrt{x} + \sqrt{y} = 4$

In Problems 25–32, find the points of intersection and sketch the graphs of the equations.

25. $x - y + 2 = 0$ **26.** $x - 2y = -1$ **27.** $x + y = 1$
     $y = x^2$ $y^2 = x + 4$ $x^2 + y^2 = 5$

28. $2x + y = 1$ **29.** $y = x^2$ **30.** $y = x^2$
     $x^2 + y^2 = 2$ $x = y^2$ $x^2 + y^2 = 2$

31. $x + 2y = 3$ **32.** $2x + 3y = 6$
     $x^2 + y^2 = 4$ $x^2 + y^2 = 16$

In computer graphics, a graph is viewed through a particular "window," or range of x and y values. Any part of the graph that is outside such a window is discarded. In Problems 33–38, find the endpoints of the portion of the graph of each of the following equations that is within the given window.

33. $x - y + 2 = 0$; window: $-4 \le x \le 4$, $-4 \le y \le 4$

34. $x + 2y - 2 = 0$; window: $-4 \le x \le 4$, $-3 \le y \le 3$

35. $2x + 3y - 1 = 0$; window: $-4 \le x \le 6$, $-3 \le y \le 5$

36. $y = x^2$; window: $-2 \le x \le 4$, $-2 \le y \le 5$

37. $y = 6 - x^2$; window: $-2 \le x \le 4$, $-3 \le y \le 5$

38. $xy = 1$; window: $-2 \le x \le 6$, $-1 \le y \le 4$

C **39.** Plot the graph of $x^2 + y^2 = 0$.

40. Plot the graph of $x^2 + y^2 = -1$.

GRAPHING CALCULATOR

41. Use a calculator to graph $x^2 - y^2 = 4$. Remember that you must first solve for y, giving $y = \pm \sqrt{x^2 - 4}$, which must be entered as $y_1 = \sqrt{(x^2 - 4)}$ and $y_2 = \sqrt{(x^2 - 4)}$. Graph $x^2 - y^2 = -4$ on the same axes.

42. Graph $y = |x|$ and $y = \sqrt{x^2}$ (a) on separate axes and (b) on the same pair of axes. What do you conclude from this?

43. Use the ZOOM In feature to find the points of intersection of $x - y + 2 = 0$ and $y = x^2$.

44. Find the points of intersection of $x + y = 1$ and $x^2 + y^2 = 5$.

45. Graph $y = x$, $y = x^3$, $y = x^7$, and $y = x^{13}$ on the same coordinate axes. What do you think the graph of $y = x^{99}$ looks like? Verify by adding its graph to the others.

46. Graph $y = x^2$, $y = x^4$, $y = x^8$, and $y = x^{24}$ using the same coordinate axes. What do you think the graph of $y = x^{100}$ looks like? Verify by adding its graph to the others.

47. Graph $y = x^2$, $y = (x - 1)^2$, and $y = (x - 3)^2$ on the same axes. What do you expect $y = (x + 2)^2$ to look like? Describe the result of replacing x by $x - k$.

48. Graph $y = x^2$, $y = x^2 + 1$, $y = x^2 + 3$ on the same axes. What do you expect $y = x^2 - 2$ to look like? Describe the result of replacing $y = f(x)$ by $y = f(x) + k$.

49. The integer and fractional parts of a number are the parts to the left and right, respectively, of the decimal point. For example, iPart $5.38 = 5$ and fPart $5.38 = 0.38$; iPart $-3.87 = -3$ and fPart $-3.87 = -0.87$. Graph $y = -$iPart x using Connected MODE and again using Dot MODE. Which is the better representation of the graph? Repeat for $y = $ fPart x.

50. Frequently, functions are defined by several equations rather than by a single one. For example, we may define $|x|$ by

$$|x| = \begin{cases} -x & \text{if } x < 0, \\ x & \text{if } x \geq 0. \end{cases}$$

This would be described on a TI by $Y_1 = (-X)(X<0) + X(X \geq 0)$; on a Casio it would be $Y_1 = -X, [-9.4, 0]$ $Y_2 = X, [0, 9.4]$. Graph

$$y = \begin{cases} 1 & \text{if } x \leq 0, \\ (x - 2)^2 & \text{if } x > 0. \end{cases}$$

51. Graph

$$y = \begin{cases} 2 & \text{if } x \leq 1, \\ (x - 2)^2 & \text{if } x > 1 \end{cases}$$

(see Problem 50 for entering into a calculator), using Connected MODE. Repeat using Dot MODE. Which gives a better representation near $x = 1$? near $x = 4$?

52. $x^2 + y^2 = 5$ and $x + y = \sqrt{5}$ have two points of intersection. Graph and use CALC, JUMP, or G SOLVE to find them. Repeat with $y = \sqrt[3]{x^2}$ and $y = -x$. What problem do we have here? (The problem will not necessarily occur on every calculator.) What seems to cause the problem?

53. Problem 24 asks for the graph of $\sqrt{x} + \sqrt{y} = 4$. If we were to graph this equation on a calculator, we would have to solve for y first, giving $y = (4 - \sqrt{x})^2$. Are the graphs of these two equations identical? If not, where are they the same, and where do they differ?

APPLICATIONS

54. In Newtonian mechanics, the mass of an object is fixed. However, the special theory of relativity says that the mass of an object varies with its velocity, with a moving body having greater mass than one that is at rest. The relationship between the mass of an object and its velocity is given by

$$m = \frac{cm_0}{\sqrt{c^2 - v^2}},$$

where m_0 is the rest mass and c is the velocity of light (2.998×10^{10} centimeters/second). Graph this expression. Use multiples of c on the x axis and multiples of m_0 on the y axis. What does this suggest about the possibility of accelerating a subatomic particle to the speed of light?

55. The profit function for the manufacture of a given item expresses the profit P in terms of the number x of items manufactured. The profit function for a certain item has been determined to be

$$P = 9x - 0.02x^2,$$

where both x and P are in thousands. The equation

$$P = 16 + 7.2x$$

was used to approximate the profit function for the range $0 \le x \le 100$. For what values of x does the approximation give the exact value of the profit? What happens if we try to extend the approximation beyond $x = 100$? [*Hint:* Graph these two equations.]

56. There is a relationship between the price p of a manufactured item and the quantity q that can be sold at price p. An equation relating these two is called a **demand equation**, and its graph is called a **demand curve**. As the price p increases, the demand q decreases. Similarly, there is a relationship between p and q from the point of view of the manufacturer. In this case, q is the quantity that the manufacturer is willing to produce when the selling price is p. An equation relating the selling price and the amount produced is called the *supply equation*. In this case, q increases with the price. When the demand exceeds the supply, there is a tendency for the price to increase; when the supply exceeds the demand, there is a tendency for the price to decrease. The point at which the supply equals the demand is called **market equilibrium**. Suppose that, for a given item, the supply equation is $q = 194p - 284$ and the demand equation is $q = 538 - 106p$, where p is the price in dollars and q is the quantity in thousands. Graph these two equations on the same set of axes and find the market equilibrium.

57. Repeat Problem 56 with supply equation $q = 20p - 70$ and demand equation $q = (200/p) - 10$.

58. The relationship between the number n of items produced and the cost C of producing them is called the **cost equation**. The relationship between the number n of items produced and the revenue R (income) realized from their production is the **revenue equation**. The point at which the cost equals the revenue is the **break-even point**. Suppose that the cost equation is $C = 0.0004n^2 - 0.8n + 1000$ and the revenue function is $R = 0.6n$. Graph these two equations on the same coordinate axes and find the break-even point(s). For what values of n does the revenue exceed the cost?

59. Repeat Problem 58 with $C = 1000(n + 100)/(n + 20)$ and $R = 22.5n$.

1.8

AN EQUATION OF A LOCUS

In the last section we considered one of the two principal problems of analytic geometry—finding the graph of an equation. Let us now consider the other—finding an equation of a locus. In other words, given a description of a curve, we want to find an equation representing that curve. Since an equation of a curve is a relationship satisfied by the x and y coordinates of each point on the curve (but by no other point), we need merely consider an arbitrary point (x, y) on the curve and give the description of the curve in terms of x and y. Let us consider some examples.

EXAMPLE 1

Find an equation for the set of all points in the xy plane which are equidistant from $(1, 3)$ and $(-2, 5)$.

SOLUTION

Let (x, y) be one such point (see Figure 1.49). Then

$$\sqrt{(x - 1)^2 + (y - 3)^2}$$
$$= \sqrt{(x + 2)^2 + (y - 5)^2}$$
$$(x - 1)^2 + (y - 3)^2$$
$$= (x + 2)^2 + (y - 5)^2$$
$$x^2 - 2x + 1 + y^2 - 6y + 9$$
$$= x^2 + 4x + 4 + y^2 - 10y + 25$$
$$6x - 4y + 19 = 0.$$

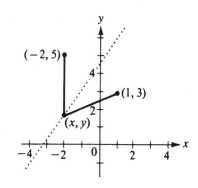

Figure 1.49

Note in the above example that the first equation given is an equation of the desired curve. Of course, we want the equation in as simple a form as possible. In carrying out the simplification, however, we must be sure that the final equation is equivalent to the original. One way to be sure of this is to simplify only by reversible steps.

It might also be noted in the example that the first step in simplifying was squaring both sides of the original equation. This is not normally a reversible operation. (If $x = 5$, then $x^2 = 25$; but if $x^2 = 25$, then $x = \pm 5$.) However, since we are dealing only with positive distances, we need consider only positive square roots in reversing the operation. Thus it is reversible. The remaining operations are clearly reversible; the first and last equations are equivalent.

There is a second way to verify that the first and last equations are equivalent. Clearly, any point (x, y) that satisfies the first equation must satisfy the subsequent ones. The only question in doubt is: Does any point satisfying the last equation also satisfy the first? In order to answer this question, let us consider a point (x, y) which satisfies the last equation. Solving for y, we have

$$y = \frac{6x + 19}{4}.$$

Let us now substitute this expression for y into both sides of the first equation. Since both sides simplify to

$$\frac{\sqrt{52x^2 + 52x + 65}}{4},$$

we see that the first equation is satisfied by any such point (x, y). In this way, we see that the first and last equations are equivalent.

EXAMPLE 2   Find an equation for the set of all points (x, y) such that the sum of its distances from $(3, 0)$ and $(-3, 0)$ is 8 (see Figure 1.50).

SOLUTION

$$\sqrt{(x - 3)^2 + y^2} + \sqrt{(x + 3)^2 + y^2} = 8$$

$$\sqrt{(x - 3)^2 + y^2} = 8 - \sqrt{(x + 3)^2 + y^2}$$

$$x^2 - 6x + 9 + y^2 = 64 - 16\sqrt{(x + 3)^2 + y^2} + x^2 + 6x + 9 + y^2$$

$$16\sqrt{(x + 3)^2 + y^2} = 64 + 12x$$

$$4\sqrt{(x + 3)^2 + y^2} = 16 + 3x$$

$$16[(x + 3)^2 + y^2] = (16 + 3x)^2$$

$$16x^2 + 96x + 144 + 16y^2 = 256 + 96x + 9x^2$$

$$7x^2 + 16y^2 = 112$$

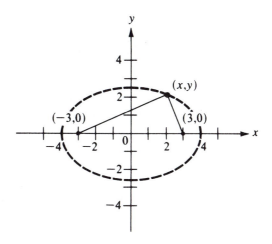

Figure 1.50

Again, the steps are reversible. Although we squared both sides twice, both sides of the equation had to be positive in each instance.

EXAMPLE 3 Find an equation for the set of all points (x, y) that are equidistant from $(-2, 4)$ and the y axis.

SOLUTION The distance from (x, y) to $(-2, 4)$ is

$$d_1 = \sqrt{(x + 2)^2 + (y - 4)^2}.$$

The distance from the point (x, y) to the y axis is the shortest (perpendicular) distance, which is the horizontal distance (see Figure 1.51). As we noted on page 4, this is the absolute value of the difference between the x coordinates of the point (x, y) and the point $(0, y)$,

$$d_2 = |x - 0| = |x|.$$

Since $d_1 = d_2$, we have

$$\sqrt{(x + 2)^2 + (y - 4)^2} = |x|$$
$$(x + 2)^2 + (y - 4)^2 = x^2$$
$$x^2 + 4x + 4 + y^2 - 8y + 16 = x^2$$
$$y^2 + 4x - 8y + 20 = 0$$

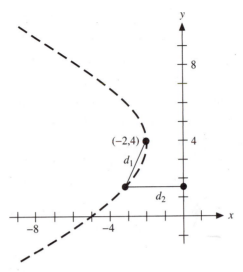

Figure 1.51

PROBLEMS

Find an equation for the set of all points (x, y) satisfying the given conditions.

A
1. It is equidistant from $(5, 8)$ and $(-2, 4)$.
2. It is equidistant from $(2, 3)$ and $(-4, 1)$.
3. Its distance from $(5, 8)$ is 3.
4. Its distance from $(3, 1)$ is 4.

B
5. It is on the line having slope 2 and containing the point $(3, -2)$.
6. It is on the line having slope -1 and containing the point $(-2, 5)$.
7. It is on the line containing $(3, -2)$ and $(5, 3)$.
8. It is on the line containing $(4, 2)$ and $(2, -1)$.
9. Its distance from $(0, 0)$ is three times its distance from $(4, 0)$.
10. Its distance from $(2, 5)$ is twice its distance from $(-3, 1)$.
11. It is the vertex of a right triangle with hypotenuse the segment from $(2, 5)$ to $(-1, 4)$.
12. It is the vertex of a right triangle with hypotenuse the segment from $(3, -2)$ to $(1, 4)$.
13. It is equidistant from $(4, 0)$ and the y axis.
14. It is equidistant from $(2, 3)$ and the x axis.
15. It is twice as far from $(3, 0)$ as it is from the y axis.

16. It is twice as far from the y axis as it is from $(3, 0)$.

17. The sum of its distances from $(0, 4)$ and $(0, -4)$ is 10.

18. The sum of its distances from $(4, -2)$ and $(2, 5)$ is 10.

19. The sum of the squares of its distances from $(3, 0)$ and $(-3, 0)$ is 50.

20. The sum of the squares of its distances from $(4, -2)$ and $(2, 5)$ is 20.

21. The difference of its distances from $(3, 0)$ and $(-3, 0)$ is 2.

22. The difference of its distances from $(4, 2)$ and $(1, -3)$ is 4.

23. The product of its distances from the coordinate axes is 4.

24. The product of its distances from $(4, 0)$ and the y axis is 4.

C 25. It is equidistant from the origin and the line $x + y + 1 = 0$. *Hint:* The distance from the point (x_1, y_1) to the line $Ax + By + C = 0$ is

$$d = \frac{|Ax_1 + By_1 + C|}{\sqrt{A^2 + B^2}}$$

REVIEW PROBLEMS

A 1. Find the point of intersection of $2x + y = 5$ and $x - 3y = 7$.

2. Use distances to determine whether or not the three points $(1, 5)$, $(-2, -1)$, and $(4, 10)$ are collinear. Check your work by using slopes.

3. Find the lengths of the medians of the triangle with vertices $(-3, 4)$, $(5, 5)$, and $(3, -2)$.

4. Use distances to determine whether or not the points $(1, 6)$, $(5, 3)$, and $(3, 1)$ are the vertices of a right triangle. Check your work by using slopes.

5. Determine x so that $(x, 1)$ is on the line joining $(0, 4)$ and $(4, -2)$.

6. Line l_1 contains the points $(4, 7)$ and $(2, 3)$, while l_2 contains $(5, 6)$ and $(-3, 4)$. Are l_1 and l_2 parallel, perpendicular, coincident, or none of these?

7. Find the slopes of the altitudes of the triangle with vertices $(-2, 4)$, $(3, 3)$, and $(-5, -2)$.

B 8. Find the points of trisection of the segment joining $(2, -5)$ and $(-3, 7)$.

9. Find an equation of the perpendicular bisector of the segment joining $(5, -3)$ and $(-1, 1)$. [*Hint:* What is the relationship between a point on the desired line and the two given points?]

10. Find the points of intersection of $x - 7y + 2 = 0$ and $x^2 + y^2 - 4x + 6y - 12 = 0$. Sketch.

11. The point $(5, -2)$ is at a distance $\sqrt{13}$ from the midpoint of the segment joining $(5, y)$ and $(-1, 1)$. Find y.

12. Prove analytically that the lines joining the midpoints of adjacent sides of a quadrilateral form a parallelogram.

13. Find the point of intersection of the medians of the triangle with vertices $(4, -3)$, $(-2, 1)$, and $(0, 5)$. (See Problem 26, Page 19.)

14. Find an equation for the set of all points (x, y) such that it is equidistant from $(0, 1)$ and the x axis.

C 15. Find the center of the circle circumscribed about the triangle with vertices $(-1, 1)$, $(6, 2)$, and $(7, -5)$.

16. Two vertices of an equilateral triangle are $(a, -a)$ and $(-a, a)$. Find the third.

17. A square has all its vertices in the first quadrant and one of its sides joins $(3, 1)$ and $(6, 3)$. Find the other two vertices.

18. A parallelogram has three vertices $(3, 4)$, $(6, 3)$, and $(1, 0)$ and the fourth vertex in the first quadrant. Find the fourth vertex.

19. Find an equation for the set of all points (x, y) such that the angle from the x axis to the line joining (x, y) and the origin equals y.

Chapter *2*   *Vectors in the Plane*

DIRECTED LINE SEGMENTS AND VECTORS

Since quantities such as force, velocity, and acceleration have direction as well as magnitude, it is convenient to represent them geometrically. To do so we use the concept of vectors, which have both magnitude and direction. Not only are vectors important in physics and engineering, but their use can considerably simplify geometric problems, especially in solid analytic geometry. One reason vectors are so useful is the wide range of interpretations they may be given. Since we are interested mainly in the geometric applications, vectors will be introduced geometrically by means of directed line segments.

Suppose A and B are points (not necessarily different) in space. The directed line segment from A to B is represented by $\overrightarrow{AB}$; B is called the **head** and A the **tail** of this segment. Two directed line segments $\overrightarrow{AB}$ and $\overrightarrow{CD}$ are **equivalent**, $\overrightarrow{AB} \equiv \overrightarrow{CD}$, (1) if both are of length zero or (2) if both have the same positive length, both lie on the same or parallel lines, and both are directed in the same way (see Figure 2.1, in which $\overrightarrow{AB} \equiv \overrightarrow{CD}$ and $\overrightarrow{EF} \equiv \overrightarrow{GH}$). With this information, we can easily prove the following theorem.

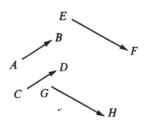

Figure 2.1

THEOREM 2.1
   (a) $\overrightarrow{AB} \equiv \overrightarrow{AB}$ *for any directed line segment* $\overrightarrow{AB}$.
   (b) *If* $\overrightarrow{AB} \equiv \overrightarrow{CD}$, *then* $\overrightarrow{CD} \equiv \overrightarrow{AB}$.
   (c) *If* $\overrightarrow{AB} \equiv \overrightarrow{CD}$ *and* $\overrightarrow{CD} \equiv \overrightarrow{EF}$, *then* $\overrightarrow{AB} \equiv \overrightarrow{EF}$.

Now let us choose an arbitrary directed line segment $\overrightarrow{AB}$. Let M_1 be the set of all directed line segments equivalent to $\overrightarrow{AB}$. Now let us choose another segment $\overrightarrow{CD}$ not in M_1 and let M_2 be the set of all directed line segments equivalent to $\overrightarrow{CD}$. Proceeding in this way, we can partition the set of all directed line segments into a collection of subsets, no two of which have any element in common. These subsets are what we call **vectors**. Thus a vector is a certain set of mutually equivalent directed line segments.

DEFINITION
*The set of all directed line segments equivalent to a given directed line segment is a **vector** v. Any member of that set is a **representative** of v. The set of all directed line segments equivalent to one of length zero is called the **zero vector** 0.*

It might be noted that a vector has magnitude (length) and direction, but not position. Any representative of a given vector has not only magnitude and direction but also position. Let us now consider how vectors may be combined.

DEFINITION
*Suppose u and v are vectors. Let $\overrightarrow{AB}$ be a representative of u. Let $\overrightarrow{BC}$ be that representative of v with tail at B. The **sum u + v** of u and v is the vector w, having $\overrightarrow{AC}$ as a representative.*

This is represented geometrically in Figure 2.2. Since the sum of two vectors is given in terms of representatives of those vectors, the question remains, "Is the sum well defined—that is, is it independent of the representatives used?" Theorem 2.1 and the congruence of triangles easily show that the sum is well defined.

It might be noted that this definition is equivalent to the well-known parallelogram law for the addition of vectors (see Figure 2.2). Let us observe that the figures given here represent vectors graphically by means of representative directed line segments. In Figure 2.2, the vector **u** is represented by two equivalent directed line segments, both of which are labeled **u**.

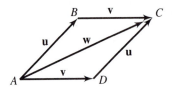

Figure 2.2

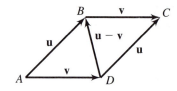

Figure 2.3

We can now use the sum of two vectors in order to define the difference.

DEFINITION   *If **u** and **v** are vectors, then **u** − **v** is the vector **w** such that **u** = **v** + **w**.*

Geometrically, **u** − **v** can again be given by a parallelogram law; but **u** − **v** is represented by the other diagonal. This is shown in Figure 2.3. Note that the direction on **u** − **v** is from the head of **v** to the head of **u**. It is easily verified that **u** − **v** is a vector **w** such that **u** = **v** + w.

When dealing with vectors, we use the term *scalar* for a real number. A scalar has magnitude but not direction. In order to distinguish a scalar from a vector, we use **boldface roman type** for vectors and *lightface italic* for scalars. This use of boldface type for vectors is generally followed in printed works; however, in handwritten work, a vector is usually written with an arrow over it.

DEFINITION   *If **v** is a vector, then |**v**| is the length of any representative of **v**. It is called the **absolute value**, or **length**, of **v**.*

Note that the absolute value of a vector is not a vector, but a scalar.

DEFINITION   *If k is a scalar and **v** a vector, then k**v** is a vector whose length is |k| |**v**| and whose direction is the same as or opposite to the direction of **v**, according to whether k is positive or negative. It is called a **scalar multiple** of **v**.*

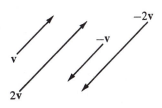

Figure 2.4

Figure 2.4 gives several examples of scalar multiples of the vector **v**.

Let us take note of the fact that we are not adding and multiplying ordinary numbers; thus it is not obvious that the rules of ordinary arithmetic hold—they must be proved from the given definitions.

THEOREM 2.2   *The following properties hold for arbitrary vectors **u**, **v**, and **w** and arbitrary scalars a and b.*

(a) **u** + **v** = **v** + **u**

(b) **u** + (**v** + **w**) = (**u** + **v**) + **w**

(c) (ab)**v** = a(b**v**)

(d) (a + b)**v** = a**v** + b**v**

(e) **v** + **0** = **v**

(f) 0**v** = **0**

(g) a**0** = **0**

(h) |a**v**| = |a| |**v**|

(i) |**u** + **v**| ≤ |**u**| + |**v**|

(j) a(**u** + **v**) = a**u** + a**v**

The proofs of these properties are quite simple. For example, Figure 2.2 can be used to prove (a). From triangle ABC and the definition of the sum of two vectors, it follows that $\overrightarrow{AC}$ is a representative of $\mathbf{u} + \mathbf{v}$. But triangle ADC shows that $\overrightarrow{AC}$ is also a representative of $\mathbf{v} + \mathbf{u}$. Thus $\mathbf{u} + \mathbf{v} = \mathbf{v} + \mathbf{u}$.

The remaining proofs are left for the student (see Problem 44). Be careful not to confuse the scalar 0 with the zero vector **0**.

If we form the scalar multiple of the nonzero vector $\mathbf{v}$ and scalar $1/|\mathbf{v}|$, the result is easily seen to be the *unit vector* (that is, the vector of length 1) in the direction of $\mathbf{v}$. It is usually written

$$\frac{\mathbf{v}}{|\mathbf{v}|}.$$

Of special interest are the unit vectors along the axes.

DEFINITION *If $O = (0, 0)$, $X = (1, 0)$, and $Y = (0, 1)$, then the vectors represented by $\overrightarrow{OX}$ and $\overrightarrow{OY}$ are denoted by* **i** *and* **j**, *respectively, and are called* **basis vectors**.

THEOREM 2.3 *Every vector in the xy plane can be written in the form*

$$a\mathbf{i} + b\mathbf{j}$$

in one and only one way. The numbers a and b are called the **components** *of the vector.*

PROOF Suppose we have a vector **v** in the plane. Let us consider the representative of **v** with its tail at the origin O (see Figure 2.5). The head is at $P = (a, b)$. Let us project P onto both axes, giving points $A = (a, 0)$ and $B = (0, b)$. Since $\overrightarrow{OA}$ represents a vector of length $|a|$ that is either in the direction of **i** or in the opposite direction, depending upon whether a is positive or negative, it represents $a\mathbf{i}$. Similarly $\overrightarrow{OB}$ represents $b\mathbf{j}$. It is clear that $\mathbf{v} = a\mathbf{i} + b\mathbf{j}$.

Since the point P can be represented in rectangular coordinates by a pair (a, b) of numbers in one and only one way, the vector **v** has one and only one representation in component form. ∎

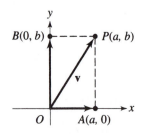

Figure 2.5

It might be noted that this implies that a pair of numbers a and b defines a vector in the plane. Thus the vector $a\mathbf{i} + b\mathbf{j}$ is sometimes represented by the ordered pair (a, b). In fact, some authors define a vector as an ordered pair of numbers and derive their geometric properties from this. It might be argued that the point (a, b) and the vector (a, b) are easily confused. However, since the vector (a, b) is represented by a directed line segment from the origin to the point (a, b), we may use the point as an interpretation of the vector. One of the reasons that vectors are so useful is their wide range of interpretations.

THEOREM 2.4   *If $\overrightarrow{AB}$, where $A = (x_1, y_1)$ and $B = (x_2, y_2)$, represents a vector $\mathbf{v}$ in the xy plane, then*

$$\mathbf{v} = (x_2 - x_1)\mathbf{i} + (y_2 - y_1)\mathbf{j}.$$

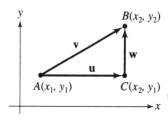

Figure 2.6

PROOF   Let C be the point (x_2, y_1) (see Figure 2.6). Then

$$\mathbf{v} = \mathbf{u} + \mathbf{w},$$

where $\mathbf{u}$ is represented by $\overrightarrow{AC}$ and $\mathbf{w}$ by $\overrightarrow{CB}$. Since $\mathbf{u}$ is of length $|x_2 - x_1|$ and is in either the same or the opposite direction as $\mathbf{i}$, depending upon whether $x_2 - x_1$ is positive or negative, it follows that $\mathbf{u} = (x_2 - x_1)\mathbf{i}$. Similarly $\mathbf{w} = (y_2 - y_1)\mathbf{j}$.

Thus

$$\mathbf{v} = (x_2 - x_1)\mathbf{i} + (y_2 - y_1)\mathbf{j}. \quad \blacksquare$$

EXAMPLE 1   Find the vector $\mathbf{v}$ in the plane represented by the directed line segment from $(3, -2)$ to $(-1, 5)$. Sketch $\mathbf{v}$.

SOLUTION
$$\begin{aligned}
\mathbf{v} &= (x_2 - x_1)\mathbf{i} + (y_2 - y_1)\mathbf{j} \\
&= (-1 - 3)\mathbf{i} + (5 + 2)\mathbf{j} \\
&= -4\mathbf{i} + 7\mathbf{j}
\end{aligned}$$

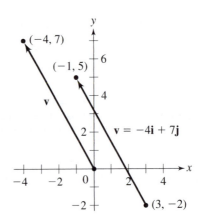

Figure 2.7

The vector **v** may now be sketched as a directed line segment from $(3, -2)$ to $(-1, 5)$ or from $(0, 0)$ to $(-4, 7)$ as shown in Figure 2.7. Of course any other parallel displacement of either of these directed line segments is an equally valid representation of **v**.

EXAMPLE 2   Find the unit vector in the direction of $\mathbf{v} = 3\mathbf{i} - \mathbf{j}$.

SOLUTION

$$|\mathbf{v}| = \sqrt{3^2 + (-1)^2} = \sqrt{10}$$

$$\mathbf{u} = \frac{\mathbf{v}}{|\mathbf{v}|} = \frac{3\mathbf{i} - \mathbf{j}}{\sqrt{10}} = \frac{3}{\sqrt{10}}\mathbf{i} - \frac{1}{\sqrt{10}}\mathbf{j}$$

THEOREM 2.5
(a) $(a_1\mathbf{i} + b_1\mathbf{j}) + (a_2\mathbf{i} + b_2\mathbf{j}) = (a_1 + a_2)\mathbf{i} + (b_1 + b_2)\mathbf{j}$
(b) $(a_1\mathbf{i} + b_1\mathbf{j}) - (a_2\mathbf{i} + b_2\mathbf{j}) = (a_1 - a_2)\mathbf{i} + (b_1 - b_2)\mathbf{j}$
(c) $d(a\mathbf{i} + b\mathbf{j}) = da\mathbf{i} + db\mathbf{j}$
(d) $|a\mathbf{i} + b\mathbf{j}| = \sqrt{a^2 + b^2}$

The proof is left to the student.

EXAMPLE 3   If $\mathbf{u} = 3\mathbf{i} - 2\mathbf{j}$ and $\mathbf{v} = -\mathbf{i} + 5\mathbf{j}$, find $\mathbf{u} + \mathbf{v}$, $\mathbf{u} - \mathbf{v}$, $|\mathbf{v}|$, and $2\mathbf{v}$.

$$\mathbf{u} + \mathbf{v} = (3 - 1)\mathbf{i} + (-2 + 5)\mathbf{j} = 2\mathbf{i} + 3\mathbf{j}$$
$$\mathbf{u} - \mathbf{v} = (3 - (-1))\mathbf{i} + (-2 - 5)\mathbf{j} = 4\mathbf{i} - 7\mathbf{j}$$
$$|\mathbf{v}| = \sqrt{(-1)^2 + 5^2} = \sqrt{26}$$
$$2\mathbf{v} = 2(-\mathbf{i} + 5\mathbf{j}) = -2\mathbf{i} + 10\mathbf{j}$$

EXAMPLE 4 Using the vectors of Example 3, give graphical representations of $\mathbf{u}$, $\mathbf{v}$, $\mathbf{u} + \mathbf{v}$, $\mathbf{u} - \mathbf{v}$, and $2\mathbf{u}$.

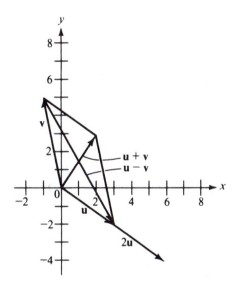

Figure 2.8

SOLUTION $\mathbf{u} = 3\mathbf{i} - 2\mathbf{j}$ is represented by the directed line segment from the origin to $(3, -2)$ (see Figure 2.8). Similarly, $\mathbf{v} = -\mathbf{i} + 5\mathbf{j}$ is represented by the directed line segment from the origin to the point $(-1, 5)$. Now we complete the parallelogram with $\mathbf{u}$ and $\mathbf{v}$ as two sides. The two diagonals are $\mathbf{u} + \mathbf{v}$ and $\mathbf{u} - \mathbf{v}$, $\mathbf{u} + \mathbf{v}$ being directed from the origin to the opposite vertex and $\mathbf{u} - \mathbf{v}$ from the head of $\mathbf{v}$ to the head of $\mathbf{u}$. Finally, $2\mathbf{u}$ is the directed line segment from the origin, in the same direction as $\mathbf{u}$, but twice as long as $\mathbf{u}$.

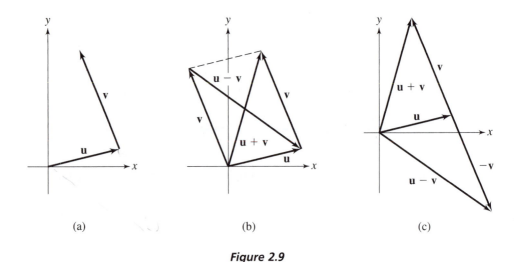

(a)   (b)   (c)

Figure 2.9

EXAMPLE 5   Given the vectors of Figure 2.9a, sketch $\mathbf{u} + \mathbf{v}$ and $\mathbf{u} - \mathbf{v}$.

SOLUTION   One way to handle this problem is to take a parallel displacement of $\mathbf{v}$ with its tail at the origin, and then use the parallelogram law. This was done in Figure 2.9b. However, let us first note that when the head of $\mathbf{u}$ and the tail of $\mathbf{v}$ correspond, then $\mathbf{u} + \mathbf{v}$ goes from the tail of $\mathbf{u}$ to the head of $\mathbf{v}$. Thus there is no need to draw the entire parallelogram; we can simply draw $\mathbf{u} + \mathbf{v}$ directly. Finally, it is easily seen that $\mathbf{u} - \mathbf{v} = \mathbf{u} + (-\mathbf{v})$. But $-\mathbf{v}$ is a vector whose length is the same as the length of $\mathbf{v}$ and whose direction is opposite that of $\mathbf{v}$. By sketching $-\mathbf{v}$ with the same tail as $\mathbf{v}$, we can then use $\mathbf{u} - \mathbf{v} = \mathbf{u} + (-\mathbf{v})$ to get the result we want as shown in Figure 2.9c.

It might be noted that the two different methods gave $\mathbf{u} - \mathbf{v}$ in two different positions. Nevertheless, they have the same length and direction; they are both $\mathbf{u} - \mathbf{v}$. It is sometimes convenient to have the tail of $\mathbf{u} - \mathbf{v}$ at the origin, and the second method above puts it in that position directly.

EXAMPLE 6   Find the endpoint A of $\mathbf{v}$, represented by $\overrightarrow{AB}$, if $\mathbf{v} = 4\mathbf{i} - 2\mathbf{j}$ and $B = (-2, 1)$.

SOLUTION   Since $\mathbf{v} = (x_2 - x_1)\mathbf{i} + (y_2 - y_1)\mathbf{j}$, where $A = (x_1, y_1)$ and $B = (x_2, y_2)$, it follows that

$$4 = -2 - x_1 \qquad -2 = 1 - y_1$$
$$x_1 = -6 \qquad\qquad y_1 = 3.$$

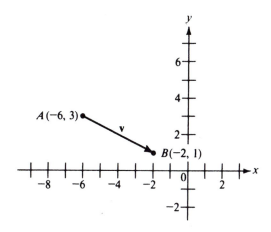

Figure 2.10

Thus $A = (-6, 3)$. See Figure 2.10.

EXAMPLE 7   If $\mathbf{v} = \overrightarrow{AB} = 4\mathbf{i} + 6\mathbf{j}$ and the midpoint of AB is $(3, -1)$, find A and B.

SOLUTION   Let $A = (x_1, y_1)$ and $B = (x_2, y_2)$. Since

$$\mathbf{v} = (x_2 - x_1)\mathbf{i} + (y_2 - y_1)\mathbf{j},$$
$$x_2 - x_1 = 4 \qquad \text{and} \qquad y_2 - y_1 = 6$$

Since the midpoint of AB is $\left(\dfrac{x_1 + x_2}{2}, \dfrac{y_1 + y_2}{2}\right)$,

$$\frac{x_1 + x_2}{2} = 3 \qquad \text{and} \qquad \frac{y_1 + y_2}{2} = -1$$

or

$$x_1 + x_2 = 6 \qquad \text{and} \qquad y_1 + y_2 = -2.$$

Solving

$$x_2 - x_1 = 4 \qquad \qquad y_2 - y_1 = 6$$
$$\text{and}$$
$$x_2 + x_1 = 6 \qquad \qquad y_2 + y_1 = -2,$$

we have $x_1 = 1$, $x_2 = 5$ and $y_1 = -4$, $y_2 = 2$. Thus $A = (1, -4)$ and $B\,(5, 2)$.

PROBLEMS

A   *In Problems 1–8, give in component form the vector* **v** *that is represented by* $\overrightarrow{AB}$.

1. $A = (4, 3), B = (-2, 1)$ **2.** $A = (2, 5), B = (0, 1)$

3. $A = (-3, 2), B = (4, 3)$ **4.** $A = (-2, 4), B = (0, 4)$

5. $A = (1, -2), B = (0, 3)$ **6.** $A = (0, 2), B = (1, 4)$

7. $A = (-3, 2), B = (1, -1)$ **8.** $A = (4, -3), B = (0, 5)$

In Problems 9–16, give the unit vector in the direction of **v**.

9. $\mathbf{v} = 3\mathbf{i} - \mathbf{j}$ **10.** $\mathbf{v} = 2\mathbf{i} + 4\mathbf{j}$

11. $\mathbf{v} = -\mathbf{i} + 2\mathbf{j}$ **12.** $\mathbf{v} = 3\mathbf{j}$

13. $\mathbf{v} = \mathbf{i} + 2\mathbf{j}$ **14.** $\mathbf{v} = 3\mathbf{i} - \mathbf{j}$

15. $\mathbf{v} = -4\mathbf{i} + 3\mathbf{j}$ **16.** $\mathbf{v} = 4\mathbf{j}$

In Problems 17–24, find the endpoints of the representative $\overrightarrow{AB}$ *of* **v** *from the given information.*

17. $\mathbf{v} = 3\mathbf{i} - \mathbf{j}, A = (1, 4)$ **18.** $\mathbf{v} = 2\mathbf{i} + 3\mathbf{j}, A = (-1, 3)$

19. $\mathbf{v} = -\mathbf{i} + 2\mathbf{j}, B = (4, 2)$ **20.** $\mathbf{v} = 2\mathbf{i} - 4\mathbf{j}, B = (0, 3)$

21. $\mathbf{v} = \mathbf{i} - \mathbf{j}, A = (5, 1)$ **22.** $\mathbf{v} = 2\mathbf{i} + \mathbf{j}, A = (-2, 0)$

23. $\mathbf{v} = 3\mathbf{i} - \mathbf{j}, B = (4, 2)$ **24.** $\mathbf{v} = 2\mathbf{i} + 2\mathbf{j}, B = (1, 1)$

In Problems 25–32, find **u** + **v** *and* **u** − **v**. *Draw a diagram showing* **u**, **v**, **u** + **v**, *and* **u** − **v**.

25. $\mathbf{u} = 3\mathbf{i} - \mathbf{j}, \mathbf{v} = \mathbf{i} + 2\mathbf{j}$ **26.** $\mathbf{u} = 2\mathbf{i} + 3\mathbf{j}, \mathbf{v} = 2\mathbf{i} - \mathbf{j}$

27. $\mathbf{u} = \mathbf{i} - \mathbf{j}, \mathbf{v} = 2\mathbf{i} + 2\mathbf{j}$ **28.** $\mathbf{u} = 3\mathbf{i} + \mathbf{j}, \mathbf{v} = 2\mathbf{i}$

29. $\mathbf{u} = \mathbf{i} - 3\mathbf{j}, \mathbf{v} = 2\mathbf{i} + 4\mathbf{j}$ **30.** $\mathbf{u} = 4\mathbf{i} + \mathbf{j}, \mathbf{v} = \mathbf{i} + 2\mathbf{j}$

31. $\mathbf{u} = 3\mathbf{i} - \mathbf{j}, \mathbf{v} = 2\mathbf{i} - 2\mathbf{j}$ **32.** $\mathbf{u} = \mathbf{i} - \mathbf{j}, \mathbf{v} = 2\mathbf{i} + \mathbf{j}$

33. Given the vectors of Figure 2.11, sketch **u** + **v**.

34. Given the vectors of Figure 2.11, sketch **u** − **v**.

B   *In Problems 35–38, find the endpoints of* $\mathbf{v} = \overrightarrow{AB}$ *from the given information.*

35. $\mathbf{v} = 3\mathbf{i} + 5\mathbf{j}$, (4, 1) is the midpoint of AB

36. $\mathbf{v} = 4\mathbf{i} - 6\mathbf{j}$, (2, 5) is the midpoint of AB

37. $\mathbf{v} = \mathbf{i} + \mathbf{j}$, (2, 0) is the midpoint of AB

38. $\mathbf{v} = 2\mathbf{i} - \mathbf{j}$, (0, −2) is the midpoint of AB

In Problems 39–42, find the vector indicated and sketch **u**, **v**, *and the new vector.*

39. $\mathbf{u} = 5\mathbf{i} + \mathbf{j}, \mathbf{v} = -2\mathbf{i} + 3\mathbf{j}$; find $2\mathbf{u} + \mathbf{v}$.

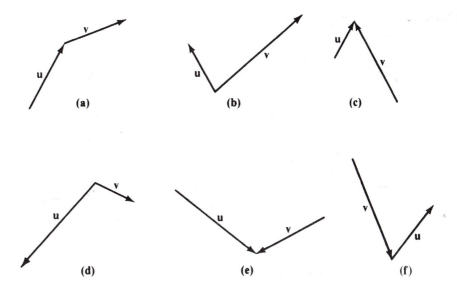

Figure 2.11

40. $\mathbf{u} = -2\mathbf{i} + 5\mathbf{j}$, $\mathbf{v} = 3\mathbf{i} - \mathbf{j}$; find $\mathbf{u} - 3\mathbf{v}$.

41. $\mathbf{u} = 3\mathbf{i} + \mathbf{j}$, $\mathbf{v} = \mathbf{i} - \mathbf{j}$; find $-2\mathbf{u} + \mathbf{v}$.

42. $\mathbf{u} = -\mathbf{i} + 2\mathbf{j}$, $\mathbf{v} = 3\mathbf{i} - 4\mathbf{j}$; find $3\mathbf{u} + 2\mathbf{v}$.

C **43.** Prove Theorem 2.1.

44. Prove Theorem 2.2.

45. Prove Theorem 2.5.

46. Show that the sum of two vectors is well defined (see the paragraph following the definition of the sum).

47. Show that $a\mathbf{i} + b\mathbf{j} = c\mathbf{i} + d\mathbf{j}$ if and only if $a = c$ and $b = d$.

48. Show that if neither $\mathbf{u}$ nor $\mathbf{v}$ is a scalar multiple of the other and $a\mathbf{u} + b\mathbf{v} = \mathbf{0}$, then $a = b = 0$.

2.2

THE DOT PRODUCT

We have considered the sums and differences of pairs of vectors, but the only product considered so far is the scalar multiple—the product of a scalar and a vector. We shall now consider the product of two vectors. There are two different product operations for a pair of vectors, the dot product and the cross product. We shall take up the dot product in this section but defer a discussion of the cross product to Chapter 9 (since it requires three dimensions). In this case, we shall give a purely algebraic definition of the dot product. Its geometric significance will be considered later.

DEFINITION   If $\mathbf{u} = a_1\mathbf{i} + b_1\mathbf{j}$ and $v = a_2\mathbf{i} + b_2\mathbf{j}$, then the **dot product (scalar product, inner product)** of $\mathbf{u}$ and $\mathbf{v}$ is

$$\mathbf{u} \cdot \mathbf{v} = a_1 a_2 + b_1 b_2.$$

Note that the dot product of two vectors is *not* another vector; it is a scalar.

EXAMPLE 1   Find the dot product of $\mathbf{u} = 3\mathbf{i} - 2\mathbf{j}$ and $\mathbf{v} = \mathbf{i} + \mathbf{j}$.

SOLUTION
$$\mathbf{u} \cdot \mathbf{v} = (3)(1) + (-2)(1) = 1$$

THEOREM 2.6   If $\mathbf{u}$, $\mathbf{v}$, and $\mathbf{w}$ are vectors and k is a scalar, then
(a) $\mathbf{u} \cdot \mathbf{v} = \mathbf{v} \cdot \mathbf{u}$,
(b) $(\mathbf{u} + \mathbf{v}) \cdot \mathbf{w} = \mathbf{u} \cdot \mathbf{w} + \mathbf{v} \cdot \mathbf{w}$,
(c) $k(\mathbf{u} \cdot \mathbf{v}) = (k\mathbf{u}) \cdot \mathbf{v} = \mathbf{u} \cdot (k\mathbf{v})$,
(d) $\mathbf{0} \cdot \mathbf{u} = 0$,
(e) $\mathbf{u} \cdot \mathbf{u} = |\mathbf{u}|^2$.

PROOF   (a) Suppose $\mathbf{u} = a_1\mathbf{i} + b_1\mathbf{j}$ and $\mathbf{v} = a_2\mathbf{i} + b_2\mathbf{j}$. Then $\mathbf{u} \cdot \mathbf{v} = a_1 a_2 + b_1 b_2$ and $\mathbf{v} \cdot \mathbf{u} = a_2 a_1 + b_2 b_1$, which are clearly equal. ∎

The remaining statements can be proved in much the same way. These proofs are left to the student (see Problems 43–46).

Several observations may be made here. Note that in (c) we have two vectors and a scalar. This statement is not true if the scalar is replaced by a third vector. Since the dot product of two vectors is a scalar, there is some question concerning the meaning of $\mathbf{w} \cdot \mathbf{u} \cdot \mathbf{v}$. Once we have taken the dot product of two of them, we no longer have a second dot product—we have a scalar multiple. Even if we interpret the second product as a scalar multiple, it can easily be shown that $\mathbf{w}(\mathbf{u} \cdot \mathbf{v})$ and $(\mathbf{w} \cdot \mathbf{u})\mathbf{v}$ are not identical. If $\mathbf{w} = \mathbf{i}$, $\mathbf{u} = \mathbf{j}$, and $\mathbf{v} = \mathbf{i} + \mathbf{j}$, then $\mathbf{w}(\mathbf{u} \cdot \mathbf{v}) = \mathbf{i}$ while $(\mathbf{w} \cdot \mathbf{u})\mathbf{v} = \mathbf{0}$.

Do not confuse the statements of Theorems 2.6(d) and 2.2(f). In the one case we have the dot product of two vectors $\mathbf{0}$ and $\mathbf{v}$, giving the scalar 0,

$$\mathbf{0} \cdot \mathbf{v} = 0.$$

In the other case we have the scalar multiple of the scalar 0 and the vector $\mathbf{v}$ to give the vector $\mathbf{0}$,

$$0\mathbf{v} = \mathbf{0}.$$

Finally we note that we can use Theorem 2.6(a), (b), and (c) to show that vector expressions can be multiplied (dot product) in much the same way as polynomials. For example,

$$(k_1\mathbf{v}_1 + k_2\mathbf{v}_2) \cdot (k_3\mathbf{v}_3 + k_4\mathbf{v}_4)$$
$$= (k_1\mathbf{v}_1) \cdot (k_3\mathbf{v}_3 + k_4\mathbf{v}_4) + (k_2\mathbf{v}_2) \cdot (k_3\mathbf{v}_3 + k_4\mathbf{v}_4)$$
$$= (k_1\mathbf{v}_1) \cdot (k_3\mathbf{v}_3) + (k_1\mathbf{v}_1) \cdot (k_4\mathbf{v}_4) + (k_2\mathbf{v}_2) \cdot (k_3\mathbf{v}_3) + (k_2\mathbf{v}_2) \cdot (k_4\mathbf{v}_4)$$
$$= k_1k_3\mathbf{v}_1 \cdot \mathbf{v}_3 + k_1k_4\mathbf{v}_1 \cdot \mathbf{v}_4 + k_2k_3\mathbf{v}_2 \cdot \mathbf{v}_3 + k_2k_4\mathbf{v}_2 \cdot \mathbf{v}_4.$$

Now let us use these properties to consider the geometric significance of the dot product. First we consider the angle between two vectors.

DEFINITION   *The **angle** between two nonzero vectors* $\mathbf{u}$ *and* $\mathbf{v}$ *is the smaller angle between the representatives of* $\mathbf{u}$ *and* $\mathbf{v}$ *having their tails at the origin.*

In Figure 2.12, we see that there are two angles, θ_1 and θ_2, formed by the vectors $\mathbf{u}$ and $\mathbf{v}$. By the above definition, θ_1 is taken to be the angle between them.

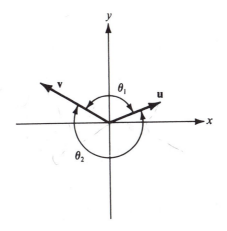

Figure 2.12

Note that the angle between two vectors is nondirected. That is, we do not consider the angle from one vector to another, which would imply a preferred direction, but rather the angle between two vectors. If θ is the angle between two vectors, then

$$0° \le \theta \le 180°.$$

THEOREM 2.7   *If* $\mathbf{u}$ *and* $\mathbf{v}$ *are vectors and* θ *is the angle between them, then*

$$\mathbf{u} \cdot \mathbf{v} = |\mathbf{u}| \cdot |\mathbf{v}| \cos \theta \qquad \cos \theta = \frac{\mathbf{u} \cdot \mathbf{v}}{|\mathbf{u}||\mathbf{v}|}$$

and

$$\mathbf{v} \cdot \mathbf{v} = |\mathbf{v}|^2.$$

PROOF   By the law of cosines (see Figure 2.13),

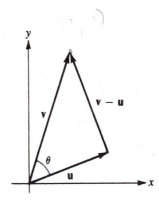

Figure 2.13

$$|v - u|^2 = |u|^2 + |v|^2 - 2|u|\,|v|\cos\theta.$$

But

$$(v - u) \cdot (v - u) = v \cdot v + u \cdot u - 2u \cdot v$$

or

$$|v - u|^2 = |v|^2 + |u|^2 - 2u \cdot v.$$

Comparing this with our first expression for $|v - u|^2$, we conclude that

$$u \cdot v = |u|\,|v|\cos\theta. \quad \blacksquare$$

We see that we can use the dot product to find the angle between two vectors.

EXAMPLE 2   Find the cosine of the angle between $u = 3i - 4j$ and $v = 5i + 12j$ (see Figure 2.14).

SOLUTION
$$\cos\theta = \frac{a_1a_2 + b_1b_2}{|u||v|}$$
$$= \frac{(3)(5) + (-4)(12)}{\sqrt{9 + 16}\sqrt{25 + 144}}$$
$$= -\frac{33}{65}$$
$$= -0.5077$$
$$\theta = 120.5°$$

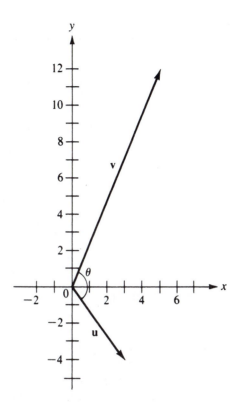

Figure 2.14

EXAMPLE 3   Find $\mathbf{u} \cdot \mathbf{v}$ for the vectors of Figure 2.15.

SOLUTION
$$\mathbf{u} \cdot \mathbf{v} = |\mathbf{u}|\,|\mathbf{v}|\,\cos\theta$$
$$= 6 \cdot 10 \cos 60°$$
$$= 6 \cdot 10 \cdot \frac{1}{2}$$
$$= 30$$

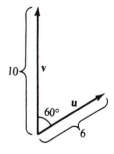

Figure 2.15

EXAMPLE 4   If $\mathbf{u} = \mathbf{i} + 2\mathbf{j}$, $\mathbf{v} = a\mathbf{i} - \mathbf{j}$, and the angle between $\mathbf{u}$ and $\mathbf{v}$ is $\pi/3$, find a.

SOLUTION

$$\cos\theta = \frac{\mathbf{u}\cdot\mathbf{v}}{|\mathbf{u}||\mathbf{v}|}$$

$$\cos\frac{\pi}{3} = \frac{a-2}{\sqrt{5}\sqrt{a^2+1}}$$

$$\frac{1}{2} = \frac{a-2}{\sqrt{5}\sqrt{a^2+1}}$$

$$\sqrt{5}\sqrt{a^2+1} = 2a-4$$

$$5(a^2+1) = 4a^2-16a+16$$

$$a^2+16a-11 = 0$$

$$a = \frac{-16\pm\sqrt{256+44}}{2} = \frac{-16\pm\sqrt{300}}{2} = \frac{-16\pm10\sqrt{3}}{2}$$

$$= -8\pm5\sqrt{3}$$

Since we squared both sides of the equation

$$\sqrt{5}\sqrt{a^2+1} = 2a-4$$

we must check our results. This equation requires that $2a-4>0$ or $a>2$. Since neither value of a in our final result satisfies this inequality, it follows that there is no number a satisfying the conditions of this problem.

We can use the dot product to determine when two vectors are *orthogonal* (perpendicular). As we shall see later, orthogonality of vectors is an important concept.

THEOREM 2.8   *The vectors* $\mathbf{u}$ *and* $\mathbf{v}$ *(not both* $\mathbf{0}$*) are* **orthogonal** (**perpendicular**) *if and only if*

$$\mathbf{u}\cdot\mathbf{v} = \mathbf{0}$$

(the zero vector is taken to be orthogonal to every other vector).

PROOF   If $\mathbf{u}$ and $\mathbf{v}$ are orthogonal, then $\theta = 90°$ and $\cos\theta = 0$. Thus

$$\mathbf{u}\cdot\mathbf{v} = |\mathbf{u}||\mathbf{v}|\cos\theta = 0.$$

Conversely, let us suppose that $\mathbf{u}\cdot\mathbf{v} = 0$. Then

$$|\mathbf{u}||\mathbf{v}|\cos\theta = 0,$$

implying that $|\mathbf{u}| = 0$, $|\mathbf{v}| = 0$, or $\cos\theta = 0$. If $|\mathbf{u}| = 0$ or $|\mathbf{v}| = 0$, then $\mathbf{u} = \mathbf{0}$ or $\mathbf{v} = \mathbf{0}$, respectively. Since the zero vector is orthogonal to every other vector, $\mathbf{u}$ and $\mathbf{v}$ are orthogonal. If $\cos\theta = 0$, $\theta = 90°$ and the vectors are orthogonal. ∎

EXAMPLE 5

Determine whether or not $\mathbf{u} = 2\mathbf{i} - \mathbf{j}$ and $\mathbf{v} = \mathbf{i} + 2\mathbf{j}$ are orthogonal.

SOLUTION

$$\mathbf{u} \cdot \mathbf{v} = (2)(1) + (-1)(2) = 0$$

They are orthogonal, since $\mathbf{u} \cdot \mathbf{v} = 0$.

Another special case of Theorem 2.7 occurs when we have parallel vectors.

DEFINITION

The nonzero vectors $\mathbf{u}$ and $\mathbf{v}$ are parallel if $\mathbf{u} = k\mathbf{v}$ for some scalar k.

If $\mathbf{u}$ and $\mathbf{v}$ are orthogonal, $\theta = 90°$ and $\mathbf{u} \cdot \mathbf{v} = 0$, as we have seen before. If $\mathbf{u}$ and $\mathbf{v}$ are parallel, $\theta = 0°$ or $\theta = 180°$ and $\mathbf{u} \cdot \mathbf{v} = \pm\, |\mathbf{u}|\,|\mathbf{v}|$.

The projection of one vector upon another is determined by the angle between them or the dot product.

DEFINITION

The projection of $\mathbf{u}$ on $\mathbf{v}$ ($\mathbf{v} \neq \mathbf{0}$) is a vector $\mathbf{w}$ such that if $\overrightarrow{AB}$ is a representative of $\mathbf{u}$ and $\overrightarrow{AC}$ is a representative of $\mathbf{v}$, then a representative of $\mathbf{w}$ is a directed line segment $\overrightarrow{AD}$ lying on the line determined by AC with BD perpendicular to that line (see Figure 2.16).

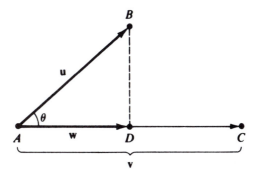

Figure 2.16

The projection of $\mathbf{u}$ on $\mathbf{v}$ is defined in terms of representatives of these vectors. Thus we again have the question of whether or not the projection is well defined. Theorem 2.1 and the congruence of triangles show that the projection of $\mathbf{u}$ on $\mathbf{v}$ is independent of the representatives considered.

THEOREM 2.9

If **w** *is the projection of* **u** *on* **v** *and* θ *is the angle between* **u** *and* **v**, *then*

$$|\mathbf{w}| = \frac{|\mathbf{u} \cdot \mathbf{v}|}{|\mathbf{v}|} \quad \text{and} \quad \mathbf{w} = \left(\frac{\mathbf{u} \cdot \mathbf{v}}{|\mathbf{v}|}\right)\frac{\mathbf{v}}{|\mathbf{v}|} = \frac{\mathbf{u} \cdot \mathbf{v}}{|\mathbf{v}|^2}\mathbf{v}.$$

PROOF

We can see from Figure 2.16 that

$$\begin{aligned}
|\mathbf{w}| &= |\mathbf{u}|\,|\cos\theta| \\
&= \frac{|\mathbf{u}|\,|\mathbf{v}|\,|\cos\theta|}{|\mathbf{v}|} \\
&= \frac{|\mathbf{u} \cdot \mathbf{v}|}{|\mathbf{v}|} \quad \text{(by Theorem 2.7).}
\end{aligned}$$

Now we can easily find **w**. Its length is $|\mathbf{u} \cdot \mathbf{v}|/|\mathbf{v}|$, and its direction is determined by **v**, since the projection is upon **v**. Thus

$$\mathbf{w} = \pm\left(\frac{|\mathbf{u} \cdot \mathbf{v}|}{|\mathbf{v}|}\right)\frac{\mathbf{v}}{|\mathbf{v}|}.$$

Now if the angle θ between **u** and **v** is less than $\pi/2$, $\mathbf{u} \cdot \mathbf{v} > 0$. But **w** and **v** have the same direction in this case and

$$\mathbf{w} = +\left(\frac{|\mathbf{u} \cdot \mathbf{v}|}{|\mathbf{v}|}\right)\frac{\mathbf{v}}{|\mathbf{v}|} = \left(\frac{\mathbf{u} \cdot \mathbf{v}}{|\mathbf{v}|}\right)\frac{\mathbf{v}}{|\mathbf{v}|}.$$

If $\theta > \pi/2$, $\mathbf{u} \cdot \mathbf{v} < 0$ and **w** and **v** have opposite directions. This leads to

$$\mathbf{w} = -\left(\frac{|\mathbf{u} \cdot \mathbf{v}|}{|\mathbf{v}|}\right)\frac{\mathbf{v}}{|\mathbf{v}|} = \left(\frac{\mathbf{u} \cdot \mathbf{v}}{|\mathbf{v}|}\right)\frac{\mathbf{v}}{|\mathbf{v}|}. \quad \blacksquare$$

EXAMPLE 6

Find the projection **w** of $\mathbf{u} = 3\mathbf{i} + \mathbf{j}$ on $\mathbf{v} = 3\mathbf{i} + 4\mathbf{j}$.

SOLUTION

$$\mathbf{u} \cdot \mathbf{v} = (3)(3) + (1)(4) = 13,$$

$$|\mathbf{v}| = \sqrt{3^2 + 4^2} = 5.$$

Thus

$$\begin{aligned}
\mathbf{w} &= \left(\frac{\mathbf{u} \cdot \mathbf{v}}{|\mathbf{v}|}\right)\frac{\mathbf{v}}{|\mathbf{v}|} \\
&= \frac{13}{5}\frac{3\mathbf{i} + 4\mathbf{j}}{5} = \frac{39}{25}\mathbf{i} + \frac{52}{25}\mathbf{j}.
\end{aligned}$$

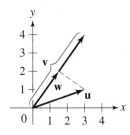

Figure 2.17 gives a graphical representation of **u**, **v**, and **w**.   ***Figure 2.17***

PROBLEMS

A   *In Problems 1–8, find the angle θ between the given vectors.*

1. $u = 3i - j, v = i + 2j$

2. $u = 4i + j, v = i + 2j$

3. $u = -i + 2j, v = 2i + j$

4. $u = i + j, v = 2i - j$

5. $u = 2i - j, v = i + 2j$

6. $u = 3i + 2j, v = i - j$

7. $u = 2i - 2j, v = 4i + j$

8. $u = i + j, v = 2i + 4j$

In Problems 9–16, find **u** · **v** *and indicate whether or not* **u** *and* **v** *are orthogonal.*

9. $u = i - j, v = 2i + j$

10. $u = 2i + j, v = i - 2j$

11. $u = 3i + 2j, v = 2i - j$

12. $u = 2i - 4j, v = 2i + j$

13. $u = i - j, v = 3i + 4j$

14. $u = i + j, v = 2i - 3j$

15. $u = 2i - 3j, v = 3i + j$

16. $u = 4i, v = i + j$

B   *In Problems 17–24, find the projection of* **u** *on* **v**.

17. $u = 2i - j, v = i + j$

18. $u = i - 3j, v = 2i + j$

19. $u = 2i + 4j, v = i - 2j$

20. $u = 4i + j, v = 2i + j$

21. $u = i - j, v = 2i + j$

22. $u = 2i - 3j, v = 3i + 2j$

23. $u = 2i + j, v = 4i - 2j$

24. $u = 3i - j, v = 2i + 2j$

25. Find **u** · **v** for the vectors of Figure 2.18.

26. Find **u** · **v** for the vectors of Figure 2.19.

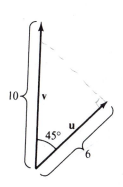

Figure 2.18

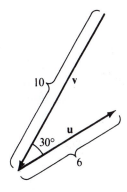

Figure 2.19

In Problems 27–38, determine the value(s) of a so that the given conditions are satisfied.

27. $u = 3i - j, v = i + aj$, **u** and **v** are perpendicular.

28. $u = i + j, v = 3i - aj$, **u** and **v** are perpendicular.

29. $\mathbf{u} = 4\mathbf{i} + \mathbf{j}$, $\mathbf{v} = 2\mathbf{i} + a\mathbf{j}$, $\mathbf{u}$ and $\mathbf{v}$ are perpendicular.

30. $\mathbf{u} = 2\mathbf{i} - \mathbf{j}$, $\mathbf{v} = a\mathbf{i} + \mathbf{j}$, $\mathbf{u}$ and $\mathbf{v}$ are perpendicular.

31. $\mathbf{u} = \mathbf{i} - 2\mathbf{j}$, $\mathbf{v} = a\mathbf{i} + \mathbf{j}$, $\mathbf{u}$ and $\mathbf{v}$ are parallel.

32. $\mathbf{u} = a\mathbf{i} - \mathbf{j}$, $\mathbf{v} = 2\mathbf{i} + a\mathbf{j}$, $\mathbf{u}$ and $\mathbf{v}$ are parallel.

33. $\mathbf{u} = \mathbf{i} + \mathbf{j}$, $\mathbf{v} = a\mathbf{i} - \mathbf{j}$, $\mathbf{u}$ and $\mathbf{v}$ are parallel.

34. $\mathbf{u} = a\mathbf{i} + 3\mathbf{j}$, $\mathbf{v} = 2\mathbf{i} + \mathbf{j}$, $\mathbf{u}$ and $\mathbf{v}$ are parallel.

35. $\mathbf{u} = a\mathbf{i} + 2\mathbf{j}$, $\mathbf{v} = \mathbf{i} - \mathbf{j}$, the angle between $\mathbf{u}$ and $\mathbf{v}$ is $\pi/3$.

36. $\mathbf{u} = 3\mathbf{i} - a\mathbf{j}$, $\mathbf{v} = 2\mathbf{i} + \mathbf{j}$, the angle between $\mathbf{u}$ and $\mathbf{v} = \pi/4$.

37. $\mathbf{u} = 2\mathbf{i} + \mathbf{j}$, $\mathbf{v} = a\mathbf{i} - \mathbf{j}$, the angle between $\mathbf{u}$ and $\mathbf{v}$ is $2\pi/3$.

38. $\mathbf{u} = \mathbf{i} - \mathbf{j}$, $\mathbf{v} = 4\mathbf{i} + a\mathbf{j}$, the angle between $\mathbf{u}$ and $\mathbf{v}$ is $\pi/4$.

In Problems 39–42, let $\mathbf{u}$ *be represented by* $\overrightarrow{AB}$, $\mathbf{v}$ *by* $\overrightarrow{AC}$, *and* $\mathbf{w}$ *by* $\overrightarrow{BC}$. *Find the projections of* $\mathbf{v}$ *and* $\mathbf{w}$ *on* $\mathbf{u}$.

39. $A = (0, 0)$, $B = (1, 4)$, $C = (2, -1)$

40. $A = (2, 3)$, $B = (-3, -1)$, $C = (4, 2)$

41. $A = (4, 1)$, $B = (3, -1)$, $C = (0, 2)$

42. $A = (2,0)$, $B = (5, 5)$, $C = (3, 5)$

C 43. Prove Theorem 2.6(b).

44. Prove Theorem 2.6(c).

45. Prove Theorem 2.6(d).

46. Prove Theorem 2.6(e).

47. Show that $|\mathbf{u} \cdot \mathbf{v}| \leq |\mathbf{u}|\,|\mathbf{v}|$.

2.3

APPLICATIONS OF VECTORS

Let us now consider some applications of vectors. Vector methods can be used in some of the proofs of elementary geometry. Sometimes the resulting proof is much shorter than either an analytic or a synthetic argument.

EXAMPLE 1

Using vector methods, prove that the line joining the midpoints of two sides of a triangle is parallel to and one-half the length of the third side.

SOLUTION

Suppose D and E are the midpoints of AB and BC, respectively. Let us give directions to the line segments involved and consider them to be representatives of vectors as shown in Figure 2.20.

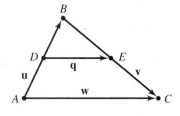

Figure 2.20

Then

$$\mathbf{u} + \mathbf{v} = \mathbf{w},$$

$$\frac{1}{2}\mathbf{u} + \mathbf{q} + \frac{1}{2}\mathbf{v} = \mathbf{w},$$

and

$$\mathbf{q} = \mathbf{w} - \frac{1}{2}(\mathbf{u} + \mathbf{v}) = \mathbf{w} - \frac{1}{2}\mathbf{w} = \frac{1}{2}\mathbf{w}.$$

Since vectors have both magnitude and direction, we have proved that $\mathbf{q}$ has the same direction as $\mathbf{w}$ and that it is half the length of $\mathbf{w}$.

One reason for the brevity of the foregoing solution is that we were interested in both the direction and magnitude of $\overrightarrow{DE}$. Thus a representation by vectors was very efficient. When perpendicularity is involved, we consider the dot product. Since this product is defined in terms of the components, it is sometimes convenient to give component representations of the vectors.

EXAMPLE 2   Prove by vector methods that the diagonals of a rhombus are perpendicular.

SOLUTION   Suppose two sides of the rhombus represent the vectors $\mathbf{u}$ and $\mathbf{v}$ as shown in Figure 2.21. Since a rhombus is a parallelogram, the diagonals are $\mathbf{u} + \mathbf{v}$ and $\mathbf{u} - \mathbf{v}$ as we have indicated on pages 52 and 53. Now let us consider the dot product.

$$(\mathbf{u} + \mathbf{v}) \cdot (\mathbf{u} - \mathbf{v}) = \mathbf{u}^2 - \mathbf{v}^2 = |\mathbf{u}|^2 - |\mathbf{v}|^2$$

Since we are dealing with a rhombus, the sides are of equal length; that is, $|\mathbf{u}| = |\mathbf{v}|$. Thus the above dot product is 0, implying that the diagonals are perpendicular.

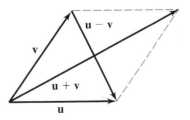

Figure 2.21

EXAMPLE 3   Prove the Pythagorean Theorem by vector methods.

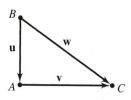

Figure 2.22

SOLUTION   Taking A in Figure 2.22 to be the right angle, we have $\mathbf{u} \cdot \mathbf{v} = 0$.

$$\mathbf{w} = \mathbf{u} + \mathbf{v}$$
$$|\mathbf{w}|^2 = |\mathbf{u} + \mathbf{v}|^2$$
$$= |\mathbf{u}|^2 + |\mathbf{v}|^2 + 2\mathbf{u} \cdot \mathbf{v}$$
$$= |\mathbf{u}|^2 + |\mathbf{v}|^2 \quad (\text{since } \mathbf{u} \cdot \mathbf{v} = 0)$$

EXAMPLE 4   Prove by vector methods that the diagonals of a square are perpendicular and equal in length.

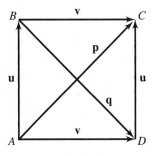

Figure 2.23

SOLUTION   Since we have a square, $|\mathbf{u}| = |\mathbf{v}|$; and $\mathbf{u} \cdot \mathbf{v} = 0$ (see Figure 2.23). Because $\mathbf{p} = \mathbf{v} + \mathbf{u}$, it follows that

$$|\mathbf{p}|^2 = |\mathbf{v} + \mathbf{u}|^2$$
$$= |\mathbf{v}|^2 + |\mathbf{u}|^2 + 2\mathbf{v} \cdot \mathbf{u}$$
$$= |\mathbf{v}|^2 + |\mathbf{u}|^2.$$

Similarly, $\mathbf{q} = \mathbf{v} - \mathbf{u}$ implies that

$$|\mathbf{q}|^2 = |\mathbf{v} - \mathbf{u}|^2$$
$$= |\mathbf{v}|^2 + |\mathbf{u}|^2 - 2\mathbf{v} \cdot \mathbf{u}$$
$$= |\mathbf{v}|^2 + |\mathbf{u}|^2.$$

Thus, $|\mathbf{p}| = |\mathbf{q}|$.

$$\mathbf{p} \cdot \mathbf{q} = (\mathbf{v} + \mathbf{u}) \cdot (\mathbf{v} - \mathbf{u})$$
$$= \mathbf{v}^2 - \mathbf{u}^2$$
$$= |\mathbf{v}|^2 - |\mathbf{u}|^2 = 0,$$

showing that the diagonals are perpendicular.

One of the most common uses of vectors is in analyzing the forces on an object. If there are several forces acting on a body in different directions, each force can be represented by a vector. A single equivalent force (resultant force) is one that has the same effect on the body as the given forces all acting together; it is simply the vector sum of all of the given forces.

Sometimes this problem must be worked in the other direction; that is, given a single force, find a pair of forces satisfying certain conditions that are equivalent to the single given force. This is often done to find the horizontal and vertical components of a given force.

EXAMPLE 5

A force of 20 pounds is directed 60° from the horizontal. What is the vector representation for this force?

SOLUTION

In effect we are to find the vectors $\mathbf{v}_1 = a\mathbf{i}$ and $\mathbf{v}_2 = b\mathbf{j}$ such that $\mathbf{v} = \mathbf{v}_1 + \mathbf{v}_2 = a\mathbf{i} + b\mathbf{j}$. It is easily seen from Figure 2.24 that $a = |\mathbf{v}| \cos 60°$ and $b = |\mathbf{v}| \sin 60°$. Thus

$$\mathbf{v} = 20 \cos 60° \mathbf{i} + 20 \sin 60° \mathbf{j}$$
$$= 10\mathbf{i} + 10\sqrt{3}\mathbf{j}.$$

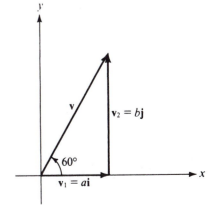

Figure 2.24

Now let us go back to the problem of finding a single force equivalent to several forces acting on a body. As we can see in the following example, the earlier problem is sometimes needed within this one.

EXAMPLE 6

The following forces are exerted on an object: 5 pounds to the right, 10 pounds upward, 2 pounds upward and to the right, inclined to the horizontal at an angle of 30°. What single force is equivalent to them?

SOLUTION

The three given forces are first represented by vectors; the equivalent force is their sum.

$$\mathbf{v}_1 = 5\mathbf{i} \quad \text{and} \quad \mathbf{v}_2 = 10\mathbf{j}$$

For the third vector, we use the method of Example 5 to see that

$$\mathbf{v}_3 = |\mathbf{v}_3| \cos 30° \, \mathbf{i} + |\mathbf{v}_3| \sin 30° \, \mathbf{j}$$
$$= 2 \cdot \frac{\sqrt{3}}{2} \mathbf{i} + 2 \cdot \frac{1}{2}\mathbf{j}$$
$$= \sqrt{3}\mathbf{i} + \mathbf{j}.$$

This gives $\mathbf{v}_3 = \sqrt{3}\mathbf{i} + \mathbf{j}$, and

$$\mathbf{u} = \mathbf{v}_1 + \mathbf{v}_2 + \mathbf{v}_3 = (5 + \sqrt{3})\mathbf{i} + 11\mathbf{j}$$

This is the vector representation of the resultant force. Now let us use it to compute the magnitude and direction of this force.

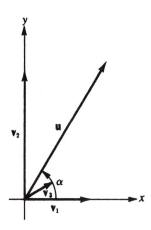

Figure 2.25

$$|\mathbf{u}| = \sqrt{(5 + \sqrt{3})^2 + 11^2}$$
$$= \sqrt{149 + 10\sqrt{3}} = 12.9$$
$$\cos\alpha = \frac{5 + \sqrt{3}}{\sqrt{149 + 10\sqrt{3}}} = 0.5220$$
$$\alpha = 58.53°$$

Thus the three given forces are equivalent to a single force of 12.9 pounds directed upward to the right at an angle of 58.53° with the x axis. See Figure 2.25.

EXAMPLE 7 Find the resultant force if there are two forces: one with magnitude 6 and directed 120° from the horizontal, and the other with magnitude 10 and directed −135° from the horizontal (see Figure 2.26).

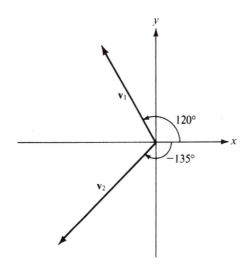

Figure 2.26

SOLUTION First we express each of the two forces in vector form as in Example 6.

$$\mathbf{v}_1 = 6\cos 120°\,\mathbf{i} + 6\sin 120°\,\mathbf{j}$$
$$= -3\mathbf{i} + 3\sqrt{3}\mathbf{j}$$
$$\mathbf{v}_2 = 10\cos(-135°)\,\mathbf{i} + 10\sin(-135°)\,\mathbf{j}$$
$$= -5\sqrt{2}\mathbf{i} - 5\sqrt{2}\mathbf{j}$$

Now the resultant force is found by adding.

$$\mathbf{v} = \mathbf{v}_1 + \mathbf{v}_2$$
$$= (-3\mathbf{i} + 3\sqrt{3}\mathbf{j}) + (-5\sqrt{2}\mathbf{i} - 5\sqrt{2}\mathbf{j})$$
$$= (-3 - 5\sqrt{2})\mathbf{i} + (3\sqrt{3} - 5\sqrt{2})\mathbf{j}$$

We also use vectors to analyze the conditions of static equilibrium for an object on which several concurrent forces act. Suppose that n concurrent forces $\mathbf{F}_1, \mathbf{F}_2, \ldots, \mathbf{F}_n$ act on an object; the condition of static equilibrium is that the sum of these forces is the zero vector.

EXAMPLE 8   Forces of 12 pounds, 18 pounds, and 24 pounds act on an object in the directions shown in Figure 2.27. What additional force (magnitude and direction) is required to produce static equilibrium?

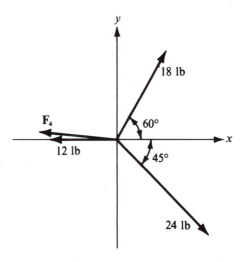

Figure 2.27

SOLUTION   We begin by writing each force in component form:

$$\mathbf{F}_1 = -12\mathbf{i};$$
$$\mathbf{F}_2 = 18\cos 60°\,\mathbf{i} + 18\sin 60°\,\mathbf{j}$$
$$= 9\mathbf{i} + 9\sqrt{3}\,\mathbf{j};$$
$$\mathbf{F}_3 = 24\cos 45°\,\mathbf{i} - 24\sin 45°\,\mathbf{j}$$
$$= \frac{24}{\sqrt{2}}\mathbf{i} - \frac{24}{\sqrt{2}}\mathbf{j}.$$

Let $\mathbf{F}_4$ be the additional force required to produce static equilibrium. The condition for this is

$$\sum_{i=1}^{4} \mathbf{F}_i = -12\mathbf{i} + 9\mathbf{i} + 9\sqrt{3}\mathbf{j} + \frac{24}{\sqrt{2}}\mathbf{i} - \frac{24}{\sqrt{2}}\mathbf{j} + \mathbf{F}_4 = \mathbf{0}.$$

Thus

$$\mathbf{F}_4 = \left(3 - \frac{24}{\sqrt{2}}\right)\mathbf{i} + \left(\frac{24}{\sqrt{2}} - 9\sqrt{3}\right)\mathbf{j}$$

$$= -14.0\mathbf{i} + 1.38\mathbf{j}.$$

The magnitude of this vector is

$$|\mathbf{F}_4| = \sqrt{(14.0)^2 + (1.38)^2} = 14.1 \text{ pounds,}$$

and the angle it makes with the positive direction of the x axis is

$$\tan^{-1}\left(\frac{1.38}{-14.0}\right) = 174.4°.$$

EXAMPLE 9

Find the tension in each of the three cords that support the 100-pound weight shown in Figure 2.28.

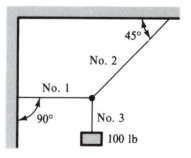

Figure 2.28

SOLUTION

The knot at the intersection of the three cords is in equilibrium under the actions of the concurrent forces of tension in the cords.

Figure 2.29 shows the directions of the forces at the knot. Note that all forces are pulling away from the knot. The tension in cord #3 is 100 pounds since it

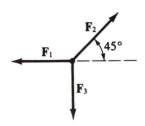

Figure 2.29

supports the 100-pound weight in equilibrium; we write the force on the knot due to this cord as

$$\mathbf{F}_3 = -100\mathbf{j}.$$

If T_1 and T_2 are the tensions in cords #1 and #2, then the forces on the knot due to them are

$$\mathbf{F}_1 = -T_1\mathbf{i}$$

and

$$\mathbf{F}_2 = T_2 \cos 45°\,\mathbf{i} + T_2 \sin 45°\,\mathbf{j} = \frac{T_2}{\sqrt{2}}\mathbf{i} + \frac{T_2}{\sqrt{2}}\mathbf{j}.$$

The condition for equilibrium requires that the sum of the forces at the knot be the zero vector:

$$-T_1\mathbf{i} + \left(\frac{T_2}{\sqrt{2}}\mathbf{i} + \frac{T_2}{\sqrt{2}}\mathbf{j}\right) - 100\mathbf{j} = \mathbf{0}$$

$$\left(-T_1 + \frac{T_2}{\sqrt{2}}\right)\mathbf{i} + \left(\frac{T_2}{\sqrt{2}} - 100\right)\mathbf{j} = \mathbf{0}.$$

Thus

$$-T_1 + \frac{T_2}{\sqrt{2}} = 0 \qquad \text{and} \qquad \frac{T_2}{\sqrt{2}} - 100 = 0.$$

Solving these equations, we get

$$T_2 = 100\sqrt{2} \text{ pounds} \qquad \text{and} \qquad T_1 = 100 \text{ pounds.}$$

EXAMPLE 10   Find the tension in each of the five cords that support the 100-pound weight shown in Figure 2.30.

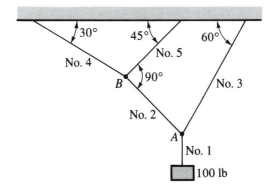

Figure 2.30

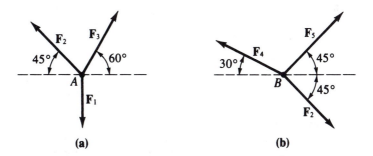

Figure 2.31

Figure 2.31a shows the direction of the forces acting upon knot A; the forces are

$$\mathbf{F}_1 = -100\mathbf{j};$$

$$\mathbf{F}_2 = T_2 \cos 135° \, \mathbf{i} + T_2 \sin 135° \, \mathbf{j} = -\frac{T_2}{\sqrt{2}}\mathbf{i} + \frac{T_2}{\sqrt{2}}\mathbf{j};$$

$$\mathbf{F}_3 = T_3 \cos 60° \, \mathbf{i} + T_3 \sin 60° \, \mathbf{j} = \frac{T_3}{2}\mathbf{i} + \frac{\sqrt{3}T_3}{2}\mathbf{j}.$$

For equilibrium, the sum of these forces must be the zero vector:

$$-100\mathbf{j} + \left(-\frac{T_2}{\sqrt{2}}\mathbf{i} + \frac{T_2}{\sqrt{2}}\mathbf{j}\right) + \left(\frac{T_3}{2}\mathbf{i} + \frac{\sqrt{3}T_3}{2}\mathbf{j}\right)$$

$$= \left(-\frac{T_2}{\sqrt{2}} + \frac{T_3}{2}\right)\mathbf{i} + \left(-100 + \frac{T_2}{\sqrt{2}} + \frac{\sqrt{3}T_3}{2}\right)\mathbf{j} = \mathbf{0}.$$

Therefore,

$$-\frac{T_2}{\sqrt{2}} + \frac{T_3}{T_2} = 0$$

$$-100 + \frac{T_2}{\sqrt{2}} + \frac{\sqrt{3}T_3}{2} = 0.$$

Adding these two equations, we have

$$-100 + \frac{(\sqrt{3}+1)T_3}{2} = 0$$

$$T_3 = \frac{200}{\sqrt{3}+1} = \frac{200(\sqrt{3}-1)}{3-1} = 100(\sqrt{3}-1).$$

Substituting this back into the first equation gives

$$-\frac{T_2}{\sqrt{2}} + \frac{100(\sqrt{3}-1)}{2} = 0$$

$$T_2 = 50\sqrt{2}(\sqrt{3}-1).$$

Now we repeat all of this at knot B. Figure 2.31b shows the direction of the forces acting upon it. Note that the tension on a cord from A to B tends to pull it apart. That is, external forces are pulling it toward A and B. The cord exerts equal and opposite forces at these two points. That is why the force $\mathbf{F}_2$ is directed in opposite directions in the two parts of Figure 2.31. From that figure, we have

$$\mathbf{F}_2 = T_2\cos(-45°)\mathbf{i} + T_2\sin(-45°)\mathbf{j} = 50(\sqrt{3}-1)\mathbf{i} - 50(\sqrt{3}-1)\mathbf{j};$$

$$\mathbf{F}_4 = T_4\cos 150°\,\mathbf{i} + T_4\sin 150°\,\mathbf{j} = -\frac{\sqrt{3}T_4}{2}\mathbf{i} + \frac{T_4}{2}\mathbf{j};$$

$$\mathbf{F}_5 = T_5\cos 45°\,\mathbf{i} + T_5\sin 45°\,\mathbf{j} = \frac{T_5}{\sqrt{2}}\mathbf{i} + \frac{T_5}{\sqrt{2}}\mathbf{j}.$$

Again, the sum of these forces must be $\mathbf{0}$:

$$[50(\sqrt{3}-1)\mathbf{i} + 50(\sqrt{3}-1)\mathbf{j}] + \left(-\frac{\sqrt{3}T_4}{2}\mathbf{i} + \frac{T_4}{2}\mathbf{j}\right) + \left(\frac{T_5}{\sqrt{2}}\mathbf{i} + \frac{T_5}{\sqrt{2}}\mathbf{j}\right)$$

$$= \left[50(\sqrt{3}-1) - \frac{\sqrt{3}T_4}{2} + \frac{T_5}{\sqrt{2}}\right]\mathbf{i} + \left[-50(\sqrt{3}-1) + \frac{T_4}{2} + \frac{T_5}{\sqrt{2}}\right]\mathbf{j} = \mathbf{0},$$

giving

$$50(\sqrt{3}-1) - \frac{\sqrt{3}T_4}{2} + \frac{T_5}{\sqrt{2}} = 0;$$

$$-50(\sqrt{3}-1) + \frac{T_4}{2} + \frac{T_5}{\sqrt{2}} = 0.$$

Solving these two equations for T_4 and T_5, we have

$$T_4 = 200 \quad \text{and} \quad T_5 = 50\sqrt{2}(\sqrt{3} + 1).$$

There are many other important applications of vectors that are beyond the scope of this book. For example, we may use vectors to analyze the forces on a wire suspended from two points, or on the cables of a suspension bridge. Similarly, vectors may be used to study the motion of planets, finding equations of their paths from a knowledge of the gravitational and centrifugal forces on them.

PROBLEMS

A *In Problems 1–6, the given forces are acting on a body. What single force is equivalent to them?*

1. $\mathbf{f}_1 = 4\mathbf{i} + 3\mathbf{j}$, $\mathbf{f}_2 = \mathbf{i} - 2\mathbf{j}$, $\mathbf{f}_3 = \mathbf{i} + \mathbf{j}$
2. $\mathbf{f}_1 = 2\mathbf{i} - \mathbf{j}$, $\mathbf{f}_2 = 3\mathbf{i} + 4\mathbf{j}$, $\mathbf{f}_3 = -\mathbf{i} + 2\mathbf{j}$
3. 5 pounds to the right; 10 pounds upward
4. 4 pounds downward; 6 pounds to the right; 2 pounds upward
5. 3 pounds downward; 4 pounds to the right and inclined upward at an angle of 45° with the horizontal
6. 4 pounds to the left; 5 pounds to the right and inclined upward at an angle of 60° with the horizontal

In Problems 7–12, give a vector representation of the given force.

7. A force of 6 pounds directed 30° from the horizontal
8. A force of 8 pounds directed 150° from the horizontal
9. A force of 3 pounds directed −45° from the horizontal
10. A force of 10 pounds directed −120° from the horizontal
11. A force of 2 pounds directed 35° from the horizontal
12. A force of 5 pounds directed −75° from the horizontal

B *In Problems 13–18, the given forces are acting on a body. What additional force will result in equilibrium? (A set of forces is in equilibrium if the sum of all of them is the zero vector.)*

13. $\mathbf{f}_1 = 2\mathbf{i} + \mathbf{j}$, $\mathbf{f}_2 = \mathbf{i} - 3\mathbf{j}$, $\mathbf{f}_3 = -3\mathbf{i} + \mathbf{j}$
14. $\mathbf{f}_1 = 4\mathbf{i} - \mathbf{j}$, $\mathbf{f}_2 = \mathbf{i} + 3\mathbf{j}$, $\mathbf{f}_3 = 2\mathbf{i} - 5\mathbf{j}$
15. 2 pounds to the left; 5 pounds upward
16. 4 pounds to the right; 6 pounds upward; 6 pounds to the left

17. 3 pounds to the right; 5 pounds to the right and upward inclined at an angle of 45° with the horizontal

18. 3 pounds upward; 2 pounds to the right and inclined upward at an angle of 30° with the horizontal; 4 pounds to the left and inclined downward at an angle of 60° with the horizontal

In Problems 19–24, find the resultant of the forces given.

19. A force of 8 pounds directed 45° from the horizontal; a force of 9 pounds directed 270° from the horizontal

20. A force of 3 pounds directed 150° from the horizontal; a force of 5 pounds directed −60° from the horizontal

21. A force of 6 pounds directed 45° from the horizontal; a force of 8 pounds directed 180° from the horizontal

22. A force of 10 pounds directed 120° from the horizontal; a force of 10 pounds directed −60° from the horizontal

23. A force of 10 pounds directed 120° from the horizontal; a force of 8 pounds directed −30° from the horizontal; a force of 20 pounds directed −90° from the horizontal

24. A force of 8 pounds directed 135° from the horizontal; a force of 10 pounds directed 30° from the horizontal; a force of 2 pounds directed −120° from the horizontal

In Problems 25–30, find the tension in all the cords of the arrangements shown in Fig. 2.32. In each case, the weight of the suspended object is 100 pounds.

25.   **26.**

27.   **28.**

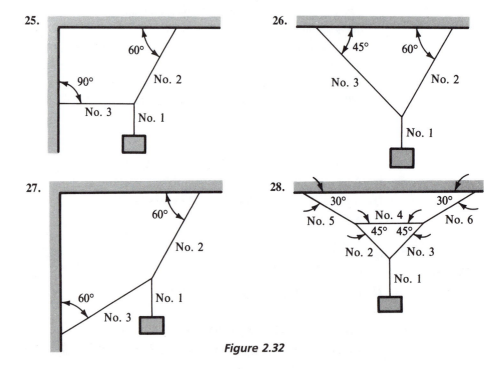

Figure 2.32

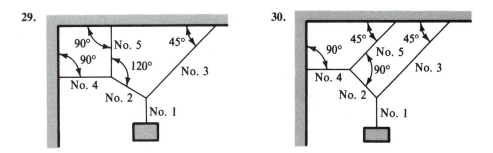

Figure 2.32 *(cont.)*

In Problems 31–39, use vector methods to prove the given theorem.

31. The segment joining the midpoints of the nonparallel sides of a trapezoid is parallel to and one-half the sum of the lengths of the parallel sides.

32. The lines joining consecutive midpoints of a quadrilateral form a parallelogram.

33. If the diagonals of a parallelogram are perpendicular, then it is a rhombus.

34. If the sum of the squares of two sides of a triangle equals the square of the third side, then the triangle is a right triangle.

35. The sum of the squares of the four sides of a parallelogram is equal to the sum of the squares of the two diagonals.

36. The diagonals of a rectangle are equal.

37. The base angles of an isosceles triangle are equal.

38. If one of the parallel sides of a trapezoid is twice the length of the other, then the diagonals intersect at a point of trisection of both of them.

39. The medians of a triangle are concurrent at a point two-thirds of the way from each vertex to the midpoint of the opposite side. [*Hint*: Let $\overrightarrow{BP} = r\,\overrightarrow{BE}$ and $\overrightarrow{AP} = s\overrightarrow{AD}$ (see Figure 2.33) and use the result of Problem 48, page 61.]

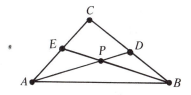

Figure 2.33

REVIEW PROBLEMS

A   **1.** Find the unit vector in the direction of $\mathbf{v} = (2, -3)$.

  2. Suppose $\overrightarrow{AB}$ is a representative of $\mathbf{v} = 2\mathbf{i} - 5\mathbf{j}$. Find B when $A = (1, 3)$.

3. Suppose $\overrightarrow{AB}$ is a representative of $\mathbf{v} = 5\mathbf{i} - 3\mathbf{j}$, and $(-2, 1)$ is the midpoint of AB. Find A and B.

4. If $\mathbf{u} = (3, -2)$ and $\mathbf{v} = (1, 6)$, find $\mathbf{u} + \mathbf{v}$, $\mathbf{u} - \mathbf{v}$, $3\mathbf{u}$, and $2\mathbf{u} + \mathbf{v}$. Sketch all the vectors.

5. If $\mathbf{u} = 2\mathbf{i} - \mathbf{j}$ and $\mathbf{v} = 3\mathbf{i} - 2\mathbf{j}$, find $\mathbf{u} \cdot \mathbf{v}$.

6. Find the angle θ between $\mathbf{u} = 2\mathbf{i} - 3\mathbf{j}$ and $\mathbf{v} = \mathbf{i} + 5\mathbf{j}$.

B   7. Find the projection $\mathbf{w}$ of $\mathbf{u} = \mathbf{i} + 4\mathbf{j}$ upon $\mathbf{v} = 2\mathbf{i} - 9\mathbf{j}$.

8. Find the projection $\mathbf{w}$ of $\mathbf{u} = 4\mathbf{i} + \mathbf{j}$ upon $\mathbf{v} = 2\mathbf{i} - 2\mathbf{j}$.

9. Determine a so that the angle between $\mathbf{u} = 5\mathbf{i} + \mathbf{j}$ and $\mathbf{v} = a\mathbf{i} - \mathbf{j}$ is $\pi/4$.

10. Determine a so that $\mathbf{u} = \mathbf{i} - 4\mathbf{j}$ and $\mathbf{v} = 3\mathbf{i} + a\mathbf{j}$ are orthogonal.

11. Find the resultant force if a force of 6 pounds is directed 90° from the horizontal, 8 pounds is directed 30° from the horizontal, and 10 pounds is directed $-135°$ from the horizontal.

12. Use vectors to prove that the diagonals of an isosceles trapezoid are equal.

13. Use vectors to prove that if the lengths of the parallel sides of a trapezoid are in the ratio $1:n$, then the point of intersection of the diagonals divides both of them in the ratio $1:(n + 1)$.

14. The following forces are exerted on an object: 30 pounds to the left, 15 pounds downward, $10\sqrt{2}$ pounds upward and to the right inclined to the horizontal at an angle of 45°. What single force is equivalent to them?

15. The handle of a lawnmower is inclined to the horizontal at an angle of 30°. If a man pushes forward and down in the direction of the handle with a force of 20 pounds, with what force is the mower being pushed forward? With what force is it being pushed into the ground?

C   16. Given $\mathbf{u} = 3\mathbf{i} + 4\mathbf{j}$ and $\mathbf{v} = 12\mathbf{i} - 5\mathbf{j}$, find the vectors $\mathbf{v}_1$ and $\mathbf{v}_2$ such that $\mathbf{v} = \mathbf{v}_1 + \mathbf{v}_2$, $\mathbf{v}_1 = k\mathbf{u}$, and $\mathbf{v}_2 \cdot \mathbf{u} = 0$.

Chapter **3** The Line

POINT-SLOPE AND TWO-POINT FORMS

The last section of Chapter 1 dealt with the problem of finding an equation of a curve from a description of it. In this chapter, as well as the next two, we shall consider this problem in more detail. Let us begin with a consideration of the line. The two simplest ways of determining a line are by a pair of points or by one point and the slope. Thus, if a line is described in either of these ways, we should be able to give an equation for it. We begin with a line described by its slope and a point on it.

THEOREM 3.1   (*Point-slope form of a line.*) *A line that has slope m and contains the point* (x_1, y_1) *has equation*

$$y - y_1 = m(x - x_1).$$

PROOF   Let (x, y) be any point different from (x_1, y_1) on the given line (see Figure 3.1). Since the line has slope, it is not vertical. Thus $x \neq x_1$, which gives

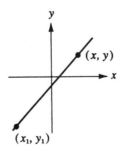

Figure 3.1

$$m = \frac{y - y_1}{x - x_1}$$

and

$$y - y_1 = m(x - x_1).$$

Although the formula was derived only for points on the line different from the given point (x_1, y_1), it is easily seen that (x_1, y_1) also satisfies the equation. Thus, every point on the line satisfies the equation. Suppose now that the point (x_2, y_2) satisfies the equation—that is,

$$y_2 - y_1 = m(x_2 - x_1).$$

If $x_2 = x_1$, then $y_2 - y_1 = 0$, or $y_2 = y_1$. In this case, $(x_2, y_2) = (x_1, y_1)$, which is on the line. If $x_2 \neq x_1$, then

$$\frac{y_2 - y_1}{x_2 - x_1} = m.$$

Thus, the slope of the line joining (x_1, y_1) and (x_2, y_2) is m, and this line has the point (x_1, y_1) in common with the given line. Thus, (x_2, y_2) is on the given line since there can be only one line with slope m containing (x_1, y_1). ∎

EXAMPLE 1   Find an equation of the line through $(-2, -3)$ with slope $1/2$ (see Figure 3.2).

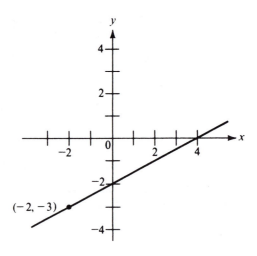

Figure 3.2

SOLUTION

$$y - y_1 = m(x - x_1)$$

$$y - (-3) = \frac{1}{2}[x - (-2)]$$

$$2y + 6 = x + 2$$

$$x - 2y - 4 = 0$$

Of course vertical lines cannot be represented by the point-slope form, since they have no slope. Again, remember that "no slope" does not mean "zero slope." A horizontal line has $m = 0$, and it can be represented by the point-slope form, which gives $y - y_1 = 0$. There is no x in the resulting equation! But the points on a horizontal line satisfy the condition that they all have the same y coordinate, no matter what the x coordinate is. Similarly, the points on a vertical line satisfy the condition that all have the same x coordinate. Thus, if (x_1, y_1) is one point on a vertical line, then $x = x_1$, or $x - x_1 = 0$ for every point (x, y) on the line.

EXAMPLE 2   Find an equation of the vertical line through $(3, -2)$ (see Figure 3.3).

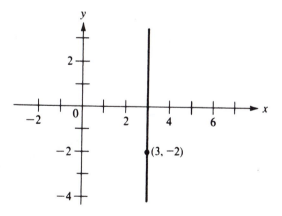

Figure 3.3

SOLUTION   Since the x coordinate of the given point is 3, all points on the line have x coordinates 3. Thus,

$$x = 3 \quad \text{or} \quad x - 3 = 0.$$

THEOREM 3.2 (*Two-point form of a line.*) *A line through* (x_1, y_1) *and* (x_2, y_2), $x_1 \neq x_2$, *has equation*

$$y - y_1 = \frac{y_2 - y_1}{x_2 - x_1}(x - x_1).$$

It might be noted that this result is often stated in the form

$$\frac{y - y_1}{x - x_1} = \frac{y_2 - y_1}{x_2 - x_1}.$$

While the symmetry of this form is appealing, the form has one serious defect—the point (x_1, y_1) is on the desired line, but it does not satisfy this equation. It does satisfy the equation of Theorem 3.2.

   The proof of Theorem 3.2 follows directly from Theorem 3.1 and the fact that $m = (y_2 - y_1)/(x_2 - x_1)$, provided $x_1 \neq x_2$. Actually, this follows so easily from Theorem 3.1 that you may prefer to use the earlier theorem after finding the slope from the two given points. Of course, the designation of the two points as "point 1" and "point 2" is quite arbitrary.

EXAMPLE 3 Find an equation of the line through $(4, 1)$ and $(-2, 3)$ (see Figure 3.4).

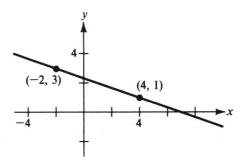

Figure 3.4

SOLUTION

$$y - y_1 = \frac{y_2 - y_1}{x_2 - x_1}(x - x_1)$$

$$y - 1 = \frac{3 - 1}{-2 - 4}(x - 4)$$

$$x + 3y - 7 = 0$$

The slope of the line is

$$m = \frac{y_2 - y_1}{x_2 - x_1} = \frac{3 - 1}{-2 - 4} = \frac{2}{-6} = -\frac{1}{3}$$

Using this with the point (4, 1) in the point-slope form, we have

$$y - y_1 = m(x - x_1)$$
$$y - 1 = -\frac{1}{3}(x - 4)$$
$$3y - 3 = -x + 4$$
$$x + 3y - 7 = 0.$$

Note that we get exactly the same result if we use the slope $-1/3$ with the other point, $(-2, 3)$.

$$y - y_1 = m(x - x_1)$$
$$y - 3 = -\frac{1}{3}(x + 2)$$
$$3y - 9 = -x - 2$$
$$x + 3y - 7 = 0$$

EXAMPLE 4 Find the perpendicular bisector of the segment joining (5, −3) and (1, 7).

SOLUTION First let us find the midpoint.

$$x = \frac{5 + 1}{2} = 3 \quad \text{and} \quad y = \frac{-3 + 7}{2} = 2.$$

The midpoint is (3, 2).
   The slope of the line joining (5, −3) and (1, 7) is

$$m = \frac{-3 - 7}{5 - 1} = -\frac{5}{2};$$

the slope of the perpendicular line is $m = 2/5$.
   Now we merely need to use the point-slope formula, using (3, 2) and $m = 2/5$.

$$y - 2 = \frac{2}{5}(x - 3)$$
$$2x - 5y + 4 = 0$$

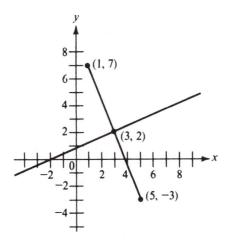

Figure 3.5

The result is shown in Figure 3.5.

In Section 1.4, we used the DRAW or PLOT facility of a graphing calculator to draw a line through two points or through a given point and with a given slope. However, this procedure has certain limitations. We cannot use a TRACE or ZOOM; indeed, we cannot change the viewing window in any way. The reason for this is that DRAW or PLOT does not generate a function which can be referred to for a TRACE or that can allow regraphing when the viewing window changes.

Now that we can use two points or a point and a slope to get an equation, we can use GRAPH. However, we must use a function rather than an equation. All that means is that we must solve our equation for y as a function of x. Except for vertical lines, whose equations have no y, this is quite straightforward.

EXAMPLE 5   Use a graphing calculator to graph the line through $(-4, 3)$ with slope $-2/3$. Use TRACE and ZOOM to verify that $(-4, 3)$ is on the line and that its slope is $-2/3$.

SOLUTION   We use a slightly altered form of our point-slope equation to have y as a function of x.

$$y = y_1 + m(x - x_1)$$

$$y = 3 - \frac{2}{3}[x - (-4)]$$

$$= 3 - \frac{2}{3}(x + 4)$$

There is no need to simplify this result; the calculator can use any form of the function. Using the normal range and $Y_1 = 3 - (2/3)(X + 4)$, we get the graph of Figure 3.6

Using TRACE, we move the cursor to the left so that $x = -4$. We can then read the y coordinate, which is seen to be 3. This is illustrated in Figure 3.6.

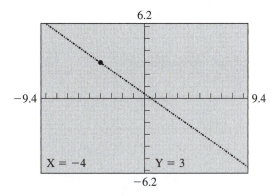

Figure 3.6

Now let us move the cursor to $x = 1.0$ and ZOOM In. This gives the graph of Figure 3.7. Again using TRACE, we identify the point on the graph with y coordinate 0. This is the point $(0.5, 0)$. Now we move the cursor 1 unit to the right to the point with $x = 1.5$. The y coordinate (-0.6666667) of this point is the slope of the line. Thus, the slope is verified to be $-2/3$.

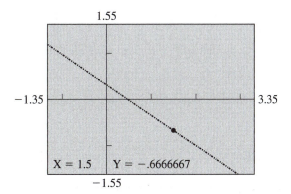

Figure 3.7

EXAMPLE 6   Find an equation of the line containing $(2, 5)$ and $(-3, 1)$. Use TRACE to verify that $(2, 5)$ and $(-3, 1)$ are on the line.

$$m = \frac{5 - 1}{2 - (-3)} = \frac{4}{5}$$

Again using our slightly altered point-slope formula with the above slope and point $(2, 5)$, we have

$$y = 5 + \frac{4}{5}(x - 2).$$

With the normal range and $Y_1 = 5 + (4/5)(X - 2)$, we get the graph shown in Figure 3.8.

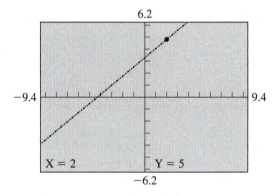

Figure 3.8

Using TRACE, we move the cursor to the right until $x = 2$. For this point, $y = 5$ as shown in Figure 3.8. A similar method can be used to verify that $(-3, 1)$ is on the graph.

PROBLEMS

A *In Problems 1–16, find an equation of the line indicated and sketch the graph.*

1. Through $(2, -4)$; $m = -2$ **2.** Through $(5, 3)$; $m = 4$

3. Through $(2, 2)$: $m = 1$ **4.** Through $(-4, 6)$; $m = 5$

5. Through $(9, 0)$; $m = 1$ **6.** Through $(0, 3)$; $m = 2$

7. Through $(4, -2)$; $m = 0$ **8.** Through $(2, 5)$; no slope

9. Through $(1, 4)$ and $(3, 5)$ **10.** Through $(2, -1)$ and $(4, 4)$

11. Through $(3, 3)$ and $(1, 1)$　　12. Through $(2, 1)$ and $(-3, 3)$

13. Through $(0, 0)$ and $(1, 5)$　　14. Through $(0, 1)$ and $(-2, 0)$

15. Through $(2, 3)$ and $(5, 3)$　　16. Through $(5, 1)$ and $(5, 3)$

B　17. Find equations of the three sides of the triangle with vertices $(1, 4)$, $(3, 0)$, and $(-1, -2)$.

18. Find equations of the medians of the triangle of Problem 17.

19. Find equations of the altitudes of the triangle of Problem 17.

20. Find the vertices of the triangle with sides $x - 5y + 8 = 0$, $4x - y - 6 = 0$, and $3x + 4y + 5 = 0$.

21. Find equations of the medians of the triangle of Problem 20.

22. Find equations of the altitudes of the triangle of Problem 20.

23. Find an equation of the chord of the circle $x^2 + y^2 = 25$ which joins $(-3, 4)$ and $(5, 0)$. Sketch the circle and its chord.

24. Find an equation of the chord of the parabola $y = x^2$ which joins $(-1, 1)$ and $(2, 4)$. Sketch the curve and its chord.

25. Find an equation of the perpendicular bisector of the segment joining $(4, 2)$ and $(-2, 6)$.

26. Find an equation of the line through the points of intersection of the circles

$$x^2 + y^2 + 2x - 19 = 0 \quad \text{and} \quad x^2 + y^2 - 10x - 12y + 41 = 0.$$

Look over your work. Is there any easier way?

27. Repeat Problem 26 for the circles

$$x^2 + y^2 + 4x + 2y + 3 = 0 \quad \text{and} \quad x^2 + y^2 - 6x - 8y + 21 = 0.$$

What is wrong?

28. Find an equation of the line through the centers of the two circles of Problem 26.

29. What condition must the coordinates of a point satisfy in order that it be equidistant from $(2, 5)$ and $(4, -1)$?

30. Find the center and radius of the circle through the points $(1, 3)$, $(4, -6)$, and $(-3, 1)$.

31. Consider the triangle with vertices $A = (3, 1)$, $B = (0, 5)$, and $C = (7, 4)$. Find equations of the altitude and the median from A. What do your results tell us about the triangle?

C　32. Show that a line through points (x_1, y_1) and (x_2, y_2) can be represented by

$$\begin{vmatrix} x & y & 1 \\ x_1 & y_1 & 1 \\ x_2 & y_2 & 1 \end{vmatrix} = 0.$$

The expression on the left-hand side of this equation is a determinant. Some authors use the notation

$$\det \begin{bmatrix} x & y & 1 \\ x_1 & y_1 & 1 \\ x_2 & y_2 & 1 \end{bmatrix}$$

for this determinant.

33. Show that the points (x_1, y_1), (x_2, y_2), (x_3, y_3) are collinear if and only if

$$\begin{vmatrix} x_1 & y_1 & 1 \\ x_2 & y_2 & 1 \\ x_3 & y_3 & 1 \end{vmatrix} = 0.$$

34. Show that if no pair of the equations

$$A_1x + B_1y + C_1 = 0$$
$$A_2x + B_2y + C_2 = 0$$
$$A_3x + B_3y + C_3 = 0$$

represent parallel lines, then the lines are concurrent if and only if

$$\begin{vmatrix} A_1 & B_1 & C_1 \\ A_2 & B_2 & C_2 \\ A_3 & B_3 & C_3 \end{vmatrix} = 0.$$

*GRAPHING
CALCULATOR*

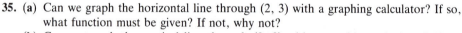

35. (a) Can we graph the horizontal line through (2, 3) with a graphing calculator? If so, what function must be given? If not, why not?
(b) Can we graph the vertical line through (2, 3) with a graphing calculator? If so, what function must be given? If not, why not?

In Problems 36–40, graph the given lines on the same coordinate axes.

36. (a) The line through (1, 3) with slope 1
(b) The line through (1, 3) with slope 2
(c) The line through (1, 3) with slope 3
(d) The line through (1, 3) with slope 4
How does the line change as the slope increases indefinitely?

37. (a) The line through (1, 3) with slope 1/2
(b) The line through (1, 3) with slope 1/3
(c) The line through (1, 3) with slope 1/4
(d) The line through (1, 3) with slope 1/5
How does the line change as the slope decreases toward zero through positive numbers?

38. (a) The line through (1, 3) with slope −1
(b) The line through (1, 3) with slope −2
(c) The line through (1, 3) with slope −3
(d) The line through (1, 3) with slope −4
How does the line change as the slope decreases indefinitely?

39. (a) The line through (1, 3) with slope 1
(b) The line through (1, 3) with slope −1
(c) The line through (1, 3) with slope 2
(d) The line through (1, 3) with slope −2

What is the relationship between a line with a positive slope and a line with the corresponding negative slope?

40. (a) The line through $(1, -3)$ with slope 2
(b) The line through $(1, 1)$ with slope 2
(c) The line through $(1, 5)$ with slope 2
(d) The line through $(1, 9)$ with slope 2
What is the relationship between these lines?

In Problems 41–44, find an equation of the line through the given point and with the given slope. Use TRACE to verify that the line contains the given point and has the proper slope.

41. $m = -3, (-2, -4)$

42. $m = 2/5, (3, -4)$

43. $m = -3/4, (2.58, 3.17)$

44. $m = 3.487, (-5.43, 2.78)$

In Problems 45–48, find an equation of the line through the given points. Use TRACE to verify that the given points are on the line.

45. $(-7, 3), (4, -4)$

46. $(2, 5), (-3, 1)$

47. $(-1.4, 5.8), (3.2, 1.6)$

48. $(-2.5, 6.1), (4.3, 1.5)$

APPLICATIONS

49. The pressure within a partially evacuated container is being measured by means of an open-end manometer. This gives the difference between the pressure in the container and atmospheric pressure. It is known that a difference of 0 mm of mercury corresponds to a pressure of 1 atmosphere and that if the pressure in the container were reduced to 0 atmospheres, a difference of 760 mm of mercury would be observed. Assuming that the difference D in mm of mercury and the pressure P in atmospheres are related by a linear relation, determine what such a relation is.

50. Knowing that water freezes at 0°C, or 32°F, that it boils at 100°C, or 212°F, and that the relation between the temperature in degrees centigrade C and in degrees Fahrenheit F is linear, find that relation.

51. The amount of a given commodity that consumers are willing to buy at a given price is called the demand for that commodity corresponding to the given price; the relationship between the price and the demand is called a demand equation. Similarly the amount that manufacturers are willing to offer for sale at a given price is called the supply corresponding to the given price, and the relationship between the price and the supply is called a supply equation. Market equilibrium exists when the supply and demand are equal. The demand and supply equations for a given commodity are

$$2p + x - 100 = 0 \quad \text{and} \quad p - x + 10 = 0,$$

respectively, where p is the price of the commodity and x is its supply or demand. At what price will there be market equilibrium? Graph both equations with p on the vertical axis. What happens to the demand as the price increases? What happens to the supply as the price increases?

52. It costs the Acme Company $9500 to manufacture 100 pairs of shoes each day and $12,250 to manufacture 150 pairs. Assuming that the cost is a linear function of the number manufactured, find the cost as a function of the number manufactured. Interpret the constants in your result.

53. Market research by the ZYZZ Electric Company has shown that the company can sell 60,000 lamps if it prices them at $20 and 20,000 if it prices them at $40. Assuming that the demand *D* (the number that can be sold) is a linear function of the price *p*, find the demand as a function of the price. For what values of the price does the result make sense?

54. The Consumer Price Index was 38.8 in 1970 and 113.6 in 1987. Assuming that this index is linearly related to time, find an equation relating the Consumer Price Index to the time given in number of years after 1970. What value for the Consumer Price Index does this predict for 1980? (The actual value was 82.4.)

3.2

SLOPE-INTERCEPT AND INTERCEPT FORMS

The *x* and *y* intercepts of a line are the points at which the line crosses the *x* and *y* axes, respectively. These points are of the form $(a, 0)$ and $(0, b)$ (see Figure 3.9), but they are usually represented simply by *a* and *b*, since the 0's are understood by their position on the axes. We shall continue using the convention that the *x* and *y* intercepts of a line are represented by the symbols *a* and *b*, respectively. It might be noted that lines parallel to the *x* axis have no *x* intercept and those parallel to the *y* axis have no *y* intercept. While a line on the *x* axis has infinitely many points in common with the *x* axis, we shall adopt the convention that it has no *x* intercept. Similarly, a line on the *y* axis has no *y* intercept. Thus no horizontal line has an *x* intercept and no vertical line has a *y* intercept. One other special case is that of a line through the origin which is neither horizontal nor vertical; it has a single point (the origin) which is both its *x* and *y* intercept. In this case $a = b = 0$. With these special points defined, we now introduce two more forms of a line.

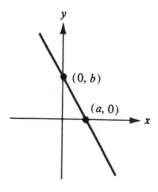

Figure 3.9

HISTORICAL NOTE

DESCARTES

René Descartes (pronounced day-CART) was a seventeenth-century French philosopher, scientist, and mathematician. He is best known as a philosopher, having founded modern philosophical rationalism.

In mathematics, he was one of the two men (the other being Pierre de Fermat) who first conceived the idea of representing points by pairs of numbers. Such numbers are often called **cartesian coordinates** after Descartes.

In an appendix of a book that he published in 1637, Descartes first noted that a curve is defined by some property that is possessed by all of its points and by no others. Furthermore, this property can be represented by an equation in *x* and *y*, and there is a correspondence between the algebraic properties of the equation and the geometric properties of the

curve. He classified algebraic equations by their degrees and noted the curves corresponding to them.

The details of his development differ in certain ways from our present-day conventions, a minor difference being that he considered only a single axis rather than two. Nevertheless, a *y* coordinate implies the existence of a *y* axis, even if it is not shown. A much more significant difference is that he considered curves only in the first quadrant. The problems associated with this restriction soon led to a widening of the scope to all quadrants.

As with many radical changes, the value of analytic geometry was not immediately perceived by the general mathematical community. However, its use in calculus soon led to its universal acceptance.

THEOREM 3.3   (*Slope-intercept form of a line.*) *A line with slope m and y intercept b has equation*

$$y = mx + b.$$

PROOF   Since the y intercept is really the point $(0, b)$, the use of the point-slope form gives

$$y - b = m(x - 0) \quad \text{or} \quad y = mx + b. \quad \blacksquare$$

THEOREM 3.4   (*Intercept form of a line.*) *A line with nonzero intercepts a and b has equation*

$$\frac{x}{a} + \frac{y}{b} = 1.$$

PROOF   Since the intercepts are the points $(a, 0)$ and $(0, b)$, the line has slope

$$m = -\frac{b}{a}.$$

Using the slope-intercept form, we have

$$y = -\frac{b}{a}x + b.$$

Dividing each term by b gives

$$\frac{y}{b} = \frac{-x}{a} + 1, \qquad \text{or} \qquad \frac{x}{a} + \frac{y}{b} = 1. \quad \blacksquare$$

It might be noted that these two forms are merely special cases of the point-slope and two-point forms; thus, the earlier forms may be used in place of these at any time. However, these forms, especially the slope-intercept form, are so convenient to use that it is well to remember them. We shall see an example of their use shortly.

EXAMPLE 1   Find an equation of the line with slope 2 and y intercept 5 (see Figure 3.10).

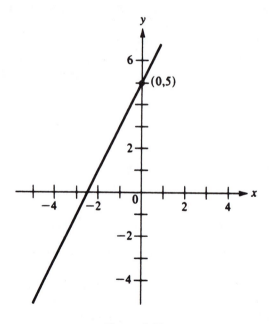

Figure 3.10

SOLUTION

$$y = mx + b$$
$$y = 2x + 5$$
$$2x - y + 5 = 0$$

There is no commonly used special form for a line with a given slope and x intercept. Although one can easily be derived, it has not proved to be as convenient as the slope-intercept form. If you know the slope and the x intercept, simply use the point-slope form with the point $(a, 0)$.

EXAMPLE 2   Find an equation of the line with x and y interceps 5 and -2, respectively (see Figure 3.11).

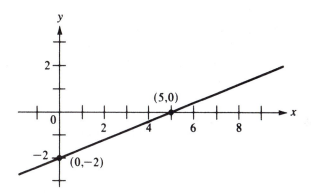

Figure 3.11

SOLUTION

$$\frac{x}{a} + \frac{y}{b} = 1$$

$$\frac{x}{5} + \frac{y}{-2} = 1$$

$$-2x + 5y = -10$$

$$2x - 5y - 10 = 0$$

Just as it was true that vertical lines could not be represented by the point-slope form, we see that vertical lines cannot be represented by the slope-intercept form, since vertical lines have neither slope nor y intercept. The intercept form is even more restrictive, accommodating neither horizontal nor vertical lines, because a horizontal line has no x intercept and a vertical line has no y intercept. Furthermore, no line through the origin can be put into the intercept form, since $a = b = 0$ gives zeros in the denominators.

In all of the examples we have considered so far, we used the special forms only as a starting point; the final form was always $Ax + By + C = 0$. The question arises whether every equation representing a line can be put into such a form and if every equation in such a form represents a line.

THEOREM 3.5   (*General form of a line.*) *Every line can be represented by an equation of the form*

$$Ax + By + C = 0,$$

where A and B are not both zero, and any such equation represents a line.

PROOF   Any line we consider either is vertical or can be put into slope-intercept form. Thus any line can be represented by either

$$x = k \quad \text{or} \quad y = mx + b.$$

Thus any line is in the form

$$x - k = 0 \quad \text{or} \quad mx - y + b = 0.$$

Both are special cases of $Ax + By + C = 0$.

Suppose we have an equation of the form $Ax + By + C = 0$, where A and B are not both 0. Let us consider two cases.

Case I: B = 0. Then

$$Ax + C = 0 \quad \text{and} \quad x = -\frac{C}{A}$$

(since $B = 0$ and A and B are not both 0, we know that $A \neq 0$ and we may divide by A). This represents an equation of a vertical line.

Case II: B ≠ 0. Solving $Ax + By + C = 0$ for y, we have

$$y = -\frac{A}{B}x - \frac{C}{B}$$

(since $B \neq 0$, we may divide by B). This represents an equation of a line with slope $-A/B$ and y intercept $-C/B$. ∎

Theorem 3.5 has the following implication for graphing: any equation of the form $Ax + By + C = 0$ represents a line, and its graph can be determined by two of its points. Since the intercepts are so easily found, finding the line through these two points (if there are two) is the quickest way of sketching a line. Of course, vertical or horizontal lines do not have two intercepts, but these are easily sketched. The only problem comes from lines through the origin. The origin is both the x and y intercept; so just find a second point in any convenient way.

EXAMPLE 3   Sketch the line $2x - 3y - 6 = 0$.

SOLUTION   When $y = 0$, $x = 3$, and when $x = 0$, $y = -2$. We did not put the equation into intercept form in order to determine the intercepts, although we might have done so; however, we can find the intercepts by inspection by setting y and x equal to zero in turn and solving for the other. Actually, this represents a convenient way of putting the line into intercept form. Since $a = 3$ and $b = -2$, the intercept form of $2x - 3y - 6 = 0$ is

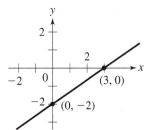

$$\frac{x}{3} + \frac{y}{-2} = 1.$$

The graph of this equation is given in Figure 3.12.

Figure 3.12

The proof of Theorem 3.5 also shows us that the slope of the line $Ax + By + C = 0$ is $-A/B$ provided $B \neq 0$. This implies that the slope of a line is determined entirely by the coefficients of x and y; the constant term has nothing to do with the slope. Thus

$$Ax + By + C_1 = 0 \quad \text{and} \quad Ax + By + C_2 = 0$$

both have the same slope, namely $-A/B$. Furthermore,

$$Ax + By + C_1 = 0 \quad \text{and} \quad Bx - Ay + C_2 = 0$$

are perpendicular since they have slopes $-A/B$ and $-B/(-A) = B/A$, respectively. This gives us an easy way to write an equation of a line parallel (or perpendicular) to a given line.

EXAMPLE 4 Find an equation of the line that is (a) parallel and (b) perpendicular to $3x + 2y - 5 = 0$ and contains the point $(3, 1)$ (see Figure 3.13).

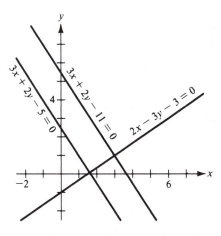

Figure 3.13

SOLUTION Any line parallel to the given line has the form $3x + 2y + C_1 = 0$ since such a line has the same slope as the given line. Now all we need to do is to determine C_1. Since the line contains the point $(3, 1)$, its equation is satisfied by $(3, 1)$. Thus we have

$$3 \cdot 3 + 2 \cdot 1 + C_1 = 0$$
$$C_1 = -11$$

and the desired line is

$$3x + 2y - 11 = 0.$$

Any line perpendicular to the given line has the form $2x - 3y + C_2 = 0$. Again the point $(3, 1)$ satisfies the equation. Thus

$$2 \cdot 3 - 3 \cdot 1 + C_2 = 0$$
$$C_2 = -3$$

and the desired line is

$$2x - 3y - 3 = 0.$$

It might be noted that, with very little practice, most of the above can be done in one's head and the answer written directly.

PROBLEMS

A In Problems 1–20, find an equation of the line described and express it in general form with integer coefficients. Sketch the line.

1. $m = 4$, $b = 2$ **2.** $m = -1$, $b = 3$

3. $m = 5$, $b = 1/2$ **4.** $m = 2/3$, $b = -1/3$

5. $m = 3/4$, $b = 2/3$ **6.** $m = -1/6$, $b = -5/4$

7. $m = 5$, $a = -2$ **8.** $m = 6$, $a = 3$

9. $m = 0$, $b = -3$ **10.** No m, $a = 2$

11. $a = 4$, $b = 2$ **12.** $a = -1$, $b = 3$

13. $a = 2$, $b = 1/2$ **14.** $a = 1/2$, $b = 1/2$

15. $a = 2/3$, $b = -2/5$ **16.** $a = -3/4$, $b = 2/3$

17. $a = b = 0$, through $(2, 5)$ **18.** $a = b = 0$, through $(-2, -3)$

19. $a = 4$, no b **20.** No a, $b = -3$

21. Find an equation of the line parallel to $2x - 5y + 1 = 0$ and containing the point $(2, 3)$.

22. Find an equation of the line perpendicular to $x + 2y - 5 = 0$ and containing the point $(4, 1)$.

B *In Problems 23–26, find an equation of the line described and express it in the general form with integer coefficients. Sketch the line.*

23. $a = b \neq 0$, through $(2, 5)$ **24.** $a = 3b \neq 0$, through $(5, -4)$

25. $a + b = 8$, through $(3, 1)$ **26.** $ab = 6$, through $(-3, 4)$

27. Find an equation of the line parallel to $4x + y + 2 = 0$ with y intercept 3.

28. Find an equation of the line perpendicular to $4x - y - 3 = 0$ with x intercept 4.

29. Find the center of the circle circumscribed about the triangle with vertices $(1, 3)$, $(4, -2)$, and $(-2, 1)$.

30. Find the center of the circle circumscribed about the triangle with sides $x + y = 2$, $x - y = 0$, and $2x - y = 4$.

31. Find the orthocenter (points of concurrency of the altitudes) of the triangle with vertices $(-10, 11)$, $(8, 2)$, and $(2, -1)$.

32. Prove analytically that the altitudes of a triangle are concurrent.

33. For what value(s) of m does the line $y = mx - 5$ have x intercept 2?

34. For what value(s) of m does the line $y = mx + 2$ contain the point $(4, 5)$?

35. For what value(s) of a does the line $(x/a) - (y/2) = 1$ have slope 2?

36. For what value(s) of b does the line $(x/3) + (y/b) = 1$ have slope -4?

37. Plot the graph of $x^2 - y^2 = 0$.

38. Plot the graph of $xy = 0$.

39. Plot the graph of $x^2 - 5x + 6 = 0$.

40. Plot the graph of $(x + y - 1)(3x - y + 2) = 0$.

41. Show that $\mathbf{v} = A\mathbf{i} + B\mathbf{j}$ is perpendicular to $Ax + By + C = 0$.

42. Show that $\mathbf{v} = B\mathbf{i} - A\mathbf{j}$ is parallel to $Ax + By + C = 0$.

C **43.** Work Problem 32 of the previous section without expanding the determinant. [*Hint:* Use Theorem 3.5.]

44. One vertex of a rectangle is $(6, 1)$; the diagonals intersect at $(2, 4)$; and one side has slope -2. Find the other three vertices.

45. One vertex of a parallelogram is $(1, 4)$; the diagonals intersect at $(2, 1)$; and the sides have slopes 1 and $-1/7$. Find the other three vertices.

GRAPHING
CALCULATOR

46. Graph the following lines.
 (a) $y = 2x - 3$ (b) $y = 2x - 1$
 (c) $y = 2x + 3$ (d) $y = 2x + 1$
 Describe the role of the constant term.

47. Graph the following lines.
 (a) $y = x + 1$ (b) $y = 2x + 1$
 (c) $y = -x + 1$ (d) $y = -2x + 1$
 Describe the role of the x coefficient.

48. Graph the following lines.
 (a) $x/2 + y/3 = 1$ (b) $x/3 - y/4 = 1$
 How must the functions be entered?

In Problems 49–52, graph the given equation and, using either TRACE or CALC, JUMP or G SOLVE, find the intercepts. Use them to rewrite the equation in intercept form.

49. $y = 5x - 3$

50. $y = 2x/3 + 5$

51. $y = 2.54x - 1.78$

52. $y = -3.87x + 2.47$

APPLICATIONS

53. If d and t represent distance from a starting point and time elapsed, respectively, give an interpretation of the slope and d intercept for $d = 60t$. Repeat for $d = 1000 - 40t$.

54. The Carolina Furniture Company buys a lathe for $8000. It is expected that the lathe will last for 12 years, at which time it will have no value. If a straight-line method of depreciation is used, give an equation for the estimated value of the lathe as a function of time. For what values of t is your equation valid?

55. Repeat Problem 54 if it is expected that the lathe will have a secondhand value of $500 after 12 years.

56. The direct cost of manufacturing a gallon of paint is $2.35. The fixed cost (which is incurred no matter how many gallons are manufactured) is $420 per day. Give the total daily cost as a function of the number of gallons of paint made. Graph the result.

57. The Ace Tennis Racket Company manufactures two models of tennis rackets, standard and deluxe. The standard model requires 15 minutes to string, and the deluxe model needs 20 minutes; the stringing machine is available for 12 hours. If x standard models and y deluxe models are made, what is the relationship between x and y, assuming that the entire 12 hours are used? Give an interpretation of the slope of this line.

58. It is estimated that an oil field contains 40 billion barrels of oil. If the oil is pumped out at the rate of 30 million barrels per year, give an equation representing the relationship between amount of oil left in the field and time. It is recommended that y represent the amount left in millions of barrels and that t represent the time in years. What is the significance of the two intercepts? If only 80% of the oil can be recovered from the field, when will the oil field be depleted?

59. The XYZ Company determines that the cost of producing x dinette sets is $C = 120x + 10,000$. What interpretation can be given to the slope of its graph? To the C intercept? These two quantities are called the **marginal cost** and **fixed cost**, respectively.

60. An encyclopedia salesperson's weekly pay is given by the equation $P = 50n + 100$, where P is the pay and n is the number of sets of encyclopedias sold. Give an interpretation for the slope and P intercept of this equation.

61. A demand equation gives the relationship between the quantity q of an item that can be sold and the price p of the item. Suppose that, for a certain item, the demand equation is given by $p = 5 - 0.01q$. Give an interpretation for both intercepts and the slope of this line.

62. It was observed that the price of videocassettes at a certain electronics store increased by 3 cents per month throughout 1990. In June, the price was $6.35. Give an equation for the price as a function of time. (Take $t = 0$ to represent the beginning of the year.) If this price increase continues, what will a cassette cost at the end of 1991? When will the price be $7.00?

In Problems 63–67, plot the given points, draw the line that best fits the data, and use the slope and/or y intercept to determine the quantity wanted.

63. Find m of $y = mx$.

x	0	1	2	3	4
y	0	4.1	8.5	12.8	16.3

64. Find m and b of $y = mx + b$.

x	1	3	5	7	10	15
y	6.4	10.9	14.5	19.8	26.0	35.2

65. Find k of $y = kx^2$. [*Hint:* Substitute $z = x^2$ and graph $y = kz$.]

x	1	2	3	4	5
y	5.3	20.5	47.0	82.9	131.0

66. The relationship between the vapor pressure P of a liquid and its absolute temperature, T, is given by the Clausius-Clapeyron equation,

$$2.303 \log_{10} P = \frac{-\Delta H}{R} \cdot \frac{1}{T} + C,$$

where ΔH is the molar heat of vaporization of the liquid and R is the ideal gas constant, 1.987 calories degree^{-1} mole^{-1}. The following data were found.

$1/T$	0.00364	0.00357	0.00341	0.00328	0.00319
$\log_{10} P$	0.0000218	0.0000230	0.0000250	0.0000272	0.0000287

What is the molar heat of vaporization of the liquid? [See Problem 65.]

67. The Freundlich equation for adsorption is

$$y = kC^{1/n},$$

where y represents the weight in grams of substance adsorbed and C is the concentration in moles/liter of the solute. In logarithmic form, the equation is

$$\log_{10} y = \log_{10} k + \frac{1}{n} \log_{10} C.$$

Experimentation with the adsorption of acetic acid from water solutions by charcoal gave the following results.

C	0.079	0.036	0.019	0.0097	0.0045
y	0.054	0.038	0.029	0.022	0.016

What are k and n? [See Problem 65].

3.3

DISTANCE FROM A POINT TO A LINE

Before considering the distance from a point to a line, let us note the result of Problem 41 of the previous section: the vector $\mathbf{v} = A\mathbf{i} + B\mathbf{j}$ is perpendicular to $Ax + By + C = 0$. This perpendicularity allows us to find the distance from any point to a given line.

THEOREM 3.6   *The distance from the point (x_1, y_1) to the line $Ax + By + C = 0$ is*

$$d = \frac{|Ax_1 + By_1 + C|}{\sqrt{A^2 + B^2}}.$$

PROOF   The distance we are considering here is the shortest, or perpendicular, distance. As noted above, the vector $\mathbf{v} = A\mathbf{i} + B\mathbf{j}$ is perpendicular to $Ax + By + C = 0$. Let (x, y) be a point on $Ax + By + C = 0$ and $\mathbf{u}$ be the vector represented by the segment from (x, y) to (x_1, y_1) (see Figure 3.14). Thus

$$\mathbf{u} = (x_1 - x)\mathbf{i} + (y_1 - y)\mathbf{j}.$$

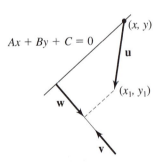

Figure 3.14

The length we seek is the length of the projection $\mathbf{w}$ of $\mathbf{u}$ upon $\mathbf{v}$. By Theorem 2.9,

$$d = |\mathbf{w}| = \frac{|\mathbf{v} \cdot \mathbf{u}|}{|\mathbf{v}|} = \frac{|A(x_1 - x) + B(y_1 - y)|}{\sqrt{A^2 + B^2}}$$

$$= \frac{|Ax_1 + By_1 - (Ax + By)|}{\sqrt{A^2 + B^2}}$$

$$= \frac{|Ax_1 + By_1 + C|}{\sqrt{A^2 + B^2}}. \quad \blacksquare$$

The following is an alternate proof that does not use vectors.

PROOF   Given the line

$$Ax + By + C = 0$$

and the point (x_1, y_1), then

$$Ax + By - (Ax_1 + By_1) = 0$$

is parallel to the given line and contains (x_1, y_1) (see Figure 3.15). Moreover, $Bx - Ay = 0$ is perpendicular to both of them. The distance we seek is the distance between the points P and Q of Figure 3.15. The point of intersection of $Bx - Ay = 0$ and $Ax + By + C = 0$ is

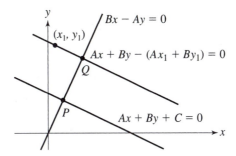

Figure 3.15

$$P = \left(\frac{-AC}{A^2 + B^2}, \frac{-BC}{A^2 + B^2} \right),$$

while the point of intersection of $Bx - Ay = 0$ and $Ax + By - (Ax_1 + By_1) = 0$ is

$$Q = \left(\frac{A(Ax_1 + By_1)}{A^2 + B^2}, \frac{B(Ax_1 + By_1)}{A^2 + B^2} \right).$$

Using the distance formula, we have

$$d = \sqrt{\left(\frac{A(Ax_1 + By_1)}{A^2 + B^2} + \frac{AC}{A^2 + B^2} \right)^2 + \left(\frac{B(Ax_1 + By_1)}{A^2 + B^2} + \frac{BC}{A^2 + B^2} \right)^2}$$

$$= \sqrt{\frac{(Ax_1 + By_1 + C)^2}{A^2 + B^2}}$$

$$= \frac{|Ax_1 + By_1 + C|}{\sqrt{A^2 + B^2}}. \quad \blacksquare$$

EXAMPLE 1   Find the distance from the point (1, 4) to the line $3x - 5y + 2 = 0$.

SOLUTION

$$d = \frac{|Ax_1 + By_1 + C|}{\sqrt{A^2 + B^2}}$$

$$= \frac{|3 \cdot 1 - 5 \cdot 4 + 2|}{\sqrt{3^2 + (-5)^2}}$$

$$= \frac{15}{\sqrt{34}}$$

This is shown graphically in Figure 3.16.

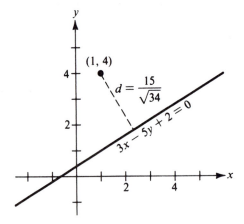

Figure 3.16

EXAMPLE 2   For what value(s) of m is the line $y - 1 = m(x + 3)$ at a distance 3 from the origin?

SOLUTION   First the given equation can be put into the general form

$$mx - y + (3m + 1) = 0.$$

Using our distance formula, we have

$$d = \frac{|Ax_1 + By_1 + C|}{\sqrt{A^2 + B^2}}$$

$$3 = \frac{m \cdot 0 - 0 + (3m + 1)}{\sqrt{m^2 + (-1)^2}}$$

$$3\sqrt{m^2 + 1} = 3m + 1$$

$$9m^2 + 9 = 9m^2 + 6m + 1$$

$$8 = 6m$$

$$m = \frac{4}{3}.$$

It might be noted that the given line has slope m and contains the point $(-3, 1)$. We have selected the slope in such a way that the line is at a distance 3 from the origin. This gives the line $4x - 3y + 15 = 0$, as shown in Figure 3.17. It might also be noted that there is a second line through $(-3, 1)$ and at a distance 3 from the origin. However, that one is a vertical line, which has no slope. It could not have been represented by our original equation.

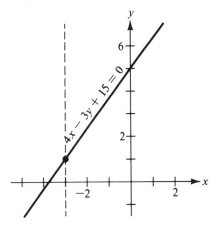

Figure 3.17

EXAMPLE 3   Find the distance between the parallel lines

$$2x - 5y - 10 = 0 \quad \text{and} \quad 2x - 5y + 4 = 0.$$

SOLUTION   First let us select a point on one of the two lines. We may select *any* point in whatever way that we find most convenient. For example, if we take $x = 0$ on the first line, then $y = -2$. Thus we have $(0, -2)$ on the first line (see Figure 3.18). Now all we have to do is find the distance from $(0, -2)$ to the other line, $2x - 5y + 4 = 0$. The result is

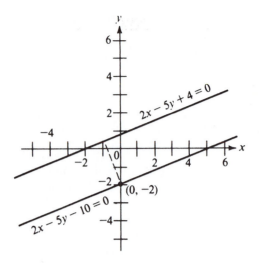

Figure 3.18

$$d = \frac{|Ax_1 + By_1 + C|}{\sqrt{A^2 + B^2}}$$

$$= \frac{|2 \cdot 0 - 5(-2) + 4|}{\sqrt{2^2 + (-5)^2}} = \frac{14}{\sqrt{29}}.$$

The absolute value in the distance formula is sometimes very inconvenient in practice. We could get rid of it if we knew whether $Ax_1 + By_1 + C$ were positive or negative. The following theorem gives us a method of determining this.

THEOREM 3.7   *If $P(x_1, y_1)$ is a point not on the line $Ax + By + C = 0$ $(B \neq 0)$, then*
(a) B and $Ax_1 + By_1 + C$ agree in sign if P is above the line.
(b) B and $Ax_1 + By_1 + C$ have opposite signs if P is below the line.

PROOF   *Case I: $B > 0$.* Let Q be the point on the given line with abscissa x_1 (see Figure 3.19). If P is above the line, then $y_1 > y$. Since $B > 0$, $By_1 > By$. Therefore,

$$Ax_1 + By_1 + C > Ax_1 + By + C.$$

Since (x_1, y) is on the line,

$$Ax_1 + By + C = 0 \quad \text{and} \quad Ax_1 + By_1 + C > 0.$$

If P is below the line, all of the above inequalities are reversed and

$$Ax_1 + By_1 + C < 0.$$

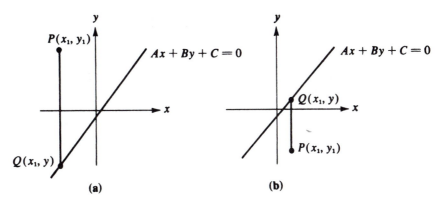

Figure 3.19

Case II: B < 0. If P is above the line, then $y_1 > y$. Since $B < 0$, $By_1 < By$. Thus,

$$Ax_1 + By_1 + C < Ax_1 + By + C.$$

Again

$$Ax_1 + By + C = 0 \quad \text{and} \quad Ax_1 + By_1 + C < 0.$$

As with Case I, all of these inequalities are reversed if P is below the line, and

$$Ax_1 + By_1 + C > 0. \quad \blacksquare$$

If $B = 0$, the line is vertical and there is no "above" nor "below." Theorem 3.7 does not apply to this case, but the distance from a point to a vertical line is easily found without using Theorem 3.6. Other methods of determining the sign of $Ax_1 + By_1 + C$ are given in Problems 34 and 35.

EXAMPLE 4 Find an equation of the line bisecting the angle from $3x - 4y - 3 = 0$ to $5x + 12y + 1 = 0$.

SOLUTION If (x, y) is any point on the desired line (see Figure 3.20), then it is equidistant from the two given lines. By Theorem 3.6,

$$\frac{|5x + 12y + 1|}{\sqrt{5^2 + 12^2}} = \frac{|3x - 4y - 3|}{\sqrt{3^2 + (-4)^2}}$$

$$5|5x + 12y + 1| = 13|3x - 4y - 3|.$$

Now let us apply Theorem 3.7. Since P is above $5x + 12y + 1 = 0$ and the coefficient of y is positive, $5x + 12y + 1$ is also positive. Similarly, since P is above $3x - 4y - 3 = 0$ and B is negative,

$$3x - 4y - 3 < 0.$$

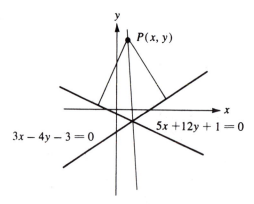

Figure 3.20

Thus

$$5(5x + 12y + 1) = -13(3x - 4y - 3) \qquad \text{or} \qquad 32x + 4y - 17 = 0.$$

Perhaps you object to the designation of P above both lines. Not every point on the bisector is above them. While this is true, the points on the bisector that are not above both are below both. Thus, we still have one expression positive and the other negative, and the result is the same.

It might be noted that we can avoid the use of Theorem 3.7 by considering both cases; that is, $5x + 12y + 1$ and $3x - 4y - 3$ either agree in signs or have opposite signs. In the one case we have $7x - 56y - 22 = 0$; the other gives $32x + 4y - 17 = 0$. Now we need to select one of them. We do so by a comparison of slopes. The first has slope $1/8$ and the second, -8. Since Figure 3.20 shows that the line we want is nearly vertical, it must have a slope that is numerically large. Thus the answer we want is $32x + 4y - 17 = 0$ with slope -8. It might be noted that the two lines above with slopes $1/8$ and -8 are perpendicular. The line $32x + 4y - 17 = 0$ bisects the angle from $3x - 4y - 3 = 0$ to $5x + 12y + 1 = 0$, while $7x - 56y - 22 = 0$ bisects the angle from the second line to the first.

PROBLEMS

A   *In Problems 1–10, find the distance from the given point to the given line.*

1. $x + y - 5 = 0$, $(2, 5)$ **2.** $2x - 4y + 2 = 0$, $(1, 3)$

3. $4x + 5y - 3 = 0$, $(-2, 4)$ **4.** $x - 3y + 5 = 0$, $(1, 2)$

5. $3x + 4y - 5 = 0$, $(1, 1)$ **6.** $5x + 12y + 13 = 0$, $(0, 2)$

7. $2x - 5y = 3$, $(3, -3)$

8. $2x + y = 5$, $(4, -1)$

9. $3x + 4 = 0$, $(2, 4)$

10. $y = 3$, $(1, 5)$

In Problems 11—16, find the distance between the given parallel lines.

11. $2x - 5y + 3 = 0$, $2x - 5y + 7 = 0$

12. $x + 2y - 2 = 0$, $x + 2y + 5 = 0$

13. $2x + y + 2 = 0$, $4x + 2y - 3 = 0$

14. $4x - y + 2 = 0$, $12x - 3y + 1 = 0$

15. $2x - y + 1 = 0$, $2x - y - 7 = 0$

16. $3x + 2y = 0$, $6x + 4y - 5 = 0$

B 17. Find the lengths of the altitudes of the triangle with vertices $(1, 2)$, $(5, 5)$, and $(-1, 7)$.

18. Find the lengths of the altitudes of the triangle with sides $x + y - 3 = 0$, $x - 2y + 4 = 0$, and $2x + 3y = 5$.

19. Find the area of the triangle of Problem 17.

20. Find the area of the triangle of Problem 18.

In Problems 21–26, find an equation of the line bisecting the angle from the first line to the second.

21. $3x - 4y - 2 = 0$, $4x - 3y + 4 = 0$

22. $8x + 15y - 5 = 0$, $5x - 12y + 1 = 0$

23. $24x - 7y + 1 = 0$, $3x + 4y - 5 = 0$

24. $12x + 35y - 4 = 0$, $15y - 8x + 3 = 0$

25. $x + y - 2 = 0$, $2x - 3 = 0$

26. $2x + y + 3 = 0$, $y + 5 = 0$

27. For what value(s) of m is the line $y = mx + 1$ at a distance 3 from $(4, 1)$?

28. For what value(s) of m is the line $y = mx + 5$ at a distance 4 from the origin?

29. For what value(s) of b is the line $(x/3) + (y/b) = 1$ at a distance 1 from the origin?

30. For what value(s) of a is the line $(x/a) + (y/2) = 1$ at a distance 2 from the point $(5, 4)$?

31. Find an equation of the set of all points equidistant from the origin and the line $x + y - 1 = 0$.

32. Find an equation for the set of all points whose distance from the origin is twice its distance from $x + y - 1 = 0$.

33. The center of the circle inscribed in a triangle is the incenter of the triangle. The center of a circle that is tangent to one side and the extensions of the other two sides is an excenter of the triangle. Find the incenter and the three excenters of the triangle with vertices $(0, 0)$, $(4, 0)$, and $(0, 3)$.

C 34. Prove that if $P = (x_1, y_1)$ is a point not on the line $Ax + By + C = 0$ $(A \neq 0)$, then
(a) A and $Ax_1 + By_1 + C$ agree in sign if P is to the right of the line.
(b) A and $Ax_1 + By_1 + C$ have opposite signs if P is to the left of the line.

35. Prove that if $P = (x_1, y_1)$ is a point not on the line $Ax + By + C = 0$ $(C \neq 0)$, then
  (a) C and $Ax_1 + By_1 + C$ agree in sign if P and the origin are on the same side of the line.
  (b) C and $Ax_1 + By_1 + C$ have opposite signs if P and the origin are on opposite sides of the line.

36. Find the center of the circle inscribed in the triangle with vertices $(0, 0)$, $(4, 0)$, and $(0, 3)$.

37. Suppose that α is the inclination of a line perpendicular (or normal) to the line l and p is the directed distance of l from the origin, p being positive if l is above the origin and negative if l is below (see Figure 3.21). Show that l can be put into the form

$$x \cos \alpha + y \sin \alpha - p = 0.$$

This is called the **normal form** of the line.

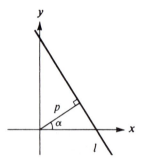

Figure 3.21

38. A board leaning against a fence makes an angle of 30° with the horizontal. If the board is 4 feet long (see Figure 3.22), what is the diameter of the largest pipe which will fit between the board, the fence, and the ground?

39. A small portion of Interstate Highway I97 is shown in Figure 3.23. FM 125 runs north-south, and FM 238 runs east-west. How far is Arcadia from I97?

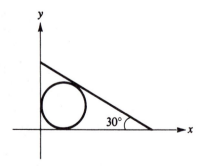

Figure 3.22

40. Basingstoke is 30 miles east of Exit 18 (see Figure 3.23). How far is it from I97?

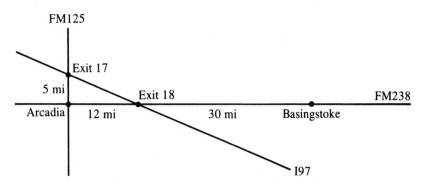

Figure 3.23

3.4

FAMILIES OF LINES

The equation

$$y = 2x + b$$

is in the form $y = mx + b$, with $m = 2$; thus, it represents a line with slope 2 and y intercept b. But what is b? Clearly we could substitute many different values for b and get equations of many different lines. It is of interest then to consider the following set, or family, of equations representing lines.

$$M = \{y = 2x + b \mid b \text{ real}\}$$

M represents a set of parallel lines all having slope 2; in fact, it represents the set of *all* lines having slope 2 (see Figure 3.24). The b in $y = 2x + b$ is called a parameter. Since the equation has a single parameter, M is called a one-parameter family of lines. Let us consider a few more examples.

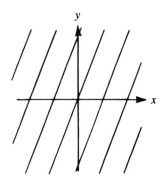

Figure 3.24

EXAMPLE 1 $\{y - 2 = m(x - 1)|m \text{ real}\}$ represents a family of lines through the point $(1, 2)$; however, it does not represent all such lines. The vertical line $x = 1$ (which has no slope) is not a member of this family (see Figure 3.25). The set of *all* lines through the point $(1, 2)$ is $\{y - 2 = m(x - 1) \mid m \text{ real}\} \cup \{x = 1\}$.

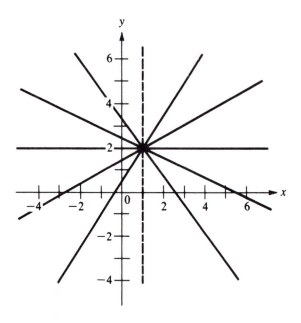

Figure 3.25

The use of a computer or graphing calculator, with its ability to graph several functions on the same set of axes, is especially useful when considering families of lines. Once again, the greatest obstacle to overcome is the requirement that the equation be solved explicitly for y as a function of x. Once this has been done, there is no need to simplify; simply replace the parameter by one value, then another, and so forth, to get several members of the family of lines. When the resulting functions are graphed, their common properties often become apparent.

To illustrate, the equation of Example 1 can be put into the form

$$y = 2 + m(x - 1).$$

By replacing m by several different values, we get

$$y_1 = 2 + 0(x - 1); \quad y_2 = 2 + 1(x - 1);$$
$$y_3 = 2 - 1(x - 1); \quad y_4 = 2 + 2(x - 1).$$

When these are graphed (see Figure 3.25), it is clear that they all contain the point $(1, 2)$.

Unfortunately, the one thing the graphs will not tell us is whether the family contains all lines through (1, 2) or whether there are exceptions. This must still be done by a consideration of the form of the given equation. In this case, the presence of the slope m should warn us that a line with no slope (a vertical line) cannot be represented.

EXAMPLE 2 $\{x/2 + y/b = 1 \mid b \text{ real}, b \neq 0\}$ represents a family of lines, all having x intercept 2 and some y intercept. It represents all such lines. However, it does not represent all lines having x intercept 2, since the line $x = 2$ is not represented, nor does it represent all lines through (2, 0), since $x = 2$ and $y = 0$ are not included (see Figure 3.26).

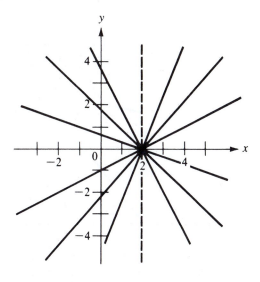

Figure 3.26

EXAMPLE 3 $\{y = mx + b \mid m, b \text{ real}\}$ is a two-parameter family of lines representing all nonvertical lines.

EXAMPLE 4 $\{x = k \mid k \text{ real}\}$ is the family of all vertical lines.

EXAMPLE 5 $\{2x + 3y - 6 + k(4x - y + 2) = 0 \mid k \text{ real}\}$ represents a family of lines (no matter what value we choose for k, the resulting equation is linear) all containing the point of intersection of

$$2x + 3y - 6 = 0 \quad\text{and}\quad 4x - y + 2 = 0$$

(because any point satisfying $2x + 3y - 6 = 0$ and $4x - y + 2 = 0$ must satisfy

$$2x + 3y - 6 + k(4x - y + 2) = 0$$

no matter what value of k we choose). Again, it does not represent *all* such lines; the line $4x - y + 2 = 0$ is not a member of this family (see Figure 3.27).

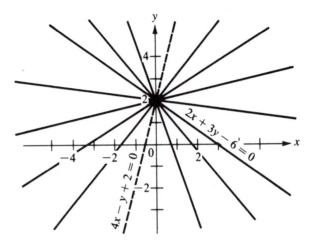

Figure 3.27

This family is more difficult to recognize than the previous ones. Once again, we can get a handle on the problem by using computer or calculator graphing. Unfortunately, solving for y is also more difficult than in the previous examples:

$$2x + 3y - 6 + k(4x - y + 2) = 0$$
$$2x + 3y - 6 + 4kx - ky + 2k = 0$$
$$(3 - k)y = -(2 + 4k)x + (6 - 2k)$$
$$y = (-(2 + 4k)x + (6 - 2k))/(3 - k).$$

The equation is left in this form rather than the "built-up" form because this is the form that must be used with a computer. Again, we may simply replace k by first one value, then another to get several examples from the family. As before, there is no need to simplify the expression; the computer can do it for us.

Upon graphing several examples, it becomes clear that they all contain the point (0, 2). However, it is not clear that this is the point of intersection of the lines

$$2x + 3y - 6 = 0 \quad \text{and} \quad 4x - y + 2 = 0.$$

Nor is it clear that the second of these two lines is not a member of the family. Thus, while a graphing calculator can help somewhat, it cannot solve this problem completely.

EXAMPLE 6 $\{Ax + By + C = 0 \mid A, B, C \text{ real}\}$ is a three-parameter family representing all lines in the plane.

Let us now consider the use of families of lines. This concept is most useful in finding an equation of a line which cannot be represented in any of the standard forms that we have seen. Suppose we consider the following example.

EXAMPLE 7   Find an equation(s) of a line(s) that contains the point (6, 0) and is a distance 5 from the point (1, 3).

SOLUTION $\{y = m(x - 6) \mid m \text{ real}\}$ represents a family of lines all containing the point (6, 0). Note that it does not represent all lines containing the point (6, 0); the only one not represented is the vertical line with equation $x = 6$. Thus, the family of all lines containing (6, 0) is (see Figure 3.28)

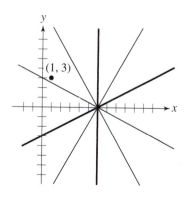

Figure 3.28

$$\{y = m(x - 6) \mid m \text{ real}\} \cup \{x = 6\}.$$

Now we must choose those members of the family that are at a distance 5 from (1, 3). We first consider those lines of the form $y = m(x - 6)$, which can be rewritten in the form

$$mx - y - 6m = 0.$$

The distance from this line to the point (1, 3) is

$$\frac{|m - 3 - 6m|}{\sqrt{m^2 + 1}} = 5.$$

Multiply both sides by $\sqrt{m^2 + 1}$ and square.

$$|-3 - 5m| = 5\sqrt{m^2 + 1}$$
$$9 + 30m + 25m^2 = 25m^2 + 25$$
$$m = \frac{8}{15}$$

Substituting this value back into the original equation, we get

$$y = \frac{8}{15}(x - 6)$$

$$8x - 15y - 48 = 0.$$

Now we must consider the line $x = 6$, which is a distance 5 from the point (1, 3). Thus, the two lines we want are

$$8x - 15y - 48 = 0 \qquad \text{and} \qquad x - 6 = 0.$$

EXAMPLE 8   Find an equation(s) of the line(s) parallel to $3x - 5y + 2 = 0$ and containing the point (3, 8).

SOLUTION   The family of all lines parallel to $3x - 5y + 2 = 0$ (including the given line) is $\{3x - 5y = k \mid k \text{ real}\}$ (see Figure 3.29). The member of the family which contains (3, 8) satisfies the condition

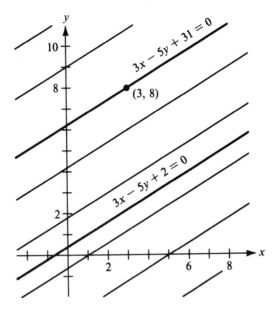

Figure 3.29

$$3 \cdot 3 - 5 \cdot 8 = k$$
$$k = -31.$$

The equation desired is $3x - 5y + 31 = 0$. The above procedure is simple enough to do mentally, and a similar procedure can be used for perpendicular lines.

EXAMPLE 9

Find an equation(s) of the line(s) perpendicular to $3x - 5y + 2 = 0$ and containing the point $(3, 8)$.

SOLUTION

The family of all lines perpendicular to $3x - 5y + 2 = 0$ is $\{5x + 3y = k \mid k \text{ real}\}$ (see Figure 3.30 on page 122). The member that contains $(3, 8)$ satisfies the conditions

$$5 \cdot 3 + 3 \cdot 8 = k$$
$$k = 39.$$

The desired equation is $5x + 3y - 39 = 0$.

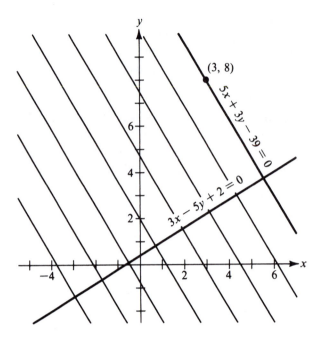

Figure 3.30

PROBLEMS

A   *In Problems 1–14, describe the family of lines given. Indicate whether or not it contains every line of that description, and, if not, give all the lines with that description which are not included in the family.*

1. $\{y - 4 = m(x + 1) \mid m \text{ real}\}$

2. $\{y = mx - 5 \mid m \text{ real}\}$

3. $\left\{\dfrac{x}{2} + \dfrac{y}{b} = 1 \mid b \text{ real}, b \neq 0\right\}$

4. $\{x = ky \mid k \text{ real}\}$

5. $\{Ax + By = 0 \mid A, B \text{ real}, A \text{ and } B \text{ not both } 0\}$

6. $\{2x - 3y = k \mid k \text{ real}\}$

7. $\left\{\dfrac{x}{a} + \dfrac{y}{b} = 1 \mid a, b \text{ real}, a \neq 0, b \neq 0\right\}$

8. $\{y = mx + b \mid m, b \text{ real}\}$

9. $\{2x + 3y + 1 + k(4x + 2y - 5) = 0 \mid k \text{ real}\}$

10. $\{x = k \mid k \text{ real}\}$

11. $\left\{\dfrac{x}{a} + \dfrac{y}{2a} = 1 \mid a \text{ real, } a \neq 0\right\}$

12. $\left\{\dfrac{x}{a} + \dfrac{y}{3 - a} = 1 \mid a \text{ real, } a \neq 0, a \neq 3\right\}$

13. $\{y = mx + m \mid m \text{ real}\}$

14. $\{y - a = m(x - a) \mid a, m \text{ real}\}$

In Problems 15–24, give, in set notation, the family described.

15. All lines parallel to $3x - 5y - 7 = 0$

16. All lines perpendicular to $3x - 5y - 7 = 0$

17. All lines containing $(2, 5)$

18. All lines with x intercept twice the y intercept

19. All lines containing the point of intersection of $3x - 5y + 1 = 0$ and $2x + 3y - 7 = 0$

20. All horizontal lines

21. All lines containing the origin

22. All lines at a distance 3 from the origin

23. All lines at a distance 5 from $(6, 0)$

24. All lines which form with the coordinate axes a triangle of area 4

In Problems 25–28, find the lines satisfying the given condition that are (a) parallel and (b) perpendicular, respectively, to the given line.

25. Containing $(5, 8)$; $3x - 5y + 1 = 0$ 26. Containing $(3, 2)$; $2x + 3y - 7 = 0$

27. y intercept 5; $4x + 2y - 5 = 0$ 28. x intercept 2; $3x + y + 2 = 0$

B 29. Find an equation(s) of the line(s) with slope 5 at a distance 3 from the origin.

30. Find an equation(s) of the line(s) perpendicular to $3x - 4y + 1 = 0$ and at a distance 4 from $(2, 3)$.

31. Find an equation(s) of the line(s) containing $(5, 4)$ and at a distance 2 from $(-1, -3)$.

32. Find an equation(s) of the line(s) containing $(3, -1)$ and at a distance 4 from $(-1, 3)$.

33. Find an equation(s) of the line(s) containing $(7, 1)$ and at a distance 5 from $(2, -5)$.

34. Find an equation(s) of the line(s) containing $(-4, 3)$ and at a distance 5 from $(-2, 2)$.

35. Find an equation(s) of the line(s) containing the point of intersection of $3x - y - 5 = 0$ and $2x + 2y - 3 = 0$ and having slope 2.

36. Find an equation(s) of the line(s) containing the point of intersection of $4x + 5y - 1 = 0$ and $3x - 2y + 1 = 0$ and the point $(1, 1)$.

37. Find an equation(s) of the line(s) containing $(4, -3)$, such that the sum of the intercepts is 5.

38. Find an equation(s) of the line(s) with slope 3 such that the sum of the intercepts is 12.

39. Find an equation(s) of the line(s) containing $(2, 3)$ and forming a triangle of area 16 with the coordinate axes.

40. Prove analytically that the bisector of an exterior angle determined by the two equal sides of an isosceles triangle is parallel to the third side.

41. An isosceles right triangle is circumscribed about the circle with center (2, 2) and radius 2. The coordinate axes are two of the sides. What is the third?

42. An isosceles right triangle is circumscribed about the circle with center (4, 2) and radius 2. The x axis is the hypotenuse. What are the other two sides?

43. An equilateral triangle is circumscribed about the circle with center (4, 2) and radius 2. The x axis is one side. What are the other two?

**GRAPHING
CALCULATOR**

44. Graph $3x + 2y - 1 = 0$ and $x - y + 3 = 0$. Graph $3x + 2y - 1 + k(x - y + 3) = 0$ for $k = 1, 2, -1, -2$. Describe the above family of lines. What value of k (if any) gives the graph of $3x + 2y - 1 = 0$? $x - y + 3 = 0$?

REVIEW PROBLEMS

A **1.** Write an equation (in general form with integer coefficients) for each of the following lines.
(a) The line through (1, 5) and (−2, 3)
(b) The line with slope 2 and x intercept 3
(c) The line with inclination 135° and y intercept 1/3
(d) The line through (2, 3) and (2, 8)

2. Write an equation (in general form with integer coefficients) for each of the following lines.
(a) The line through (4, 2) and parallel to $3x - y + 4 = 0$
(b) The line with x intercept 1/2 and y intercept −5/4
(c) The horizontal line through (3, −2)

3. Find the slope and intercepts of each of the following lines.
(a) $x - 4y + 1 = 0$ (b) $2x + 3y + 5 = 0$
(c) $5x + 2y = 0$ (d) $3x + 1 = 0$

4. Find the distance from (2, −5) to $12x + 5y + 7 = 0$.

5. Find the distance between $3x - y + 5 = 0$ and $6x - 2y - 7 = 0$.

6. Describe the family of lines given. Indicate whether or not it contains every line of that description, and, if not, give all lines with that description that are not included in the family.
(a) $\{y - 1 = m(x + 3) \mid m \text{ real}\}$
(b) $\{y = 3x + b \mid b \text{ real}\}$
(c) $\{x/a - y/3 = 1 \mid a \text{ real}, a \neq 0\}$

7. Give, in set notation, the family described.
(a) All lines containing (5, −1)
(b) All lines perpendicular to $3x + 2y - 6 = 0$
(c) All lines at a distance 3 from (2, 5)

8. Find k of $u = kv$ by graphing the given data and approximating by a line.

v	-1	2	4	7	8
u	-1.45	2.90	5.71	9.89	11.32

B

9. Find an equation of the perpendicular bisector of the segment joining (4, 1) and (0, -3).

10. A triangle has vertices (1, 5), (-2, 3), and (4, -1). Find equations for the three altitudes.

11. Find the lengths of the medians of the triangle of Problem 10.

12. Find equations for the three medians of the triangle of Problem 10.

13. If the line l has slope 3 and contains the point (-1, 1), at what points does it cross the coordinate axes?

14. Find an equation of the line bisecting the angle from $x + y - 5 = 0$ to $x - 7y + 3 = 0$.

15. Find an equation(s) of the line(s) containing (5, 1) and at a distance 1 from the origin.

16. Find an equation(s) of the line(s) with slope 3 and containing the point of intersection of $2x + 3y - 5 = 0$ and $3x - 7y + 5 = 0$.

17. Sketch $x^2 - xy + 3x - 3y = 0$.

18. Find m and b of $y = mx + b$ by graphing the given data and approximating by a line.

x	1	3	7	10	12	15
y	7.5	13.3	26.5	36.0	42.8	51.6

Chapter **4** *The Circle*

THE STANDARD FORM FOR AN EQUATION OF A CIRCLE

The standard form for an equation of a circle is a direct consequence of the definition and the length formula.

DEFINITION *A circle is the set of all points in a plane at a fixed positive distance (radius) from a fixed point (center).*

THEOREM 4.1 *A circle with center (h, k) and radius r has equation*

$$(x - h)^2 + (y - k)^2 = r^2.$$

PROOF If (x, y) is any point on the circle, then the distance from the center (h, k) to (x, y) is r (see Figure 4.1).

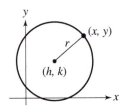

Figure 4.1

$$r = \sqrt{(x - h)^2 + (y - k)^2}$$

Squaring, we have

$$(x - h)^2 + (y - k)^2 = r^2.$$

Since the steps above are reversible, we see that every point satisfying the equation of Theorem 4.1 is on the circle described. ∎

EXAMPLE 1 Give an equation for the circle with center $(3, -5)$ and radius 2 (see Figure 4.2).

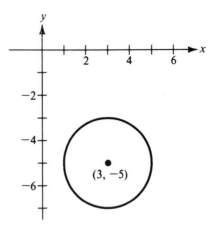

Figure 4.2

SOLUTION From Theorem 4.1, an equation is

$$(x - 3)^2 + [y - (-5)]^2 = 2^2,$$

or

$$(x - 3)^2 + (y + 5)^2 = 4.$$

Although the above form is a convenient one, in that it shows at a glance the center and radius of the circle, another form is usually used. Called the general form, it is comparable to the general form of a line. Let us first illustrate this form with the result of Example 1. Squaring the two binomials and combining similar terms, we have

$$(x - 3)^2 + (y + 5)^2 = 4$$
$$x^2 - 6x + 9 + y^2 + 10y + 25 = 4$$
$$x^2 + y^2 - 6x + 10y + 30 = 0.$$

Normally an equation of a circle will be given in this form. Let us now repeat the manipulation, starting with the standard form of Theorem 4.1.

$$(x - h)^2 + (y - k)^2 = r^2$$
$$x^2 - 2hx + h^2 + y^2 - 2ky + k^2 = r^2$$
$$x^2 + y^2 - 2hx - 2ky + (h^2 + k^2 - r^2) = 0$$

The last equation is in the form

$$x^2 + y^2 + D'x + E'y + F' = 0.$$

Upon multiplication by a nonzero constant, A, we have

$$Ax^2 + Ay^2 + Dx + Ey + F = 0 \qquad (A \neq 0),$$

as the following theorem states.

THEOREM 4.2   *Every circle can be represented in the general form*

$$Ax^2 + Ay^2 + Dx + Ey + F = 0 \qquad (A \neq 0).$$

It is a simple matter to take an equation of a circle in the standard form and reduce it to the general form. We have already seen an example of this. However, it is somewhat more difficult to go from the general form to the standard form. The latter is accomplished by the process of "completing the square." To see how this is accomplished, suppose we consider

$$(x + a)^2 = x^2 + 2ax + a^2.$$

The constant term a^2 and the coefficient of x have a definite relationship; namely, the constant term is the square of one-half the coefficient of x. Thus,

$$a^2 = \left[\frac{1}{2}(2a) \right]^2.$$

Note, however, that this relationship holds only when the coefficient of x^2 is 1.

This relationship suggests the following procedure. If the coefficients of x^2 and y^2 are not one, make them one by division. Group the x terms and the y terms on one side of the equation and take the constant to the other side. Then complete the square on both the x and the y terms. Remember that whatever is added to one side of an equation must be added to the other in order to maintain equality.

EXAMPLE 2   Express $2x^2 + 2y^2 - 2x + 6y - 3 = 0$ in the standard form. Sketch the graph of the equation.

SOLUTION

$$2x^2 + 2y^2 - 2x + 6y - 3 = 0$$

$$x^2 + y^2 - x + 3y - \frac{3}{2} = 0$$

$$(x^2 - x \quad) + (y^2 + 3y \quad) = \frac{3}{2}$$

$$\left(x^2 - x + \frac{1}{4}\right) + \left(y^2 + 3y + \frac{9}{4}\right) = \frac{3}{2} + \frac{1}{4} + \frac{9}{4}$$

$$\left(x - \frac{1}{2}\right)^2 + \left(y + \frac{3}{2}\right)^2 = 4$$

Thus, the original equation represents a circle with center $(1/2, -3/2)$ and radius 2. The graph is given in Figure 4.3.

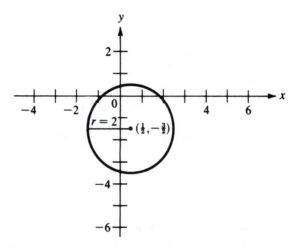

Figure 4.3

The next two examples show that the converse of Theorem 4.2 is not true: that is, an equation of the form

$$Ax^2 + Ay^2 + Dx + Ey + F = 0$$

does not necessarily represent a circle.

EXAMPLE 3   Express $x^2 + y^2 + 4x - 6y + 13 = 0$ in standard form. Sketch the graph of the equation.

SOLUTION

$$x^2 + y^2 + 4x - 6y + 13 = 0$$
$$(x^2 + 4x \quad) + (y^2 - 6y \quad) = -13$$
$$(x^2 + 4x + 4) + (y^2 - 6y + 9) = -13 + 4 + 9$$
$$(x + 2)^2 + (y - 3)^2 = 0$$

Since neither of the two expressions on the left-hand side of the last equation can be negative, their sum can be zero only if both expressions are zero. This is possible only when $x = -2$ and $y = 3$. Thus, the point $(-2, 3)$ is the only point in the plane that satisfies the original equation. The graph is given in Figure 4.4.

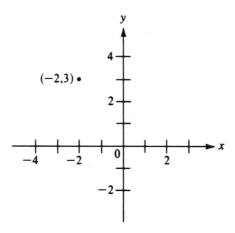

Figure 4.4

EXAMPLE 4

Express $x^2 + y^2 + 2x + 8y + 19 = 0$ in standard form.

SOLUTION

$$x^2 + y^2 + 2x + 8y + 19 = 0$$
$$(x^2 + 2x \quad) + (y^2 + 8y \quad) = -19$$
$$(x^2 + 2x + 1) + (y^2 + 8y + 16) = -19 + 1 + 16$$
$$(x + 1)^2 + (y + 4)^2 = -2$$

Again, since neither expression on the left-hand side of the last equation can be negative, their sum cannot possibly be negative. There is no point in the plane satisfying this equation. It has no graph.

The results illustrated by the last three examples are stated in the next theorem.

THEOREM 4.3

The graph of every equation of the form

$$Ax^2 + Ay^2 + Dx + Ey + F = 0 \qquad (A \neq 0)$$

HISTORICAL NOTE

FERMAT

Pierre de Fermat (pronounced fair-MAH) was an outstanding French mathematician of the seventeenth century. Nevertheless, he made his living as a lawyer. His legal responsibilities were sufficiently light that they left him with enough free time to pursue mathematics as a hobby.

Although Descartes is generally credited with the invention of analytic geometry, it is now recognized that Fermat had developed it quite independently of Descartes. In fact, Fermat had had it quite well developed by 1629, eight years before Descartes had published his work. However, Fermat had not shared his work with anyone until 1636; and it was not published until 1679, fourteen years after his death.

Fermat's ideas were quite similar to those of Descartes except that he used two axes instead of one, and his development was more systematic and complete than that of Descartes. By 1629, he had investigated the general equation of a line, a circle with center at the origin, and conic sections.

By 1638, he had developed a method of finding tangents which is similar to that used today and substantially better than that of Descartes. Furthermore, he used it to find the maximum and minimum values of functions, thus setting the stage for the development of calculus.

In spite of Fermat's geometric advances, his major contributions to mathematics were in the field of number theory. A conjecture that still bears his name is called **Fermat's Last Theorem**. It is based upon the idea of a **Pythagorean** triple, which is a set of three positive integers x, y, and z such that $x^2 + y^2 = z^2$. There are many known Pythagorean triples—for example, 3, 4, 5 and 5, 12, 13. The question arose concerning the possibility of triples satisfying the equation $x^n + y^n = z^n$, where n is larger than 2. Fermat said no. In a book, he indicated that he had an ingenious proof that was too long to be given in the margin. His reputation as a mathematician was such that most people accepted that he really had a proof although many now doubt it. In spite of the many attempts by outstanding mathematicians, this problem remained unsolved for over 350 years. In 1993, Andrew J. Wiles of Princeton University presented an outline of a proof that there are no such triples for n greater than 2. Although he had not yet (as of the summer of 1994) published a more detailed proof, many who have seen his outline believe him to have a proof.

In any case, the numerous attempts to prove Fermat's Last Theorem have resulted in a great deal of number theoretic research.

is either a circle or a point, or contains no points.
*(The last two cases are called the **degenerate cases** of a circle.)*

The use of a computer or calculator for the graphing of circles has several problems associated with it. The first is the old problem of having to solve for y as a function of x. Since an equation of a circle is a quadratic in y (as well as in x), this requires the use of the quadratic formula. Suppose we consider this with the equation of Example 2. We begin by rearranging the terms in the following order: y^2 terms, y terms, and constant terms (at least constant with respect to y—there may be x terms here). Then, looking upon y as the only variable and x as a constant, we solve by the quadratic formula.

$$2x^2 + 2y^2 - 2x + 6y - 3 = 0$$

$$2y^2 + 6y + (2x^2 - 2x - 3) = 0$$

$$y = \frac{-6 \pm \sqrt{36 - 4 \cdot 2(2x^2 - 2x - 3)}}{4}$$

If the equation is in the standard form (or if we first complete the square to put it into that form), we can easily solve for y without using the quadratic formula. For example, starting with the result of Example 2—that is, with the equation of the circle in the standard form—we have

$$(x - 1/2)^2 + (y + 3/2)^2 = 4$$

$$(y + 3/2)^2 = 4 - (x - 1/2)^2$$

$$y + 3/2 = \pm\sqrt{4 - (x - 1/2)^2}$$

$$y = -3/2 \pm \sqrt{4 - (x - 1/2)^2}.$$

By either method, we have two functions because of the $\pm$. They must be entered individually in the form

$Y_1 = (-6 + \sqrt{(36 - 8(2X^2 - 2X - 3)))/4}$
$Y_2 = (-6 - \sqrt{(36 - 8(2X^2 - 2X - 3)))/4}$

or as

$Y_1 = -3/2 + \sqrt{(11/2 - (X - 1/2)^2)}$
$Y_2 = -3/2 - \sqrt{(11/2 - (X - 1/2)^2)}.$

Once again, as we have seen in Section 1.7, the two halves of the circle may not meet to give a closed curve. In this connection, see Problem 47.

This is our first example of a second-degree equation. The general second-degree equation has the form

$$Ax^2 + Bxy + Cy^2 + Dx + Ey + F = 0.$$

A curve with an equation in this form represents a conic section: that is, the figure formed by the intersection of a plane and a right circular cone (see Section 5.1). In the case of a circle, $B = 0$ and $A = C$.

In the introduction to the graphing calculator, a program is given to graph any second-degree equation. It relieves us of the tediousness of having to solve a quadratic equation in y and entering the two resulting functions.

We now have three methods of graphing an equation in the form of a circle using a calculator: by use of the quadratic formula, by completing the square and solving for y, and by use of the graphing program. Whereas completing the square puts the equation into standard form giving us the center and radius, the other two methods do not lead us to them naturally. Nevertheless, we can find them from the graph by using CALC, JUMP, or G SOLVE. To do so, we merely find the highest point on the upper half of the circle and the lowest point on the lower half. The midpoint of the vertical line segment joining these points is the center, and the distance from the center to either of these points is the radius.

EXAMPLE 5 Using a graphing calculator, graph the equation $3.27x^2 + 3.27y^2 - 18.19x + 7.03y - 21.35 = 0$ using each of the three methods we have discussed. Find the center and the radius.

SOLUTION Since we want the center and the radius, let us solve by completing the square.

$$3.27x^2 + 3.27y^2 - 18.19x + 7.03y - 21.35 = 0$$
$$x^2 + y^2 - 5.5627x + 2.1498y - 6.5291 = 0$$
$$x^2 - 5.5627x + y^2 + 2.1498y = 6.5291$$
$$x^2 - 5.5627x + 7.7359 + y^2 + 2.1498y + 1.1554 = 15.4204$$
$$(x - 2.7813)^2 + (y + 1.0749)^2 = 15.4204$$

This tells us that the center is $(2.7813, -1.0749)$ and $r = \sqrt{15.4204} = 3.9269$. Now we solve this equation for y.

$$(y + 1.0749)^2 = 15.4204 - (x - 2.7813)^2$$
$$y + 1.0749 = \pm\sqrt{15.4204 - (x - 2.7813)^2}$$
$$y = -1.0749 \pm \sqrt{15.4204 - (x - 2.7813)^2}$$

Using these two functions, we have the result shown in Figure 4.5.

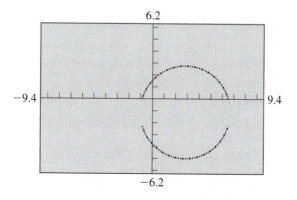

Figure 4.5

ALTERNATE SOLUTION We can get the graph of Figure 4.5 by using the graphing program with

$$A = 3.27 \qquad B = 0 \qquad C = 3.27$$
$$D = -18.19 \qquad E = 7.03 \qquad F = -21.35.$$

Alternately, we can use the quadratic formula to solve for y.

$$3.27x^2 + 3.27y^2 - 18.19x + 7.03y - 21.35 = 0$$
$$3.27y^2 + 7.03y + (3.27x^2 - 18.19x - 21.35) = 0$$
$$y = \frac{-7.03 \pm \sqrt{7.03^2 - 4 \cdot 3.27(3.27x^2 - 18.19x - 21.35)}}{2 \cdot 3.27}$$

This results in the entries

$$Y_1 = (-7.03 + \sqrt{(7.03^2 - 4*3.27(3.27X^2 - 18.19X - 21.35)))}/(2*3.27)$$
$$Y_2 = (-7.03 - \sqrt{(7.03^2 - 4*3.27(3.27X^2 - 18.19X - 21.35)))}/(2*3.27)$$

and gives the graph of Figure 4.5.

Using CALC, JUMP, or G SOLVE with the first curve generated, we can easily determine the maximum to be (2.7813, 2.8520). Using CALC, JUMP, or G SOLVE on the second curve, we find the minimum of (2.7813, −5.0018). Thus the center has x coordinate 2.7813 and y coordinate

$$(2.8520 - 5.0018)/2 = -1.0749.$$

The radius is

$$2.8520 - (-1.0749) = 3.9269.$$

PROBLEMS

A　*In Problems 1–16, write an equation of the circle described in both the standard form and the general form. Sketch the graph of each equation.*

1. Center (1, 3); radius 5　　　　　**2.** Center (0, 0); radius 1

3. Center (5, −2); radius 2　　　　　**4.** Center (0, 3); radius 1/2

5. Center (1/2, −3/2); radius 2　　　**6.** Center (−2/3, −1/2); radius 3/2

B　**7.** Center (4, −2); (3, 3) on the circle　　**8.** Center (−1, 0); (4, −3) on the circle

9. (2, −3) and (−2, 0) are the endpoints of a diameter.

10. (−3, 5) and (2, 4) are the endpoints of a diameter.

11. Radius 3; in the first quadrant and tangent to both axes

12. Radius 5; in the fourth quadrant and tangent to both axes
13. Radius 2; tangent to $x = 2$ and $y = -1$ and above and to the right of these lines
14. Radius 3; tangent to $x = -3$ and $y = 4$ and below and to the left of these lines
15. Tangent to both axes at $(4, 0)$ and $(0, -4)$
16. Tangent to $x = -2$ and $y = 2$ at $(-2, 0)$ and $(-4, 2)$

In Problems 17–28, express the equation in standard form. Sketch if the graph is nonempty, either by hand or by using a graphing calculator.

17. $x^2 + y^2 - 2x - 4y + 1 = 0$ 18. $x^2 + y^2 + 4x - 6y - 3 = 0$

19. $x^2 + y^2 + 6x - 16 = 0$ 20. $x^2 + y^2 - 10x + 4y + 29 = 0$

21. $4x^2 + 4y^2 - 4x - 12y + 1 = 0$ 22. $9x^2 + 9y^2 - 12x - 24y - 13 = 0$

23. $5x^2 + 5y^2 - 8x - 4y - 121 = 0$ 24. $9x^2 + 9y^2 - 18x - 12y - 23 = 0$

25. $9x^2 + 9y^2 - 6x + 18y + 11 = 0$ 26. $36x^2 + 36y^2 - 36x + 24y - 23 = 0$

27. $36x^2 + 36y^2 - 48x - 36y + 25 = 0$ 28. $8x^2 + 8y^2 + 24x - 4y + 19 = 0$

29. Find the point(s) of intersection of

$$x^2 + y^2 - x - 3y - 6 = 0 \quad \text{and} \quad 4x - y - 9 = 0.$$

30. Find the point(s) of intersection of

$$x^2 + y^2 + 4x - 12y + 6 = 0 \quad \text{and} \quad 3x - 5y + 2 = 0.$$

31. Find the point(s) of intersection of

$$x^2 + y^2 + 5x + y - 26 = 0 \quad \text{and} \quad x^2 + y^2 + 2x - y - 15 = 0.$$

32. Find the point(s) of intersection of

$$x^2 + y^2 + x + 12y + 8 = 0 \quad \text{and} \quad 2x^2 + 2y^2 - 4x + 9y + 4 = 0.$$

33. What happens when we try to solve

$$x^2 + y^2 - 2x + 4y + 1 = 0 \quad \text{and} \quad x - 2y + 2 = 0$$

simultaneously? Interpret geometrically.

34. What happens when we try to solve

$$x^2 + y^2 - 4x - 2y + 1 = 0 \quad \text{and} \quad x^2 + y^2 + 6x - 6y + 14 = 0$$

simultaneously? Interpret geometrically.

35. Find the line through the points of intersection of

$$x^2 + y^2 - x + 3y - 10 = 0 \quad \text{and} \quad x^2 + y^2 - 2x + 2y - 11 = 0.$$

36. For what value(s) of k is the line $x + 2y + k = 0$ tangent to the circle

$$x^2 + y^2 - 2x + 4y + 1 = 0?$$

p.134-135 (4.6)

37. Prove analytically that if P_1 and P_2 are the ends of a diameter of a circle and Q is any point on the circle, then $\angle P_1 Q P_2$ is a right angle.

38. A set of points in the plane has the property that every point in it is twice as far from $(1, 1)$ as it is from $(5, 3)$. What equation must be satisfied by every point (x, y) in the set?

39. Given two circles, how many lines are tangent to both of them? [*Note:* There is more than one answer, depending upon the relative positions of the circles.]

C **40.** Find equations of the lines tangent to $(x - 4)^2 + y^2 = 4$ and $(x + 2)^2 + y^2 = 1$. [*Hint:* See the previous problem for the number of tangent lines.]

41. Find the relation between A, D, E, and F of Theorem 4.2 in order that the equation represent
(**a**) a circle   (**b**) a point   (**c**) no graph.
If the equation represents a circle, find h, k, and r in terms of A, D, E, and F.

42. In general, squaring both sides of an equation is not reversible (if $x = 2$, then $x^2 = 4$; but if $x^2 = 4$, then $x = \pm 2$). Yet, in the proof of Theorem 4.1, the argument was declared to be reversible even though both sides of an equation were squared. Why?

GRAPHING CALCULATOR

In Problems 43–47, use a graphing calculator to graph and determine the center and the radius using (a) the quadratic formula, (b) completing the square and solving for y, and (c) the graphing program.

43. $2x^2 + 2y^2 - 6x - 2y - 27 = 0$

44. $x^2 + y^2 - 10x + 7y + 25 = 0$

45. $4.17x^2 + 4.17y^2 + 15.95x - 25.38y - 12.43 = 0$

46. $3.05x^2 + 3.05y^2 + 25.87x - 10.54y + 35.17 = 0$

47. $1.32x^2 + 1.32y^2 + 8.15x - 7.58y + 13.43 = 0$

48. Graph $x^2 + y^2 + 6x + 27y - 15 = 0$ using the "standard" window. Change the range to $-10 \le x \le 10$. How does the graph change?

In Problems 49–52, use CALC, JUMP, or G SOLVE to find the points of intersection of the given curves.

49. $x - y + 2 = 0$, $x^2 + y^2 + 4x - 6y - 3 = 0$

50. $2x - y + 2 = 0$, $x^2 + y^2 - 4x - 6y - 3 = 0$

51. $x^2 + y^2 + 6x - 2y - 6 = 0$, $x^2 + y^2 - 2x - 6y + 6 = 0$

52. $x^2 + y^2 - 4 = 0$, $x^2 + y^2 - 10x - 4y + 13 = 0$

APPLICATIONS

53. A mosque has a "keyhole" entry consisting of a rectangle surmounted by a circle, as shown in Figure 4.6. Find an equation of the circle with the given placement of the axes.

54. The dome of a church has a cross section as shown in Figure 4.7. The circular arcs all have a radius of 5 feet, and the upper circles are tangent to each other at a point on the center line and tangent to the lower circle. Find equations of the three circles.

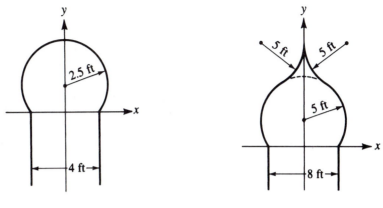

Figure 4.6   Figure 4.7

55. One nice property of round sewer covers is that they cannot fall into the sewer no matter how they are oriented. Contrast this with a square, rectangular, or triangular sewer cover. In spite of appearances to the contrary, circles are not the only possible design for sewer covers that cannot fall in. Starting with the equilateral triangle of Figure 4.8 and using arcs of circles, design a sewer cover that is not round but still cannot fall in.

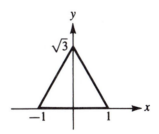

Figure 4.8

56. Given the round-headed screw shown in Figure 4.9, find an equation of the circular arc with the given placement of axes.

57. A Gothic door has the shape shown in Figure 4.10. The triangle at the top is equilateral, and each of the two arcs is circular with center at the opposite vertex. Find equations of the arcs with the given placement of axes.

58. A belt is to go around two pulleys of radii 1 inch and 2 inches with centers 6 inches apart. How long should the belt be? [*Hint:* See Problem 40.]

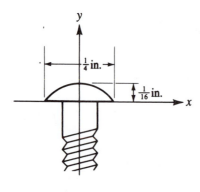

Figure 4.9

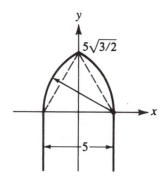

Figure 4.10

CONDITIONS TO DETERMINE A CIRCLE

We have seen two forms for equations of a circle: the standard form,

$$(x - h)^2 + (y - k)^2 = r^2,$$

with the three parameters h, k, and r, and the general form,

$$Ax^2 + Ay^2 + Dx + Ey + F = 0 \qquad (A \neq 0),$$

with the parameters A, D, E, and F. However, since $A \neq 0$, we can divide through by A to obtain

$$x^2 + y^2 + D'x + E'y + F' = 0,$$

which, like the standard form, has only three parameters. Thus we need three equations in h, k, and r or in D', E', and F' in order to determine these parameters and give the equation desired. Since each condition on a circle determines one such equation, three conditions are required to determine a circle.

As already noted above, the values of h, k, and r in the standard form for a circle are three "conditions" that determine the circle. This is illustrated in Figure 4.11.

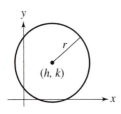

Figure 4.11

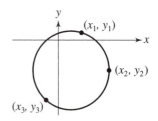

Figure 4.12

Similarly, three noncollinear points determine a circle as shown in Figure 4.12. The three points are the three conditions in this case. Knowing them gives three equations in D', E', and F' in the (slightly altered) general form of a circle. Note that one point (two coordinates) on a circle is a single "condition," while each coordinate of the center is a condition. More generally, knowing that the center is on a given line can be counted as a "condition" to determine a circle; knowing h and k is equivalent to knowing that the center is on the lines $x = h$ and $y = k$.

Finally, let us consider the following three conditions: the equation of a tangent line, the point of tangency, and another point on the circle (see Figure 4.13). The center is on the perpendicular to the tangent at the point of tangency. It is also on the perpendicular bisector of the segment joining any two points of the circle. These two lines determine the center of the circle; the radius is now easily found.

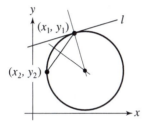

Figure 4.13

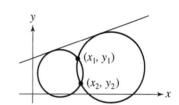

Figure 4.14

While three conditions often determine a single circle, there are situations in which they determine two circles. For example, a tangent line and a pair of points on a circle determine two circles—not one (see Figure 4.14).

EXAMPLE 1   Find an equation of the circle through points $(1, 5)$, $(-2, 3)$, and $(2, -1)$.

SOLUTION   The desired equation is

$$x^2 + y^2 + D'x + E'y + F' = 0$$

for suitable choices of D', E', and F'. Since the three given points are on the circle, they satisfy this equation. Thus

$$1 + 25 + D' + 5E' + F' = 0,$$
$$4 + 9 - 2D' + 3E' + F' = 0,$$
$$4 + 1 + 2D' - E' + F' = 0$$

or

$$D' + 5E' + F' = -26,$$
$$-2D' + 3E' + F' = -13,$$
$$2D' - E' + F' = -5 .$$

Solving simultaneously, we have $D' = -9/5$, $E' = -19/5$, and $F' = -26/5$. Thus the circle is

$$x^2 + y^2 - \frac{9}{5}x - \frac{19}{5}y - \frac{26}{5} = 0$$

or

$$5x^2 + 5y^2 - 9x - 19y - 26 = 0.$$

The example above illustrates the use of the general form to find the desired equation. The general form is rarely used because the constants D', E', and F' have no easily discernible geometric significance. The problem of finding an equation of a circle through three given points is the only one using this form. Even this problem can be solved using the standard form if we recall that the perpendicular bisector of a chord of a circle contains the center. Let us use this on the preceding problem.

ALTERNATE SOLUTION

Since the points $(1, 5)$ and $(-2, 3)$ are on the circle, the segment from one to the other is a chord of the desired circle (see Figure 4.15).

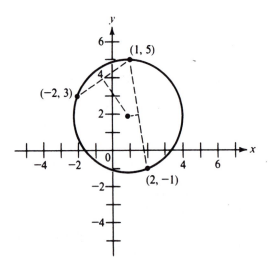

Figure 4.15

For the chord from $(1, 5)$ to $(-2, 3)$:

$$\text{midpoint} = \left(-\frac{1}{2}, 4\right)$$

$$m = \frac{5 - 3}{1 + 2} = \frac{2}{3}$$

and the perpendicular bisector is

$$y - 4 = -\frac{3}{2}\left(x + \frac{1}{2}\right)$$

$$4y - 16 = -6x - 3$$

$$6x + 4y = 13.$$

Repeating for the chord from $(1, 5)$ to $(2, -1)$:

$$\text{midpoint} = \left(\frac{3}{2}, 2\right)$$

$$m = \frac{5 + 1}{1 - 2} = -6$$

and the perpendicular bisector is

$$y - 2 = \frac{1}{6}\left(x - \frac{3}{2}\right)$$

$$12y - 24 = 2x - 3$$

$$2x - 12y = -21.$$

Thus the center is on $6x + 4y = 13$ and $2x - 12y = -21$. Solving simultaneously, we see that the center is $(9/10, 19/10)$. The radius is the distance from the center to any of the given points, say $(1, 5)$.

$$r = \sqrt{\frac{1}{100} + \frac{961}{100}} = \sqrt{\frac{962}{100}}$$

Thus the desired equation is

$$\left(x - \frac{9}{10}\right)^2 + \left(y - \frac{19}{10}\right)^2 = \frac{962}{100}$$

or

$$5x^2 + 5y^2 - 9x - 19y - 26 = 0.$$

EXAMPLE 2

Find an equation(s) of the circle(s) of radius 4 with center on the line $4x + 3y + 7 = 0$ and tangent to $3x + 4y + 34 = 0$.

SOLUTION

The three conditions lead to the following three relations involving h, k and r (see Figure 4.16).

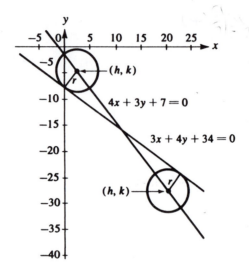

Figure 4.16

(1) $$r = 4$$

(2) $$4h + 3k + 7 = 0$$

(3) $$\frac{|3h + 4k + 34|}{\sqrt{3^2 + 4^2}} = r$$

The first and third give

$$|3h + 4k + 34| = 20.$$

Solving the second for k, we have

$$k = -\frac{4h + 7}{3},$$

and substituting into

$$|3h + 4k + 34| = 20,$$

we have

$$\left| 3h - \frac{16h + 28}{3} + 34 \right| = 20$$

$$|74 - 7h| = 60$$

$$74 - 7h = \pm 60$$

$$h = 2 \quad \text{or} \quad h = \frac{134}{7};$$

and $k = -5$ or $k = -195/7$, respectively. Thus the two solutions are

$$(x - 2)^2 + (y + 5)^2 = 16 \quad \text{and} \quad \left(x - \frac{134}{7} \right)^2 + \left(y + \frac{195}{7} \right)^2 = 16,$$

or

$$x^2 + y^2 - 4x + 10y + 13 = 0$$

and

$$49x^2 + 49y^2 - 1876x + 2730y + 55{,}197 = 0.$$

ALTERNATE SOLUTION

This problem can also be solved in the following way. Since the desired circle has radius 4 and is tangent to $3x + 4y + 34 = 0$, its center is on a line parallel to $3x + 4y + 34 = 0$ and at a distance 4 from it. There are two such lines (see Figure 4.17) given by

$$\frac{|3x + 4y + 34|}{5} = 4$$

$$3x + 4y + 34 = \pm 20$$

$$3x + 4y + 14 = 0 \quad \text{or} \quad 3x + 4y + 54 = 0.$$

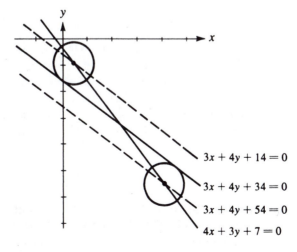

Figure 4.17

Since the center is also on $4x + 3y + 7 = 0$, we can find its coordinates by solving this equation simultaneously with each of the two equations above. From

$$3x + 4y + 14 = 0 \quad \text{and} \quad 4x + 3y + 7 = 0,$$

we get center $(2, -5)$; from

$$3x + 4y + 54 = 0 \quad \text{and} \quad 4x + 3y + 7 = 0,$$

we get center $(134/7, -195/7)$. Using these centers with the given radius, 4, we have the desired circles.

EXAMPLE 3   Find an equation(s) of the circle(s) tangent to both axes and containing the point $(-8, -1)$.

SOLUTION   The three conditions give

(1) $$h = -r$$
(2) $$k = -r$$
(3) $$(-8 - h)^2 + (-1 - k)^2 = r^2$$

(see Figure 4.18). Substituting (1) and (2) into (3), we have

$$(-8 + r)^2 + (-1 + r)^2 = r^2$$
$$r^2 - 18r + 65 = 0$$
$$(r - 5)(r - 13) = 0$$
$$r = 5 \quad \text{or} \quad r = 13.$$

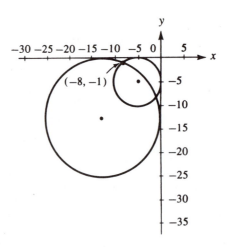

Figure 4.18

Thus we have the circle with radius 5 and center $(-5, -5)$ with equation

$$(x + 5)^2 + (y + 5)^2 = 25$$

or

$$x^2 + y^2 + 10x + 10y + 25 = 0;$$

or we have the circle with radius 13 and center $(-13, -13)$ with equation

$$(x + 13)^2 + (y + 13)^2 = 169$$

or

$$x^2 + y^2 + 26x + 26y + 169 = 0.$$

EXAMPLE 4　Find an equation(s) of the circle(s) tangent to $3x - 4y - 4 = 0$ at $(0, -1)$ and containing the point $(-1, -8)$.

SOLUTION　The center of the desired circle is on the line perpendicular to the tangent line at $(0, -1)$ (see Figure 4.19). An equation of this perpendicular is

$$4x + 3y = 4 \cdot 0 + 3(-1) \qquad \text{or} \qquad 4x + 3y + 3 = 0.$$

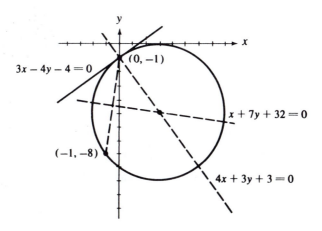

Figure 4.19

Thus, for center (h, k) we have

(1) $$4h + 3k + 3 = 0.$$

The center is also on the perpendicular bisector of the line joining $(0, -1)$ and $(-1, -8)$ (see Figure 4.19). The slope of the line joining $(0, -1)$ and $(-1, -8)$ is

7; thus the slope of a perpendicular line is $-1/7$. The midpoint of the segment from $(0, -1)$ to $(-1, -8)$ is $(-1/2, -9/2)$. By the point-slope formula, the perpendicular bisector is

$$y + \frac{9}{2} = -\frac{1}{7}\left(x + \frac{1}{2}\right)$$

or

$$x + 7y + 32 = 0.$$

Thus,

(2) $$h + 7k + 32 = 0.$$

Solving (1) and (2) simultaneously, we have

$$h = 3 \quad \text{and} \quad k = -5.$$

Using this point with $(0, -1)$, we find the radius

$$r = \sqrt{(3 - 0)^2 + (-5 + 1)^2} = 5.$$

Thus, the desired equation is

$$(x - 3)^2 + (y + 5)^2 = 25$$

or

$$x^2 + y^2 - 6x + 10y + 9 = 0.$$

PROBLEMS

B   *In Problems 1–25, find an equation(s) of the circle(s) described.*

1. Through $(-1, 2)$, $(3, 4)$, and $(2, -1)$
2. Through $(-2, -1)$, $(0, 3)$, and $(2, 0)$
3. Circumscribed about the triangle with vertices $(2, 3)$, $(0, 5)$, and $(1, -1)$
4. Circumscribed about the triangle with vertices $(1, 1)$, $(-2, 1)$, and $(1, 4)$
5. Circumscribed about the triangle with sides $x - y = 0$, $x + 2y = 0$, and $4x + y = 35$
6. Through $(2, 1)$, $(-4, 4)$, and $(6, -1)$ [*Watch out!*]
7. Tangent to the x axis; center on $2x + y - 1 = 0$; radius 5
8. Tangent to $2x + 3y + 13 = 0$ and $2x - 3y - 1 = 0$; contains $(0, 4)$
9. Tangent to $3x + 4y - 15 = 0$ at $(5, 0)$; contains $(-2, -1)$
10. Tangent to $5x - 12y + 89 = 0$ at $(-1, 7)$; contains $(16, 0)$
11. Tangent to $x + y = 0$ and $x - y - 6 = 0$; center on $3x - y + 3 = 0$
12. Tangent to $x - 3y - 7 = 0$ and $3x + y - 21 = 0$; center on $x - 3y + 3 = 0$

13. Tangent to $x - 3y = 0$ at $(0, 0)$; center on $2x + y + 1 = 0$

14. Tangent to $x - y = 0$ at $(2, 2)$; center on $2x + 3y - 7 = 0$

15. Contains $(-1, 4)$ and $(3, 2)$; center on $3x - y + 3 = 0$

16. Contains $(5, 2)$ and $(-1, 6)$; center on $x = y$

17. Tangent to $2x + 3y - 5 = 0$ at $(1, 1)$; tangent to $2x + 3y + 10 = 0$

18. Tangent to $y = 0$ at $(4, 0)$; tangent to $3x - 4y - 17 = 0$

19. Tangent to both axes; radius 3

20. Tangent to $x = 0$; center on $x + y = 10$; contains $(2, 9)$

21. Tangent to $3x - 4y + 3 = 0$ at $(-1, 0)$; radius 7

22. Tangent to $x^2 + y^2 - 22x + 20y + 77 = 0$ at $(91/17, 10/17)$; containing $(0, 1)$

23. Tangent to $x^2 + y^2 - 8x - 22y + 112 = 0$ and $3x + 4y + 19 = 0$; radius 5

24. Tangent to $2x - 3y + 6 = 0$ at $(3, 4)$; center on $3x + 2y - 17 = 0$

25. Tangent to $x + 2y - 7 = 0$ at $(5, 1)$; center on $2x - y + 5 = 0$

In Problems 26–30, use the result of Problem 31 to solve.

26. Find an equation(s) of the circle(s) containing $(1, -4)$ and the points of intersection of

$$x^2 + y^2 + 2x - 4y + 1 = 0 \quad \text{and} \quad x^2 + y^2 + 4x + 6y - 3 = 0.$$

27. Find an equation(s) of the circle(s) with center $(3, -1)$ and containing the points of intersection of

$$x^2 + y^2 - 4x - 6y + 9 = 0 \quad \text{and} \quad x^2 + y^2 - 2x - 14y + 15 = 0.$$

28. Find an equation(s) of the circle(s) with center on $x + y - 2 = 0$ and containing the points of intersection of

$$x^2 + y^2 + 4x + 6y - 3 = 0 \quad \text{and} \quad x^2 + y^2 + 2x + 2y - 2 = 0.$$

29. Find an equation(s) of the circle(s) with radius 2 and containing the points of intersection of

$$x^2 + y^2 + 2x - 2y - 3 = 0 \quad \text{and} \quad x^2 + y^2 - x - 1 = 0.$$

30. Find an equation(s) of the line(s) containing the points of intersection of

$$x^2 + y^2 - 2x - 8y + 8 = 0 \quad \text{and} \quad x^2 + y^2 + 2x - 3 = 0.$$

C 31. Suppose that

$$Ax^2 + Ay^2 + Dx + Ey + F = 0$$

and

$$A'x^2 + A'y^2 + D'x + E'y + F' = 0$$

represent two circles which intersect at the points P_1 and P_2. Show that the family

$$M = \{Ax^2 + Ay^2 + Dx + Ey + F$$
$$+ k(A'x^2 + A'y^2 + D'x + E'y + F') = 0 \mid k \text{ real}\}$$

consists of circles containing P_1 and P_2, together with the line containing these two points (when $k = -A/A'$).

32. Show that if (x_1, y_1), (x_2, y_2), and (x_3, y_3) are three noncollinear points, then the circle containing these three points has equation

$$\begin{vmatrix} x^2 + y^2 & x & y & 1 \\ x_1^2 + y_1^2 & x_1 & y_1 & 1 \\ x_2^2 + y_2^2 & x_2 & y_2 & 1 \\ x_3^2 + y_3^2 & x_3 & y_3 & 1 \end{vmatrix} = 0$$

33. Show that if (x_1, y_1), (x_2, y_2), and (x_3, y_3) are three collinear points, then the determinant of Problem 32 is linear.

34. Find an equation(s) of the line(s) tangent to $x^2 + y^2 + 4x - 10y + 4 = 0$ from the point $(3, 2)$.

35. Find an equation(s) of the line(s) tangent to $x^2 + y^2 - 8x + 2y - 152 = 0$ and having slope 1/3.

REVIEW PROBLEMS

A *In Problems 1–5, put the equation into standard form and identify it as a circle, a point, or having no graph. If it is a circle, give its center and radius. If it is a point, give its coordinates.*

1. $x^2 + y^2 - 10x + 4y + 13 = 0$

2. $x^2 + y^2 + 6x - 2y + 10 = 0$

3. $36x^2 + 36y^2 - 24x + 108y + 85 = 0$

4. $4x^2 + 4y^2 - 4x + 12y - 15 = 0$

5. $36x^2 + 36y^2 - 36x + 24y + 49 = 0$

6. Find an equation of the line through the points of intersection of $x^2 + y^2 - 4x + 2y + 1 = 0$ and $x^2 + y^2 - 8x - 2y + 8 = 0$.

B 7. Find an equation of the circle containing $(0, 0)$, $(-1, 1)$, and $(3, 3)$.

8. Find an equation of the circle tangent to $x + y = 3$ at $(2, 1)$ and with center on $3x - 2y - 6 = 0$.

9. Find an equation of the circle with center $(4, 1)$ and tangent to $3x + 4y - 2 = 0$.

10. Find an equation of the circle of radius 4 that is tangent to the x axis and has its center on $x - 2y = 2$.

11. Find an equation of the circle inscribed in the triangle with vertices $(5, 4)$, $(-15, -1)$, and $(23/3, -20/3)$.

12. Prove analytically that the perpendicular bisector of a chord of a circle contains the center.

13. Find an equation of the line tangent to $x^2 + y^2 = 25$ at $(4, -3)$.

Chapter **5** Conic Sections

5.1

INTRODUCTION

In Chapter 3 we saw that an equation of the first degree always represents a line, and every line can always be represented by an equation of the first degree. Now let us consider second-degree equations and their geometric representation. The general equation of the second degree has the form

$$Ax^2 + Bxy + Cy^2 + Dx + Ey + F = 0,$$

where A, B, and C are not all zero. We shall see that equations of the second degree represent (with two trivial exceptions) conic sections: that is, curves formed by the intersection of a plane with a right circular cone. Conversely, all conic sections are represented by second-degree equations.

The conic sections are shown in Figure 5.1 on page 150. Furthermore, a cone has two portions, or *nappes*, separated from each other by the vertex. Note that a cone has no base or end; it extends infinitely far in both directions. Thus some of the conic sections are unbounded. The traditional conic sections are the parabola, ellipse, and hyperbola; a circle is a special case of the ellipse. The remaining situations are called *degenerate* conics. In addition, there are two other situations represented by second-degree equations: a pair of parallel lines, and no graph at all. No matter what the cone or the plane, there must be some intersection; and it cannot be a pair of parallel lines. There are the two exceptions mentioned above.

5.2

THE PARABOLA

DEFINITION   *A **parabola** is the set of all points in a plane equidistant from a fixed point (**focus**) and a fixed line (**directrix**) not containing the focus.*

Suppose we choose the focus to be the point $(c, 0)$ and we choose the directrix to be $x = -c$, $c \neq 0$ (see Figure 5.2 on page 151). Let us choose a point (x, y) on the parabola and see what condition must be satisfied by x and y. From the definition, we have

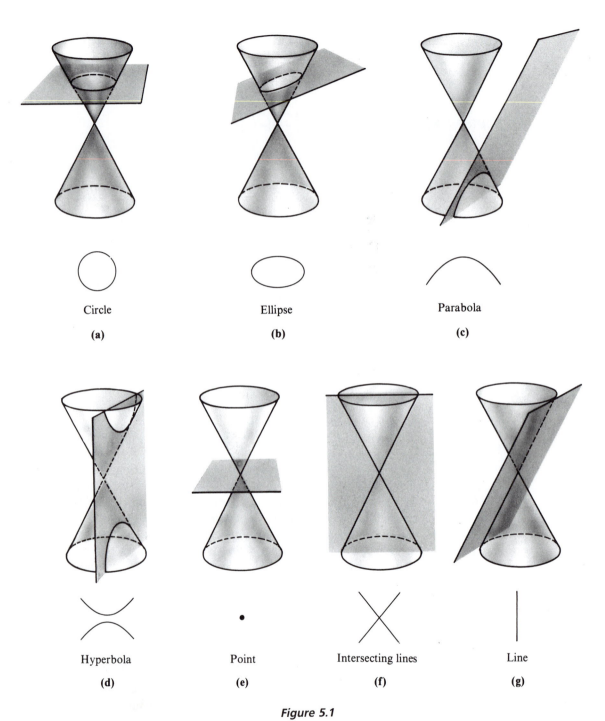

Circle

(a)

Ellipse

(b)

Parabola

(c)

Hyperbola

(d)

Point

(e)

Intersecting lines

(f)

Line

(g)

Figure 5.1

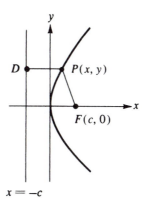

Figure 5.2

$$\overline{PF} = \overline{PD}$$
$$\sqrt{(x - c)^2 + y^2} = |x + c| \qquad \text{(See Note 1)}$$
$$(x - c)^2 + y^2 = (x + c)^2 \qquad \text{(See Note 2)}$$
$$x^2 - 2cx + c^2 + y^2 = x^2 + 2cx + c^2$$
$$y^2 = 4cx.$$

Note 1: Since $\overline{PD}$ is a horizontal distance,

$$\overline{PD} = |x - (-c)| = |x + c|.$$

You might feel that we should drop the absolute value signs, since it is clear from Figure 5.2 that $x + c$ must be positive. However, we did not insist that c be positive (although Figure 5.2 is given for a positive value of c). If c is negative, x is also negative and $x + c$ is negative.

Note 2: When we square both sides of an equation, there is a possibility of introducing extraneous roots. For instance, $(0, 1)$ is not a root of $x + y = x - y$, but it is a root of $(x + y)^2 = (x - y)^2$. The reason here is that $x + y = 1$, while $x - y = -1$ for $(0, 1)$, and $1^2 = (-1)^2 = 1$. This situation cannot occur when $(x + y)^2 = (x - y)^2$ and $x + y$ and $x - y$ are either both positive, both negative, or both zero. Since $\sqrt{(x - c)^2 + y^2}$ and $|x + c|$ must both be positive in any case, we have introduced no extraneous roots; that is, any point satisfying

$$(x - c)^2 + y^2 = (x + c)^2$$

must also satisfy

$$\sqrt{(x - c)^2 + y^2} = |x + c|.$$

We see that if a point is on the parabola with focus $(c, 0)$ and directrix $x = -c$, it must satisfy the equation $y^2 = 4cx$. Furthermore, since Note 2 indicates that all steps in the above argument are reversible, any point satisfying the equation $y^2 = 4cx$ is on the given parabola.

THEOREM 5.1   *A point (x, y) is on the parabola with focus $(c, 0)$ and directrix $x = -c$ if and only if it satisfies the equation*

$$y^2 = 4cx.$$

Let us observe some properties of this parabola before considering others. First of all, the x axis is a line of symmetry: that is, the portion below the x axis is the mirror image of the portion above. This line is called the **axis** of the parabola. It is perpendicular to the directrix and contains the focus (see Figure 5.3). The

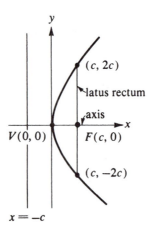

Figure 5.3

point of intersection of the axis and the parabola is the **vertex**. The vertex of the parabola $y^2 = 4cx$ is the origin. Finally, the line segment through the focus, perpendicular to the axis and having both ends on the parabola, is the **latus rectum** (literally, straight side). Since the latus rectum of $y^2 = 4cx$ must be vertical and since it contains $(c, 0)$, the x coordinate of both ends is c. Substituting $x = c$ into $y^2 = 4cx$, we have

$$y^2 = 4c^2$$
$$y = \pm 2c.$$

Thus one end of the latus rectum is $(c, 2c)$ and the other $(c, -2c)$; its length is $4|c|$.

Finally the role of the x and y may be reversed throughout, as the next theorem states.

THEOREM 5.2 *A point (x, y) is on the parabola with focus $(0, c)$ and directrix $y = -c$ if and only if it satisfies the equation*

$$x^2 = 4cy.$$

EXAMPLE 1 Sketch and discuss $y^2 = 8x$.

SOLUTION The equation is of the form

$$y^2 = 4cx,$$

with $c = 2$. Thus, it represents a parabola with vertex at the origin and axis on the x axis. The focus is at $(2, 0)$, and the directrix is $x = -2$. Finally, the length of the latus rectum is 8. This length may be used to determine the ends, $(2, \pm 4)$, of the latus rectum, which helps in sketching the curve (see Figure 5.4).

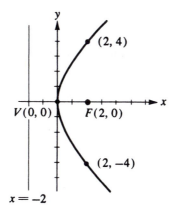

Figure 5.4

EXAMPLE 2 Sketch and discuss $x^2 = -12y$.

SOLUTION This equation is in the form $x^2 = 4cy$, with $c = -3$. Thus, it is a parabola with vertex at the origin and axis on the y axis. The focus is $(0, -3)$, the length of the latus rectum is 12, and the equation of the directrix is $y = 3$ (see Figure 5.5).

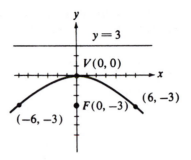

Figure 5.5

It might be observed from these two examples that the sign of *c* gives the direction in which the parabola opens. If *c* is positive, then the parabola opens in the positive direction (to the right or upward); if *c* is negative, then the parabola opens in the negative direction (to the left or downward).

EXAMPLE 3   Find an equation(s) of the parabola(s) with vertex at the origin and focus $(-4, 0)$.

SOLUTION   Since the focus and vertex are on the *x* axis, the *x* axis is the axis of the parabola (see Figure 5.6). Thus the equation is in the form $y^2 = 4cx$. Since the focus is $(-4, 0)$, $c = -4$ and the equation is $y^2 = -16x$.

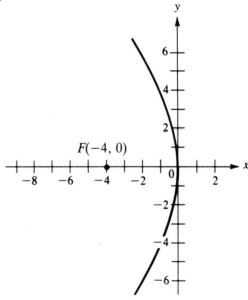

Figure 5.6

It is quite simple to use a graphing calculator for these parabolas. Although we may use the conic graphing program, solving for y is the simpler alternative.

EXAMPLE 4

Use a calculator to graph (a) $x^2 = -4.3y$ and (b) $y^2 = 7.49x$.

SOLUTION

Solving each of these equations for y gives

$$y = -x^2/4.3$$
$$y = \pm \sqrt{7.49x}.$$

Graphing with the standard window gives the graph of Figure 5.7.

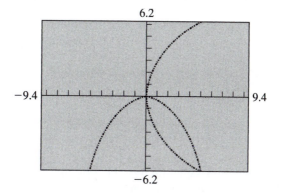

Figure 5.7

A frequently encountered problem is that of finding the tangent to a curve at a given point. This problem is easily solved by the use of calculus. We can solve this problem here without calculus by using a property of conic sections: namely, a tangent to a nondegenerate conic section has only one point in common with it. Let us use this property to solve the following problem.

EXAMPLE 5

Find an equation of the line tangent to $y^2 = -8x$ at the point $(-2, 4)$ (see Figure 5.8).

SOLUTION

The line with slope m through the point $(-2, 4)$ is

$$y - 4 = m(x + 2).$$

This is the desired line. Now all we need to do is determine m. Since the tangent line and the parabola have only one point in common, let us solve simultaneously. Solving the equation of the line for y and substituting into the equation of the parabola, we have

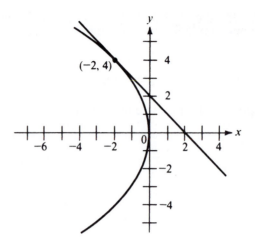

Figure 5.8

$$(4 + mx + 2m)^2 = -8x$$
$$16 + m^2x^2 + 4m^2 + 8mx + 16m + 4m^2x = -8x$$
$$m^2x^2 + (4m^2 + 8m + 8)x + (4m^2 + 16m + 16) = 0.$$

This quadratic equation has only one solution provided $B^2 - 4AC = 0$, where A, B, and C are the coefficients of x^2 and x, and the constant term, respectively, of the equation.

$$(4m^2 + 8m + 8)^2 - 4m^2(4m^2 + 16m + 16) = 0$$
$$64m^2 + 128m + 64 = 0$$
$$64(m + 1)^2 = 0$$
$$m = -1$$

Thus the desired line is

$$y - 4 = -(x + 2)$$
$$x + y - 2 = 0.$$

It might be noted that there is one line through the point $(-2, 4)$ that has only $(-2, 4)$ in common with the parabola, and that line is the horizontal line $y = 4$. Thus it is not true that *every* line through $(-2, 4)$ having only one point in common with the parabola is a tangent line. See Problem 31 for more on this line.

___

Parabolas have several interesting applications. One of the most important is their reflective property. While all three of the conic sections have special reflective properties, the one with the widest application is the parabola. If we have a

HISTORICAL NOTE

CONIC SECTIONS

The conic sections were invented by the Greek scholar Menaechmus around 350 B.C. He considered them in an attempt to "double the cube." This is one of the classical Greek problems (subsequently proved to be impossible) of constructing, using only a straightedge and compass, a cube having twice the volume of a given cube. Menaechmus described the conic sections as the intersection of a cone with a plane at a right angle to a side of the cone. He obtained the three types of conic sections by varying the vertex angle of the cone. We get a parabola, an ellipse, or a hyperbola if the vertex of the cone is a right angle, an acute angle, or an obtuse angle, respectively.

Most of his work on conic sections has been lost, although from the fragments that we have we can conclude that he investigated their properties quite extensively. Nevertheless, it is not known how he constructed them as plane figures.

Apollonius of Perga summarized the previous work and extended it in his eight books titled *Conic Sections.* Instead of using planes at right angles to three different types of cones, he used a single cone with planes intersecting it at different angles, much as we do today. It was he who gave the conic sections their names: *hyperbolē* ("excess"), *elleipsis* ("defect"), and *parabolē* ("application"). Although he did not represent them by equations, we can relate them to the equation $y^2 = 2px + ax^2$. In this equation, a hyperbola, an ellipse, and a parabola correspond to values of a that are positive (excess), negative (defect), and zero (application), respectively.

Finally, Pappus proved the focus-directrix property of conic sections in A.D. 320. This provided a unifying concept that did not require a reference to solid geometry.

Although the conic sections had been largely ignored for several centuries, interest in them was revived when Kepler showed the orbits of the planets to be elliptic.

parabolic mirror, a ray of light from the focus of the parabola is reflected by the mirror along a line that is parallel to the axis. While stated in terms of light, the same can be said of infrared radiation, radio waves, and microwaves, as well as sound waves. Thus, the reflectors on searchlights and automobile headlights are parabolic, with the light source at the focus. Actually, they are paraboloids, which are parabolas rotated about their axes (see page 366). The same principle is used (with the source and destination reversed) in reflecting telescopes, dish antennas, solar furnaces, and "spy" listening devices.

Another (though minor) application of the parabola is in the orbits or comets. At a given distance from the sun, there is a threshold velocity called the **escape velocity**. At this or greater velocity, the comet escapes entirely from the solar system; at a lesser velocity, it remains within the sun's gravitational field. The comet's path is elliptic, hyperbolic, or parabolic, depending upon whether its velocity is, respectively, less than, greater than, or equal to the escape velocity. In the latter two situations, the comet approaches the sun once and flies off into space, never to return.

All of this assumes that we are dealing with only two bodies: the sun and the comet. The gravitational forces of the planets can also have an effect upon the orbit. A near approach to a large planet can either increase or decrease the velocity of a comet, depending upon the circumstances (see Problem 44). Thus, a comet in a long elliptic orbit can have its velocity increased to equal or exceed the escape velocity, giving it a new orbit that is either parabolic or hyperbolic. Similarly, a near approach to a large planet can also change the orbit from parabolic or hyperbolic to elliptic. In this way, a comet may be "captured" by a solar system.

When dealing with a solar system, the distances are great and constantly changing with the movement of the planets, asteroids, or comets. Thus, the gravitational forces, which depend upon the distances between the bodies, are constantly changing. If we are dealing with the trajectory of a golf ball or a jet of water, the vertical distances are so small that the force of gravity may be taken to be a constant. In that case (neglecting air resistance), the trajectory is parabolic.

Parabolas also occur in certain engineering applications. When a cable hangs of its own weight, it does not hang in a parabolic arc; it forms a **catenary** (Latin for "chain"), with a more complicated formula. However, a freely hanging cable does not have its weight uniformly distributed across its length. When a bridge is suspended from the cable, it is desirable to have the weight of the bridge uniformly distributed. When this is done, the cable then takes the form of a parabola. Similarly, parabolic arches have greater strength than other shapes. Thus, concrete arch bridges often use parabolic arches.

PROBLEMS

A   *In Problems 1–12, sketch and discuss the given parabola.*

1. $y^2 = 16x$ 2. $y^2 = -12x$ 3. $x^2 = 4y$

4. $x^2 = -8y$ 5. $y^2 = 10x$ 6. $x^2 = -7y$

7. $x^2 = 5y$ 8. $y^2 = -9x$ 9. $x^2 = -2y$

10. $y^2 = 3x$ 11. $x^2 = 6y$ 12. $y^2 = -5x$

B   *In Problems 13–20, find an equation(s) of the parabola(s) described.*

13. Vertex: $(0, 0)$; axis: x axis; contains $(1, 5)$

14. Vertex: $(0, 0)$; axis: y axis; contains $(1, 5)$

15. Vertex: $(0, 0)$; axis: x axis; length of latus rectum: 5

16. Vertex: $(0, 0)$; focus: $(0, 5)$

17. Focus: $(-3, 0)$; directrix: $x = 3$

18. Focus: $(0, 8)$; directrix: $y = -8$

19. Vertex: $(0, 0)$; contains $(2, 3)$ and $(-2, 3)$

20. Vertex: $(0, 0)$; contains $(-3, -4)$ and $(-3, 4)$

In Problems 21–24, the required parabola is not in the standard position, so that Theorems 5.1 and 5.2 cannot be used. Instead, go back to the definition of a parabola.

21. Find an equation of the parabola with focus $(4, 0)$ and directrix $x = 0$

22. Find an equation of the parabola with focus $(2, 4)$ and directrix $y = -2$.

23. Find an equation of the parabola with focus $(0, 0)$ and directrix $x + y = 4$.

24. Find an equation of the parabola with focus $(1, 1)$ and directrix $x + y = 0$.

25. Find an equation of the line tangent to $y = x^2$ at $(1, 1)$.

26. Find an equation of the line tangent to $x^2 = -5y$ at $(5, -5)$.

27. Find an equation of the line tangent to $y^2 = -16x$ and parallel to $x + y = 1$.

28. Find equations of the lines tangent to $y^2 = 4x$ and containing $(-2, 1)$.

C   **29.** Prove Theorem 5.2.

30. Prove that the ordinate of any point P of the parabola $y^2 = 4cx$ is the mean proportional between the length of the latus rectum and the abscissa of P.

31. Note that in Example 5 the horizontal line $y = 4$ has only one point in common with the given parabola. Why was it not found in Example 5? [*Hint:* An assumption was made at one point that prevented us from finding it.]

32. Show that the tangent line to $y^2 = 4cx$ at (x_0, y_0) is $yy_0 = 2c(x + x_0)$.

GRAPHING
CALCULATOR

33. Graph the following.
  (a) $x^2 = y$ (b) $x^2 = 4y$ (c) $x^2 = 8y$
  (d) $x^2 = -y$ (e) $x^2 = -4y$ (f) $x^2 = -8y$
  What role is played by the coefficient of y? What is the effect of the minus?

34. Graph the following.
  (a) $y^2 = x$ (b) $y^2 = 4x$ (c) $y^2 = 8x$
  (d) $y^2 = -x$ (e) $y^2 = -4x$ (f) $y^2 = -8x$
  What is the role played by the coefficient of x? What is the effect of the minus?

35. Graph the following.
  (a) $y^2 = 4x$ (b) $y^2 = 4(x - 1)$ (c) $y^2 = 4(x - 2)$
  (d) $y^2 = 4(x + 1)$ (e) $y^2 = 4(x + 2)$
  What is the effect of replacing x by $x - h$?

36. Graph the following.
  (a) $y^2 = 4x$ (b) $(y - 1)^2 = 4x$ (c) $(y - 2)^2 = 4x$
  (d) $(y + 1)^2 = 4x$ (e) $(y + 2)^2 = 4x$
  What is the effect of replacing y by $y - k$?

37. The catenary has equation $y = (e^x + e^{-x})/2$. Graph it and $y = x^2$ on the same set of coordinate axes. For what values of x are the y values within 1 of each other? Within 0.5 of each other?

APPLICATIONS

38. An automobile headlight has a parabolic reflector that is 6 inches in diameter and 3 inches deep. How far from the vertex should be light bulb be placed?

39. A lighthouse uses a parabolic reflector that is 1 meter in diameter. How deep should the reflector be if the light source is placed halfway between the vertex and the plane of the rim?

40. In order to hear what is said in a football huddle, a listening device is set up on the sideline. It consists of a parabolic dish with a microphone at its focus. The dish is 4 feet across and 16 inches deep. Give an equation for the parabola with the origin of the coordinate system at the vertex of the parabola and the parabola opening to the right. At what point should the microphone be placed?

41. A large reflecting telescope has a parabolic mirror that is 4 meters in diameter and 25 centimeters deep. It requires a cage for the observer with an eyepiece at the focus of the mirror. How far up the tube of the telescope must the cage be built? This design is called a **prime focus** telescope.

42. The prime focus design of Problem 41 is not practical for small telescopes. Instead, a 45° plane mirror is placed between the vertex and focus and the image is reflected to an eyepiece outside the telescope tube, as shown in Figure 5.9. This design, which is not practical for large telescopes, is called a **Newtonian focus** after Sir Isaac Newton. A Newtonian focus telescope has a parabolic mirror of diameter 6 inches with equation $x^2 = 120y$ and a 45° mirror 24 inches from the vertex. Where should the eyepiece be placed?

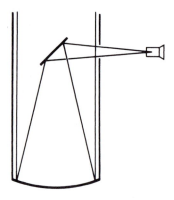

Figure 5.9

43. A comet from deep space approaches the sun along a parabolic orbit. When the comet is 100 million miles from the sun, the line joining the sun and the comet makes an angle of 60° with the axis of the parabola. How close to the sun will the comet get? [*Hint:* The point of a parabola that is closest to the focus is the vertex. Use the definition of a parabola rather than an equation in standard form.]

44. When approaching the sun, a comet passes close to the orbit of Saturn, as shown in Figure 5.10. If it crosses Saturn's orbit ahead of the planet, as in Figure 5.10a, will its speed be increased or decreased? Explain why. Repeat for the comet crossing behind Saturn, as in Figure 5.10b.

45. A concrete arch spans a width of 40 feet with a 20-foot-wide road passing under the bridge. The minimum clearance over the roadway must be 10 feet. What is the height of the smallest such arch that can be used?

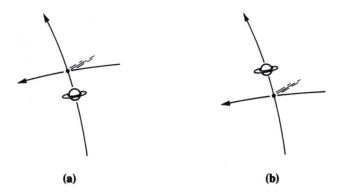

(a)   **(b)**

Figure 5.10

46. The towers of a suspension bridge are 500 feet apart and extend 100 feet above the road surface. The main cables between the towers reach to within 10 feet of the road at the center of the bridge, and there are vertical supporting cables every 10 feet. Find the lengths of those supporting cables at 50 foot intervals.

47. An arrow is shot horizontally from a point 10 feet above the ground. It hits the ground 200 feet out from the starting point. Give an equation for the arrow's trajectory, with the origin of the coordinate system at the initial position of the arrow.

48. A ball is thrown upward and outward from the top edge of a 50-foot-high building. It reaches its highest point 20 feet above and 10 feet out from the building. How far from the building is the ball when it hits the ground?

5.3

THE ELLIPSE

DEFINITION   *An ellipse is the set of all points (x, y) such that the sum of the distances from (x, y) to a pair of distinct fixed points (foci) is a fixed constant.*

Let us choose the foci to be $(c, 0)$ and $(-c, 0)$ (see Figure 5.11) and let the fixed constant be $2a$. If (x, y) represents a point on the ellipse, we have the following.

$$\sqrt{(x - c)^2 + y^2} + \sqrt{(x + c)^2 + y^2} = 2a$$

$$\sqrt{(x - c)^2 + y^2} = 2a - \sqrt{(x + c)^2 + y^2}$$

$$x^2 - 2cx + c^2 + y^2 = 4a^2 - 4a\sqrt{(x + c^2) + y^2} + x^2$$
$$+ 2cx + c^2 + y^2$$

$$4a\sqrt{(x + c)^2 + y^2} = 4a^2 + 4cx$$

$$\sqrt{(x + c)^2 + y^2} = a + \frac{cx}{a}$$

$$x^2 + 2cx + c^2 + y^2 = a^2 + 2cx + \frac{c^2 x^2}{a^2}$$

$$\frac{a^2 - c^2}{a^2} x^2 + y^2 = a^2 - c^2$$

$$\frac{x^2}{a^2} + \frac{y^2}{a^2 - c^2} = 1$$

The triangle of Figure 5.11, with vertices $(c, 0)$, $(-c, 0)$, and (x, y), has one side of length $2c$. The sum of the lengths of the other two sides is $2a$. Thus

$$2a > 2c$$
$$a > c$$
$$a^2 > c^2$$
$$a^2 - c^2 > 0.$$

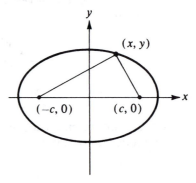

Figure 5.11

Since $a^2 - c^2$ is positive, we may replace it by another positive number, b^2. Thus

$$\frac{x^2}{a^2} + \frac{y^2}{b^2} = 1, \quad \text{where} \quad b^2 = a^2 - c^2.$$

Observe that we squared both sides of the equation at two of the steps. In both cases, both sides of the equation are nonnegative. Thus we have introduced no extraneous roots, and the steps may be reversed (see Note 2 on page 151).

Note that there are two axes of symmetry: the x axis and the y axis. Furthermore, $(\pm a, 0)$ are the x intercepts and $(0, \pm b)$ are the y intercepts, where $a > b$ (since $b^2 = a^2 - c^2$). Thus the x axis is called the **major axis** and the y axis is the **minor axis**. The points $(\pm a, 0)$ on the major axis are called the vertices, the points

$(0, \pm b)$ on the minor axis are the **covertices**, and the point of intersection of the axes, $(0, 0)$, is called the **center** (see Figure 5.12). The **foci** $(\pm c, 0)$ are on the major axis.

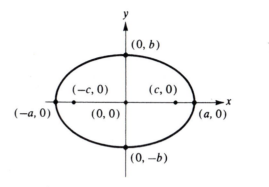

Figure 5.12

THEOREM 5.3   *A point (x, y) is on the ellipse with vertices $(\pm a, 0)$, and foci $(\pm c, 0)$ if and only if it satisfies the equation*

$$\frac{x^2}{a^2} + \frac{y^2}{b^2} = 1$$

where

$$b^2 = a^2 - c^2.$$

An ellipse has two **latera recta** (plural of latus rectum), which are chords of the ellipse perpendicular to the major axis and containing the foci. If $x = \pm c$, then

$$\frac{c^2}{a^2} + \frac{y^2}{b^2} = 1$$

$$\frac{y^2}{b^2} = \frac{a^2 - c^2}{a^2} = \frac{b^2}{a^2}$$

$$y^2 = \frac{b^4}{a^2}$$

$$y = \pm \frac{b^2}{a}$$

Thus, one latus rectum has endpoints $(c, \pm b^2/a)$, while the other has endpoints $(-c, \pm b^2/a)$. In both cases the length is $2b^2/a$. As with the parabola, this length may be used as an aid in sketching; however, the vertices and covertices allow one to make a reasonable sketch.

Again, the roles of the x and y may be reversed.

THEOREM 5.4   *A point (x, y) is on the ellipse with vertices $(0, \pm a)$ and foci $(0, \pm c)$ if and only if it satisfies the equation*

$$\frac{y^2}{a^2} + \frac{x^2}{b^2} = 1,$$

where

$$b^2 = a^2 - c^2.$$

One question that immediately arises is how we can tell whether we have

$$\frac{x^2}{a^2} + \frac{y^2}{b^2} = 1 \qquad \text{or} \qquad \frac{y^2}{a^2} + \frac{x^2}{b^2} = 1$$

The numbers in the denominator are not labeled a and b, so how do we know which is a and which is b? The answer is "size." In both cases, $a > b$. Thus the larger denominator is a^2, and the smaller is b^2.

EXAMPLE 1   Sketch and discuss $9x^2 + 25y^2 = 225$.

SOLUTION   First, we put the equation into standard form by dividing through by 225:

$$\frac{x^2}{25} + \frac{y^2}{9} = 1.$$

Now

$$a^2 = 25, \qquad b^2 = 9,$$

and

$$c^2 = a^2 - b^2 = 16.$$

This ellipse has center $(0, 0)$, vertices $(\pm 5, 0)$, covertices $(0, \pm 3)$, and foci $(\pm 4, 0)$. The latera recta have length $2b^2/a = 2 \cdot 9/5 = 3.6$ (see Figure 5.13).

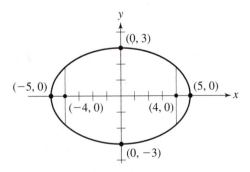

Figure 5.13

EXAMPLE 2   Sketch and discuss $25x^2 + 16y^2 = 400$.

SOLUTION   Putting the equation into standard form gives

$$\frac{x^2}{16} + \frac{y^2}{25} = 1$$

Now

$$a^2 = 25, \qquad b^2 = 16,$$

$$c^2 = a^2 - b^2 = 9.$$

This ellipse has center $(0, 0)$, vertices $(0, \pm 5)$, covertices $(\pm 4, 0)$, and foci $(0, \pm 3)$. The latera recta have length $2b^2/a = 2 \cdot 16/5 = 6.4$ (see Figure 5.14).

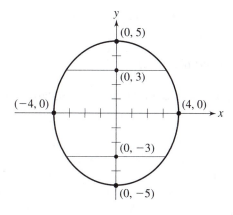

Figure 5.14

EXAMPLE 3   Find an equation of the ellipse with vertices $(0, \pm 8)$ and foci $(0, \pm 5)$.

SOLUTION   Since the vertices are on the y axis (see Figure 5.15), we have the form

$$\frac{y^2}{a^2} + \frac{x^2}{b^2} = 1.$$

Furthermore, $a = 8$ and $c = 5$; thus $b^2 = a^2 - c^2 = 64 - 25 = 39$. The final result is

$$\frac{y^2}{64} + \frac{x^2}{39} = 1.$$

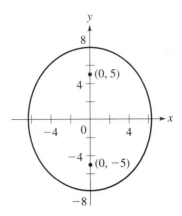

Figure 5.15

In addition to the quantities named, each ellipse is associated with a number, called the eccentricity. For any ellipse the eccentricity is

$$e = \frac{c}{a}.$$

The eccentricity of an ellipse satisfies the inequalities $0 < e < 1$. It gives a measure of the shape of the ellipse: the closer the eccentricity is to 0, the more nearly circular is the ellipse. For instance, in Example 1, $e = 4/5$, while in Example 2, $e = 3/5$. The ellipse of Example 2 is more nearly circular than the ellipse of Example 1, as can be easily seen by the sketches.

There is also a **directrix** associated with each focus of an ellipse. Associated with the focus, $(c, 0)$ of the ellipse

$$\frac{x^2}{a^2} + \frac{y^2}{b^2} = 1$$

is the directrix

$$x = \frac{a}{e} = \frac{a^2}{c}.$$

If P is any point of the ellipse, the distance from P to the focus divided by the distance from P to the directrix is equal to the eccentricity. This is sometimes used as the definition of an ellipse.

Suppose we start with focus $(c, 0)$, directrix $x = a^2/c$, and eccentricity $e = c/a$. Now let us find the set of all points $P = (x, y)$ such that the distance from P to the focus divided by the distance from P to the directrix equals the eccentricity.

$$\frac{\sqrt{(x - c)^2 + y^2}}{\frac{a^2}{c} - x} = \frac{c}{a}$$

$$\sqrt{(x - c)^2 + y^2} = a - \frac{cx}{a}$$

$$x^2 - 2cx + c^2 + y^2 = a^2 - 2cx + \frac{c^2x^2}{a^2}$$

$$\frac{a^2 - c^2}{a^2}x^2 + y^2 = a^2 - c^2$$

$$\frac{x^2}{a^2} + \frac{y^2}{a^2 - c^2} = 1$$

With $b^2 = a^2 - c^2$, this becomes

$$\frac{x^2}{a^2} + \frac{y^2}{b^2} = 1.$$

The same result can be obtained using focus $(-c, 0)$, directrix $x = -a^2/c$, and eccentricity $e = c/a$.

For a parabola, the distance from a point on the parabola to the focus divided by the distance of the point from the directrix is always 1. Thus we define $e = 1$ for every parabola.

In Section 5.1 we indicated that the circle is a special case of the ellipse. If $c = 0$, then $b = a$ in Theorems 5.3 and 5.4. Thus the equations in those theorems become $x^2 + y^2 = a^2$, an equation of a circle. With $c = 0$, the eccentricity is also zero, which agrees with our statement that an ellipse with an eccentricity near zero is nearly circular. Note, however, that there is no directrix when $c = 0$; the focus-directrix definition cannot be used in this special case.

We can again find tangent lines by using the property that a tangent to an ellipse has only one point in common with the ellipse.

EXAMPLE 4   Find equations of the lines containing $(5, 1)$ and tangent to $9x^2 + 25y^2 = 225$.

SOLUTION   Note first of all that $(5, 1)$ does not satisfy the given equation; it is not on the ellipse (see Figure 5.16). Thus we can expect not one, but two tangent lines. A line through $(5, 1)$ with slope m has equation

$$y - 1 = m(x - 5).$$

Let us solve this equation and the given equation simultaneously. Solving this equation for y and substituting into the other, we have

$$9x^2 + 25(1 + mx - 5m)^2 = 225$$

$$9x^2 + 25(1 + m^2x^2 + 25m^2 + 2mx - 10m - 10m^2x) = 225$$

$$(25m^2 + 9)x^2 + (-250m^2 + 50m)x + (625m^2 - 250m - 200) = 0$$

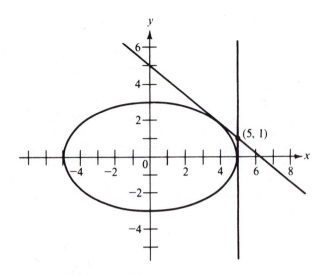

Figure 5.16

Since a tangent line and the ellipse should have only one point in common, this quadratic equation should have only one solution. Thus

$$B^2 - 4AC = 0$$
$$(-250m^2 + 50m)^2 - 4(25m^2 + 9)(625m^2 - 250m - 200) = 0$$
$$62{,}500m^4 - 25{,}000m^3 + 250m^2 - 62{,}500m^4$$
$$+ 25{,}000m^3 - 2500m^2 + 9000m + 7200 = 0$$
$$9000m + 7200 = 0$$
$$m = -\frac{4}{5}.$$

Substituting this back into
$$y - 1 = m(x - 5),$$
we have
$$y - 1 = -\frac{4}{5}(x - 5)$$
$$4x + 5y - 25 = 0.$$

Recall that we had indicated that there are two answers, but we have found only one. This is because the use of the point-slope form for the tangent line assumes that there is a slope. It is easily seen from Figure 5.16 that the other tangent line is vertical; its equation is

$$x = 5.$$

Ellipses have two applications in common with the other conic sections as well as some that are uniquely theirs. It was believed for two thousand years that the planets move in circular orbits (around the earth)—the so-called **Aristotelian model**. After all, the universe must be perfect, and the circle is the perfect figure (whatever that means). Such philosophical musings were deemed to be proof enough of the claim. It was not until Johannes Kepler showed in the seventeenth century that the orbits are elliptic, with the sun at one focus, that the Aristotelian model of the solar system was abandoned. Nevertheless, circular orbits are possible, and several planetary orbits (including the earth's) are very nearly circular. In fact, if we were to scale down the earth's orbit, making the major axis 8 inches long, the minor axis would be about 7.999 inches long! With such a small difference, we should be hard put to recognize the orbit as an ellipse rather than a circle.

The other property that ellipses have in common with the other conic sections is the reflective property of elliptic mirrors. A light source at one focus of an ellipse is reflected to the other focus. The principal application of this is in a so-called whispering gallery—a room with an elliptic dome (actually an **ellipsoid**, which is a three-dimensional ellipse). A person at one focus can whisper to someone at the other without others in the room hearing. Some examples of whispering galleries are the National Statuary Hall of the Capitol Building, the Mormon Tabernacle, in Salt Lake City, Utah, and the dome of St. Paul's Cathedral, in London. A more serious application is the use of elliptic reflectors of ultrasound to break up kidney stones. By placing the reflector in such a way that the stone is at one focus and the sound source is at the other, the sound can be concentrated on the stone, causing it to vibrate to pieces.

Another application is in aerodynamics and hydrodynamics. An elliptic wing, keel, or rudder has been shown to give less drag than other shapes. Perhaps the most widely known application of an elliptic wing was the British Spitfire of World War II fame. Gears are sometimes made elliptic in order to change constant rotational speeds to variable ones. Finally, arches, though ideally parabolic, are often elliptic.

PROBLEMS

A *In Problems* 1–10, *sketch and discuss the given ellipse.*

1. $\dfrac{x^2}{169} + \dfrac{y^2}{25} = 1$

2. $\dfrac{x^2}{144} + \dfrac{y^2}{169} = 1$

3. $\dfrac{x^2}{25} + \dfrac{y^2}{4} = 1$

4. $\dfrac{x^2}{36} + \dfrac{y^2}{16} = 1$

5. $\dfrac{x^2}{25} + \dfrac{y^2}{49} = 1$

6. $x^2 + 4y^2 = 4$

7. $9x^2 + 4y^2 = 36$

8. $9x^2 + y^2 = 9$

9. $16x^2 + 9y^2 = 144$ **10.** $4x^2 + 25y^2 = 100$

B   *In Problems 11–18, find an equation(s) of the ellipse(s) described.*

11. Center: $(0, 0)$; vertex: $(0, 13)$; focus: $(0, -5)$

12. Center: $(0, 0)$; covertex: $(0, 5)$, focus: $(-12, 0)$

13. Center: $(0, 0)$; vertex: $(5, 0)$; contains $(\sqrt{15}, 2)$

14. Center: $(0, 0)$; axes on the coordinate axes; contains $(2, 2)$ and $(-4, 1)$

15. Vertices: $(\pm 6, 0)$; length of latus rectum: 3

16. Covertices: $(\pm 2, 0)$; length of latus rectum: 2

17. Foci: $(\pm 6, 0)$; $e = 3/5$

18. Foci: $(\pm 2, 0)$; directrices: $x = \pm 8$

19. Find an equation of the line tangent to $x^2 + 4y^2 = 20$ at $(2, 2)$.

20. Find an equation of the line tangent to $2x^2 + 3y^2 = 11$ at $(2, 1)$.

21. Find an equation of the line containing $(3, -2)$ and tangent to $4x^2 + y^2 = 8$.

22. Find an equation of the line containing $(2, 4)$ and tangent to $3x^2 + 8y^2 = 84$.

C   **23.** Suppose, in Figure 5.17, that A and B are fixed pins on the arm ABP and that AP and BP have lengths a and b, respectively. Show that if A is free to side in channel XX' and B in channel YY', the point P traces an ellipse.

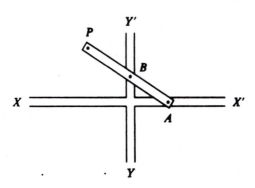

Figure 5.17

24. Given the focus $(-c, 0)$, directrix $x = -a^2/c$, and eccentricity $e = c/a$, show that these define the ellipse

$$\frac{x^2}{a^2} + \frac{y^2}{a^2 - c^2} = 1.$$

25. Prove Theorem 5.4.

26. Show that the line tangent to

$$\frac{x^2}{a^2} + \frac{y^2}{b^2} = 1$$

at (x_0, y_0) is

$$\frac{xx_0}{a^2} + \frac{yy_0}{b^2} = 1.$$

GRAPHING CALCULATOR

27. Graph $x^2/4 + y^2/9 = 1$. Remember that you must first solve for y as a pair of functions of x and then graph both functions. What is the relationship between the x and y intercepts and the numbers in the equation? Repeat for $x^2/9 + y^2/4 = 1$.

28. Graph the following.

(a) $\dfrac{x^2}{4} + \dfrac{y^2}{4} = 1$ (b) $\dfrac{x^2}{9} + \dfrac{y^2}{4} = 1$

(c) $\dfrac{x^2}{16} + \dfrac{y^2}{4} = 1$ (d) $\dfrac{x^2}{25} + \dfrac{y^2}{4} = 1$

What are the relationships between the graphs and the denominators of the x^2 terms?

29. Graph the following.

(a) $\dfrac{x^2}{4} + \dfrac{y^2}{4} = 1$ (b) $\dfrac{x^2}{4} + \dfrac{y^2}{9} = 1$

(c) $\dfrac{x^2}{4} + \dfrac{y^2}{16} = 1$ (d) $\dfrac{x^2}{4} + \dfrac{y^2}{25} = 1$

What is the relationship between the graphs and the denominators of the y^2 terms?

30. Graph the following.

(a) $\dfrac{x^2}{6} + \dfrac{y^2}{2} = 1$ (b) $\dfrac{x^2}{8} + \dfrac{y^2}{4} = 1$

(c) $\dfrac{x^2}{10} + \dfrac{y^2}{6} = 1$ (d) $\dfrac{x^2}{12} + \dfrac{y^2}{8} = 1$

What are the foci of these ellipses? Such ellipses are called **confocal**.

31. Graph the following.

(a) $\dfrac{x^2}{4} + \dfrac{y^2}{9} = 1$ (b) $\dfrac{(x-1)^2}{4} + \dfrac{y^2}{9} = 1$

(c) $\dfrac{(x-2)^2}{4} + \dfrac{y^2}{9} = 1$ (d) $\dfrac{(x+1)^2}{4} + \dfrac{y^2}{9} = 1$

(e) $\dfrac{(x+2)^2}{4} + \dfrac{y^2}{9} = 1$

What is the effect of replacing x by $x - h$?

32. Graph the following.

(a) $\dfrac{x^2}{4} + \dfrac{y^2}{9} = 1$ (b) $\dfrac{x^2}{4} + \dfrac{(y-1)^2}{9} = 1$

(c) $\dfrac{x^2}{4} + \dfrac{(y-2)^2}{9} = 1$ (d) $\dfrac{x^2}{4} + \dfrac{(y+1)^2}{9} = 1$

(e) $\dfrac{x^2}{4} + \dfrac{(y+2)^2}{9} = 1$

What is the effect of replacing y by $y - k$?

APPLICATIONS

33. The earth moves in an elliptic orbit about the sun, with the sun at one focus. The least and greatest distances of the earth from the sun are 91,446,000 miles and 94,560,000 miles, respectively. What is the eccentricity of the ellipse? How long are the major and minor axes?

34. The distance (center to center) of the moon from the earth varies from a minimum of 221,463 miles to a maximum of 252,710 miles. Find the eccentricity of the moon's orbit and the lengths of the major and minor axes.

35. The accompanying table gives the semimajor axis and eccentricity of the planetary orbits. Use this information to find the minimum and maximum distances (center to center) of Mercury from the sun.

Planet	Semimajor axis (millions of km)	Eccentricity
Mercury	57.9	0.2056
Venus	108.2	0.0068
Earth	149.6	0.0167
Mars	227.9	0.0934
Jupiter	778.3	0.0484
Saturn	1427.0	0.0560
Uranus	2869.0	0.0461
Neptune	4497.1	0.0100
Pluto	5900	0.2484

Source: Michael A. Seeds, *Foundations of Astronomy*, 2d ed. (Belmont, CA: Wadsworth Publishing Company, 1988), p. 609.

36. Using the information in the table of Problem 35, show that Pluto is sometimes closer than Neptune to the sun. Pluto revolves about the sun once every 248 years. For 20 of them (including the period from January 1979 to March 1999), it is the eighth planet from the sun.

37. The orbit of Halley's comet has an eccentricity of 0.97 and semimajor axis of 2885 million kilometers. Give an equation for the orbit with the center at the origin and major axis on the x axis. Where is (the center of) the sun on this graph?

38. A room is elliptic with vertical walls 6 feet high and an ellipsoidal ceiling. If it is 40 feet long and 20 feet wide, where should two people stand (other than next to each other) so that they can whisper to each other without being heard by others in the room?

39. Show that an ellipse can be drawn using two pins and a loop of string. Where should the pins be placed, and how long should the loop be?

40. A cabinetmaker wants to build a table with an elliptic top. It is to be 120 inches long and 50 inches wide at its widest point. He uses two pins and a string (see Problem 39) to draw the top. Where should he put the pins, and how long should the string be?

41. The ellipse in Washington, DC, is 3525 feet long and 1265 feet wide. Give an equation for it if the axes are placed with the origin at the center and major axis on the x axis. Where are the foci?

42. A hall that is 10 feet wide has a ceiling that is a semiellipse. The ceiling is 10 feet high at the sides and 12 feet high in the center. Find its equation with the x axis horizontal and the origin at the center of the ellipse.

43. A concrete arch bridge is to be built in the form of a semiellipse. It must span a width of 20 feet. In addition, the central 14 feet of it must be at least 8 feet high. Where is the highest point of the span?

5.4

THE HYPERBOLA

DEFINITION *A hyperbola is the set of all points (x, y) in a plane such that the positive difference between the distances from (x, y) to a pair of distinct fixed points (foci) is a fixed constant.*

Again, let us choose the foci to be $(c, 0)$ and $(-c, 0)$ (see Figure 5.18) and choose the fixed constant to be $2a$. If (x, y) represents a point on the hyperbola, we have the following.

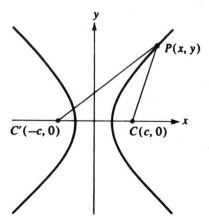

Figure 5.18

$$\sqrt{(x - c)^2 + y^2} - \sqrt{(x + c)^2 + y^2} = \pm 2a$$

$$\sqrt{(x - c)^2 + y^2} = \sqrt{(x + c)^2 + y^2} \pm 2a$$

$$x^2 - 2cx + c^2 + y^2 = x^2 + 2cx + c^2 + y^2$$
$$\pm 4a\sqrt{(x + c)^2 + y^2} + 4a^2$$

$$\mp 4a\sqrt{(x + c)^2 + y^2} = 4a^2 + 4cx$$

$$\mp\sqrt{(x + c)^2 + y^2} = a + \frac{cx}{a}$$

$$x^2 + 2cx + c^2 + y^2 = a^2 + 2cx + \frac{c^2 x^2}{a^2}$$

$$\frac{c^2 - a^2}{a^2}x^2 - y^2 = c^2 - a^2$$

$$\frac{x^2}{a^2} - \frac{y^2}{c^2 + a^2} = 1$$

In the triangle *PCC'* of Figure 5.18,

$$\overline{PC'} < \overline{PC} + \overline{CC'}$$
$$\overline{PC'} - \overline{PC} < \overline{CC'}$$
$$2a < 2c$$
$$a < c$$
$$c^2 - a^2 > 0.$$

Since $c^2 - a^2$ is positive, we may replace it by another positive number, b^2. Thus

$$\frac{x^2}{a^2} - \frac{y^2}{b^2} = 1,$$

where $b^2 = c^2 - a^2$.

Again we squared both sides of the equation at two of the steps. The first time, both sides of the equation were positive; the second time, they were either both positive or both negative. Thus we have introduced no extraneous root, and the steps may be reversed (see Note 2 on page 151).

Again, both the *x* axis and the *y* axis are axes of symmetry and again $(\pm a, 0)$ are the *x* intercepts. However, there are no *y* intercepts; when $x = 0$, we have

$$-\frac{y^2}{b^2} = 1,$$

which is not satisfied by any real number *y*. The *x* axis (containing two points of the hyperbola) is called the **transverse axis**; the *y* axis is called the **conjugate axis**. The points $(\pm a, 0)$ on the transverse axis are called the **vertices**, and the point of intersection of the axes, $(0, 0)$, is called the **center** (see Figure 5.19).

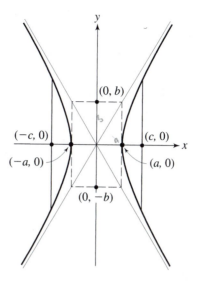

Figure 5.19

THEOREM 5.5   *A point (x, y) is on the hyperbola with vertices $(\pm a, 0)$ and foci $(\pm c, 0)$ if and only if it satisfies the equation*

$$\frac{x^2}{a^2} - \frac{y^2}{b^2} = 1,$$

where

$$b^2 = c^2 - a^2.$$

For every hyperbola there are two lines that the curve approaches more and more closely at its extremities. These two lines are called **asymptotes** (see Figure 5.19, in which the slanting dotted lines are the asymptotes). It might be noted that parabolas do not have asymptotes. Thus a hyperbola is not—as might appear from inaccurate diagrams—a pair of parabolas. The hyperbola

$$\frac{x^2}{a^2} - \frac{y^2}{b^2} = 1, \quad \text{or} \quad y = \pm\frac{b}{a}\sqrt{x^2 - a^2},$$

has asymptotes

$$y = \pm\frac{b}{a}x.$$

Let us verify this, at least for the portion in the first quadrant. Here we are dealing with

$$y = \frac{b}{a}\sqrt{x^2 - a^2} \quad \text{and} \quad y = \frac{b}{a}x$$

for positive values of x. For a given value of x, let us consider the difference d between the y coordinates of the points on the hyperbola and the line.

$$d = \frac{b}{a}x - \frac{b}{a}\sqrt{x^2 - a^2} = \frac{b}{a}(x - \sqrt{x^2 - a^2})$$

Multiplying numerator and denominator by $x + \sqrt{x^2 + a^2}$, we have

$$d = \frac{b}{a}\frac{x^2 - (x^2 - a^2)}{x + \sqrt{x^2 - a^2}} = \frac{ab}{x + \sqrt{x^2 - a^2}}.$$

Now the numerator is a constant; but, for large positive values of x, both terms of the denominator are large and positive. In fact, the larger the value of x, the larger the denominator and, therefore, the smaller d is. Thus d approaches zero as x gets larger, which shows that the line is an asymptote of the hyperbola. Of course, similar arguments can be used to show the same thing in the other three quadrants.

A convenient way of sketching the asymptotes is to plot both $(\pm a, 0)$ and $(0, \pm b)$ (even though the second pair of points is not on the hyperbola) and sketch the rectangle determined by them (see Figure 5.19). The diagonals of this rectangle are the asymptotes.

Again, two **latera recta** contain the foci and are perpendicular to the transverse axis. By using the same method as in the case of the parabola and ellipse, we can show their length to be

$$\frac{2b^2}{a}.$$

As with the parabola and the ellipse, the roles of x and y can be reversed.

THEOREM 5.6   *A point (x, y) is on the hyperbola with vertices $(0, \pm a)$ and foci $(0, \pm c)$ if and only if it satisfies the equation*

$$\frac{y^2}{a^2} - \frac{x^2}{b^2} = 1,$$

where

$$b^2 = c^2 - a^2.$$

It might be noted that a and b are determined by the sign of the term in which they appear; a^2 is always the denominator of the positive term and b^2 the denominator of the negative term. There is no requirement that a be greater than b, as there was for an ellipse.

The asymptotes of the hyperbola

$$\frac{y^2}{a^2} - \frac{x^2}{b^2} = 1$$

are

$$y = \pm \frac{a}{b} x.$$

Since the formulas for the asymptotes for the two cases are rather easy to confuse, a method that always works is to replace the 1 by 0 in the standard form and solve for y.

EXAMPLE 1   Sketch and discuss $\dfrac{x^2}{9} - \dfrac{y^2}{16} = 1$.

SOLUTION   We see that $a^2 = 9$, $b^2 = 16$, and $c^2 = a^2 + b^2 = 25$. This hyperbola has center $(0, 0)$, vertices $(\pm 3, 0)$, and foci $(\pm 5, 0)$. The asymptotes are found by replacing the 1 of the standard form by 0 and solving for y.

$$\frac{x^2}{9} - \frac{y^2}{16} = 0$$

$$y^2 = \frac{16x^2}{9}$$

$$y = \pm \frac{4}{3} x$$

The length of the latera recta is $2b^2/a = 32/3$ (see Figure 5.20).

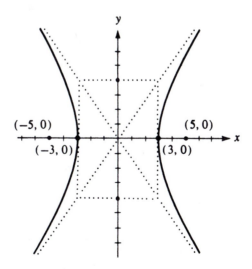

Figure 5.20

EXAMPLE 2

Sketch and discuss $16x^2 - 9y^2 + 144 = 0$.

SOLUTION

Putting this equation into standard form, we have

$$\frac{y^2}{16} - \frac{x^2}{9} = 1.$$

We see that $a^2 = 16$, $b^2 = 9$, and $c^2 = a^2 + b^2 = 25$. This hyperbola has center $(0, 0)$, vertices $(0, \pm 4)$, and foci $(0, \pm 5)$. Its asymptotes are $y = \pm 4x/3$ and the length of the latera recta is $2b^2/a = 9/2$ (see Figure 5.21).

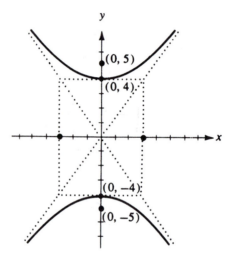

Figure 5.21

Note the relationship between the equations of these two examples when in the standard forms; the left-hand sides are simply opposite in sign. Such hyperbolas are called **conjugate hyperbolas**.

EXAMPLE 3

Find an equation of the hyperbola with foci $(\pm 4, 0)$ and vertex $(2, 0)$.

SOLUTION

Since the foci are on the transverse axis and we are given that they are on the x axis (see Figure 5.22), we must have the form

$$\frac{x^2}{a^2} - \frac{y^2}{b^2} = 1.$$

The foci tell us that $c = 4$, and the vertex gives $a = 2$; thus $b^2 = c^2 - a^2 = 12$. The resulting equation is

$$\frac{x^2}{4} - \frac{y^2}{12} = 1.$$

or

$$3x^2 - y^2 = 12.$$

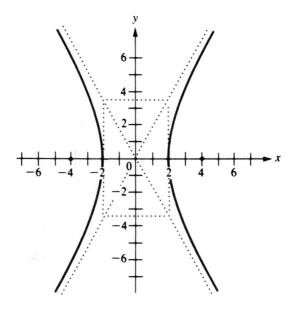

Figure 5.22

Hyperbolas, as well as the other conic sections, can be determined by a single focus, a directrix, and an eccentricity. For the hyperbola

$$\frac{x^2}{a^2} - \frac{y^2}{b^2} = 1,$$

we have **eccentricity**

$$e = \frac{c}{a}$$

and **directrices**

$$x = \pm\frac{a}{e} = \pm\frac{a^2}{c},$$

where $x = a^2/c$ is used in conjunction with the focus $(c, 0)$ and $x = -a^2/c$ with the focus $(-c, 0)$. Since $c > a$, $e = c/a > 1$. Furthermore, a single focus and directrix gives the entire hyperbola—not merely one branch. Either focus with its corresponding directrix generates a hyperbola.

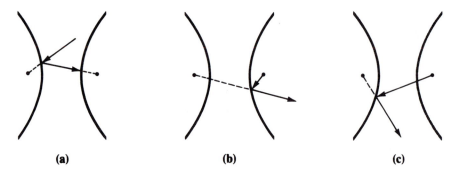

(a)   (b)   (c)

Figure 5.23

Once again, the principal applications of hyperbolas are related to their reflective properties and to celestial motion. A ray of light directed toward one focus is reflected by a hyperbolic mirror toward the other focus as in Figure 5.23a; a ray directed away from one focus is reflected away from the other as in Figures 5.23b and c. One application of this is in the construction of reflecting telescopes. As we have already seen in Section 5.2, a parabolic mirror reflects parallel rays of light to the focus of the parabola. Since the focus is in the path of the incoming light, it presents something of a problem. If the telescope is very large, a cage big enough to hold an observer with an eyepiece can be set up at the focus. This is called a **prime focus** telescope and is shown in Figure 5.24a. However, the large size of the cage cuts down on the light-gathering ability of the telescope. One solution is to put a 45° plane mirror between the parabola and its focus in order to direct the light to the side, away from the path of the incoming light, as shown in Figure 5.24b. This is called a **Newtonian focus** telescope. This presents a problem for very large telescopes because, in order to observe a portion of the sky for a long period of time, the telescope must be constantly moved to counteract the earth's movement. This requires the observer to move as well. This can be difficult with a large telescope, since the observer is well up the tube of the telescope. A third alternative is to use a secondary hyperbolic mirror to direct the light back toward the primary parabolic mirror and out through a hole in that mirror. The secondary mirror has one focus coinciding with the focus of the primary mirror and the other behind the primary. This is called a **Cassegrain focus** for its inventor and is illustrated in Figure 5.24c. A Cassegrain focus telescope still requires movement on the part of the observer, but this motion is easier to deal with because it is at the base of the telescope rather than at the top. This system is especially useful with radio telescopes, which have extremely long focal lengths.

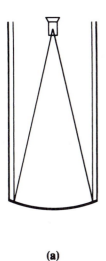
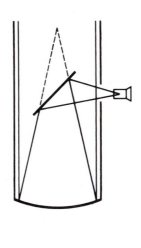
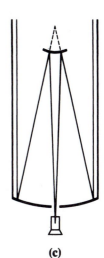

(a)   (b)   (c)

Figure 5.24

A question that may occur to you is this: Why not use a plane mirror rather than a hyperbolic one? The answer is that we want to use as small a mirror as possible to reflect the light out of the telescope. If we use a plane mirror, it must be placed much lower and therefore must be much larger.

As we have seen in Section 5.2, comets that approach the sun once and fly off, never to return, follow either a parabolic or a hyperbolic orbit. A parabolic orbit is quite unusual because it requires that the comet's velocity be exactly the escape velocity; hyperbolic orbits are much more likely.

Another application of hyperbolas is in the use of loran (long-range navigation). A radio signal is sent simultaneously from two widely separated and accurately known points. By noting the differing times of arrival of the two signals (as well as their order of arrival), one's position can be determined as being on one branch of a particular hyperbola with the two stations as foci. If a third such station is added, it can be used with either of the first two to restrict one's position to a second hyperbola. The point of intersection of these two half hyperbolas is the desired location. A ship's or plane's loran receiver has a computer component that makes the actual plotting of the hyperbolas unnecessary.

PROBLEMS

A   *In Problems 1–14, sketch and discuss each equation.*

1. $\dfrac{x^2}{16} - \dfrac{y^2}{9} = 1$

2. $\dfrac{x^2}{4} - \dfrac{y^2}{1} = 1$

3. $\dfrac{y^2}{9} - \dfrac{x^2}{4} = 1$

4. $\dfrac{y^2}{1} - \dfrac{x^2}{9} = 1$ 5. $\dfrac{x^2}{144} - \dfrac{y^2}{25} = 1$ 6. $\dfrac{y^2}{25} - \dfrac{x^2}{144} = 1$

7. $\dfrac{y^2}{25} - \dfrac{x^2}{9} = 1$ 8. $4x^2 - 9y^2 = 36$ 9. $4x^2 - y^2 = 4$

10. $4x^2 - y^2 + 16 = 0$ 11. $x^2 - y^2 = 9$ 12. $16x^2 - 9y^2 = -36$

13. $36y^2 - 100x^2 = 225$ 14. $9x^2 - 4y^2 - 9 = 0$

B   *In Problems 15–26, find an equation(s) of the hyperbola(s) described.*

15. Vertices: $(\pm 2, 0)$; focus: $(-4, 0)$

16. Foci: $(0, \pm 5)$; vertex: $(0, 2)$

17. Asymptotes: $y = \pm 2x/3$; vertex: $(6, 0)$

18. Asymptotes: $y = \pm 3x/4$; focus: $(0, -10)$

19. Asymptotes: $y = \pm 4x/3$; contains $(3\sqrt{2}, 4)$

20. Asymptotes: $y = \pm 3x/4$; length of latera recta: $9/2$

21. Vertices: $(\pm 5, 0)$; contains $(9/5, -4)$

22. Foci: $(\pm 2\sqrt{61}, 0)$; contains $(65/6, 5)$

23. Vertices: $(0, \pm 3)$; $e = 5/3$

24. Foci: $(\pm 10, 0)$; $e = 5/2$

25. Directrices: $x = \pm 9/5$; $e = 5/3$

26. Directrices: $y = \pm 25/13$; focus: $(0, -13)$

27. Find an equation of the line tangent to $x^2 - y^2 = 144$ at $(13, 5)$.

28. Find an equation of the line tangent to $x^2 - y^2 = 16$ at $(-5, 3)$.

29. Find an equation(s) of the lines(s) tangent to $x^2 - y^2 = 9$ and containing $(9, 9)$.

30. Find an equation(s) of the line(s) tangent to $4x^2 - 9y^2 = 7$ and containing $(-7, 7)$.

C   31. Show that there is a number k such that, if P is any point of a hyperbola, the product of the distances of P from the asymptotes of the hyperbola is k.

32. Show that the line tangent to

$$\frac{x^2}{a^2} - \frac{y^2}{b^2} = 1$$

at (x_0, y_0) is

$$\frac{xx_0}{a^2} - \frac{yy_0}{b^2} = 1.$$

GRAPHING CALCULATOR

33. Graph $x^2/4 - y^2/9 = 1$. What is the relationship between the x intercepts and the numbers in the equation? Repeat for $x^2/9 - y^2/4 = 1$.

34. Graph $y^2/4 - x^2/9 = 1$. What is the relationship between the y intercepts and the numbers in the equation? Repeat for $y^2/9 - x^2/4 = 1$.

35. Graph $x^2/16 - y^2/9 = 1$ and $y = \pm 3x/4$. What is the relationship between them?

36. Graph $x^2/4 - y^2/9 = 1$, $-x^2/4 + y^2/9 = 1$, and $y = \pm 3x/2$. How are they related?

37. Graph the following.

(a) $\dfrac{x^2}{4} - \dfrac{y^2}{4} = 1$

(b) $\dfrac{x^2}{9} - \dfrac{y^2}{4} = 1$

(c) $\dfrac{x^2}{16} - \dfrac{y^2}{4} = 1$

(d) $\dfrac{x^2}{25} - \dfrac{y^2}{4} = 1$

What is the relationship between the graphs and the denominators of the x^2 terms?

38. Graph the following.

(a) $\dfrac{x^2}{4} - \dfrac{y^2}{4} = 1$

(b) $\dfrac{x^2}{4} - \dfrac{y^2}{9} = 1$

(c) $\dfrac{x^2}{4} - \dfrac{y^2}{16} = 1$

(d) $\dfrac{x^2}{4} - \dfrac{y^2}{25} = 1$

What is the relationship between the graphs and the denominators of the y^2 terms?

39. Graph the following.

(a) $\dfrac{y^2}{4} - \dfrac{x^2}{4} = 1$

(b) $\dfrac{y^2}{9} - \dfrac{x^2}{4} = 1$

(c) $\dfrac{y^2}{16} - \dfrac{x^2}{4} = 1$

(d) $\dfrac{y^2}{25} - \dfrac{x^2}{4} = 1$

What is the relationship between the graphs and the denominators of the x^2 terms?

40. Graph the following.

(a) $\dfrac{y^2}{4} - \dfrac{x^2}{4} = 1$

(b) $\dfrac{y^2}{4} - \dfrac{x^2}{9} = 1$

(c) $\dfrac{y^2}{4} - \dfrac{x^2}{16} = 1$

(d) $\dfrac{y^2}{4} - \dfrac{x^2}{25} = 1$

What is the relationship between the graphs and the denominators of the y^2 terms?

41. Graph the following.

(a) $\dfrac{x^2}{4} - \dfrac{y^2}{8} = 1$

(b) $\dfrac{x^2}{6} - \dfrac{y^2}{6} = 1$

(c) $\dfrac{x^2}{8} - \dfrac{y^2}{4} = 1$

(d) $\dfrac{x^2}{10} - \dfrac{y^2}{2} = 1$

Where are the foci? These are called **confocal hyperbolas**.

42. We encountered confocal ellipses in Problem 30 of the previous section and confocal hyperbolas in Problem 41 above. The following ellipses and hyperbolas all have foci at $(\pm 2\sqrt{2}, 0)$. Graph as many of each as possible and determine how the ellipses and hyperbolas are related. Be sure to use scales on the axes that do not distort the curves.

(a) $\dfrac{x^2}{10} + \dfrac{y^2}{2} = 1$

(b) $\dfrac{x^2}{12} + \dfrac{y^2}{4} = 1$

(c) $\dfrac{x^2}{14} + \dfrac{y^2}{6} = 1$

(d) $\dfrac{x^2}{6} - \dfrac{y^2}{2} = 1$

(e) $\dfrac{x^2}{4} - \dfrac{y^2}{4} = 1$ **(f)** $\dfrac{x^2}{2} - \dfrac{y^2}{6} = 1$

APPLICATIONS

43. A Cassegrain focus telescope (see page 181) has the dimensions shown in Figure 5.25. Give an equation of the hyperbolic mirror if the axes are placed with x axis on the transverse axis and origin at the center.

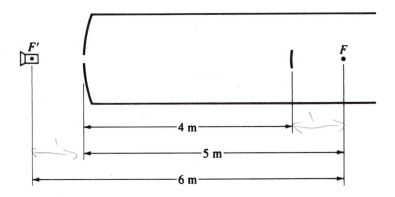

Figure 5.25

44. A radio telescope has a parabolic reflector (of radio waves) with focus 100 feet above the vertex. A small hyperbolic reflector, 90 feet above the vertex and with one focus coincident with that of the parabola, reflects the radio waves to its other focus, which is 20 feet below the vertex of the parabola. Find the lengths of the transverse and conjugate axes of the hyperbola.

45. A comet follows a hyperbolic orbit about the sun, reaching its closest point to the sun at a vertex of 43 million miles. When the line joining the sun and the comet is perpendicular to the transverse axis of the hyperbola, the comet is 137 million miles from the sun. Give an equation for the comet's orbit if the axes are placed with the x axis on the transverse axis and the origin at the center. Where is the sun?

46. It has been found that alpha particles aimed at the nucleus of an atom are repelled along a hyperbolic path. An alpha particle is shot toward the nucleus of an atom (at the origin) from a distant point on the line $y = 2x$. It is deflected along a path that approaches $y = -2x$, coming within 10 Angstrom units of the nucleus. Find an equation of the particle's path.

47. A and B are two loran stations located at $(-500, 0)$ and $(500, 0)$, respectively. A ship's receiver detects radio signals sent simultaneously from the two stations and indicates that the ship is 600 miles closer to A than to B. Similarly, it is found that the ship is 1000 miles closer to C, at $(0, 1300)$, than to D, at $(0, -1300)$. Where is the ship?

48. A man standing at a point $Q = (x, y)$ hears the crack of a rifle at point $P_1 = (1000, 0)$ and the sound of the bullet hitting the target $P_2 = (-1000, 0)$ at the same time. If the bullet travels at 2000 feet/second and sound travels at 1100 feet/second, find an equation relating x and y.

REVIEW PROBLEMS

A *In Problems 1–6, sketch and discuss.*

1. $y^2 = -6x$

2. $x^2 - 4y^2 = 4$

3. $y^2 = 16x$

4. $4x^2 + 9y^2 = 36$

5. $x^2 - y^2 + 9 = 0$

6. $4x^2 + 3y^2 = 48$

B 7. Find an equation(s) of the parabola(s) with vertex $(0, 0)$ and focus $(5, 0)$.

8. Find an equation(s) of the parabola(s) with vertex $(0, 0)$ and containing the points $(2, 4)$ and $(8, 8)$.

9. Find an equation(s) of the ellipse(s) with center $(0, 0)$, vertex $(10, 0)$, and focus $(6, 0)$.

10. Find an equation(s) of the ellipse(s) with center $(0, 0)$, vertex $(0, -3\sqrt{2})$, and containing the point $(2, -3)$.

11. Find an equation(s) of the hyperbola(s) with vertices $(0, \pm 8)$ and focus $(0, 10)$.

12. Find an equation(s) of the hyperbola(s) with asymptotes $y = \pm 3x/2\sqrt{2}$ and containing $(4, 3)$.

13. Find an equation of the line tangent to $4x^2 + 3y^2 = 16$ at $(1, 2)$.

C 14. The inner edge of a track has equation

$$\frac{x^2}{a^2} + \frac{y^2}{b^2} = 1.$$

The track is of width d. When asked to find an equation of the outer edge, a student gave the answer

$$\frac{x^2}{(a + d)^2} + \frac{y^2}{(b + d)^2} = 1.$$

Prove the student to be correct or incorrect.

Chapter **6** *Transformation of Coordinates*

TRANSLATION OF CONIC SECTIONS

The coordinate axes are something of an artificiality which we introduced on the plane in order to represent points and curves algebraically. Since the axes are of this nature, their placement is quite arbitrary. Thus we might prefer to move them in order to simplify some equation. Any change in the position of the axes may be represented by a combination of a **translation** and a **rotation**. A translation of the axes gives a new set of axes parallel to the old ones (see Figure 6.1a), while in a rotation, the axes are rotated about the origin (see Figure 6.1b).

Let us consider translation first. If the axes are translated in such a way that the origin of the new coordinate system is the point (h, k) of the old system (see Figure 6.2), then every point has two representations: (x, y) in the old coordinate system, and (x', y') in the new. The relationship between the old and new coordinate system is easily seen from Figure 6.2 to be

$$x = x' + h \qquad \text{or} \qquad x' = x - h$$

and

$$y = y' + k \qquad \text{or} \qquad y' = y - k.$$

(a)

(b)

Figure 6.1

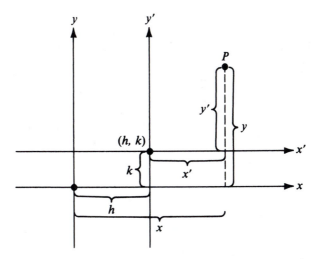

Figure 6.2

These equations (either set) are called **equations of translation**.

Furthermore, we can consider translation from a vector point of view. The vector $x\mathbf{i} + y\mathbf{j}$ can be used to represent the point (x, y). A graphical representation of this vector is a directed line segment with its tail at the origin and its head at (x, y). It is easily seen from Figure 6.3 that $\mathbf{u} = \mathbf{v} + \mathbf{w}$ or

$$x\mathbf{i} + y\mathbf{j} = (x' + h)\mathbf{i} + (y' + k)\mathbf{j}.$$

Thus

$$x = x' + h$$

and

$$y = y' + k.$$

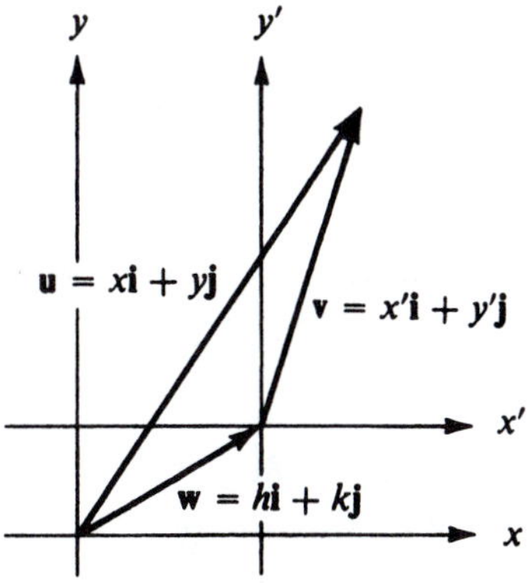

Figure 6.3

Suppose we have a parabola with vertex at (h, k) and axis $y = k$ (see Figure 6.4). Let us put in a new pair of axes, the x' and y' axes, which are parallel to and in the same directions as the original axes and have their origin at the point (h, k) of the original system. Since the parabola's vertex is now at the origin of this coordinate system, its equation is

$$y'^2 = 4cx',$$

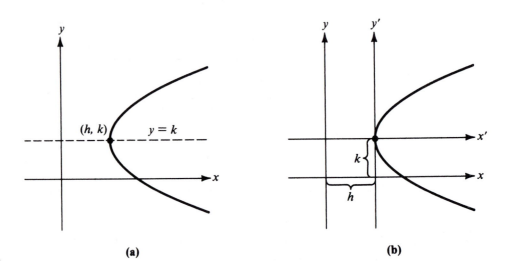

(a)   **(b)**

Figure 6.4

where $|c|$ is the distance from vertex to focus. Now the relationship between the old and new coordinates is given by the equations of translation

$$x' = x - h \quad \text{and} \quad y' = y - k.$$

Thus the equation of the parabola in the original coordinate system is

$$(y - k)^2 = 4c(x - h).$$

We can go through the same analysis for all three of our conics and obtain the following results.

THEOREM 6.1   *A point (x, y) is on the parabola with focus $(h + c, k)$ and directrix $x = h - c$ if and only if it satisfies the equation*

$$(y - k)^2 = 4c(x - h).$$

A point (x, y) is on the parabola with focus $(h, k + c)$ and directrix $y = k - c$ if and only if it satisfies the equation

$$(x - h)^2 = 4c(y - k).$$

THEOREM 6.2   A point (x, y) is on the ellipse with center (h, k), vertices $(h \pm a, k)$, and covertices $(h, k \pm b)$ if and only if it satisfies the equation

$$\frac{(x - h)^2}{a^2} + \frac{(y - k)^2}{b^2} = 1.$$

The foci are $(h \pm c, k)$, where $c^2 = a^2 - b^2$.

A point (x, y) is on the ellipse with center (h, k), vertices $(h, k \pm a)$, and covertices $(h \pm b, k)$ if and only if it satisfies the equation

$$\frac{(y - k)^2}{a^2} + \frac{(x - h)^2}{b^2} = 1.$$

The foci are $(h, k \pm c)$, where $c^2 = a^2 - b^2$.

THEOREM 6.3   A point (x, y) is on the hyperbola with center (h, k), vertices $(h \pm a, k)$, and foci $(h \pm c, k)$ if and only if it satisfies the equation

$$\frac{(x - h)^2}{a^2} - \frac{(y - k)^2}{b^2} = 1,$$

where $b^2 = c^2 - a^2$.

A point (x, y) is on the hyperbola with center (h, k), vertices $(h, k \pm a)$, and foci $(h, k \pm c)$, if and only if it satisfies the equation

$$\frac{(y - k)^2}{a^2} - \frac{(x - h)^2}{b^2} = 1,$$

where $b^2 = c^2 - a^2$.

Suppose we now consider the equation

$$(y - k)^2 = 4c(x - h)$$

and carry out the indicated multiplications. The result is

$$y^2 - 2ky + k^2 = 4cx - 4ch$$
$$y^2 - 4cx - 2ky + (k^2 + 4ch) = 0.$$

This is in the form

$$y^2 + D'x + E'y + F' = 0,$$

where $D' = -4c$, $E' = -2k$, and $F' = k^2 + 4ch$. If we now multiply through by some number C ($C \neq 0$), we have

$$Cy^2 + Dx + Ey + F = 0.$$

We can consider the other five standard forms of Theorems 6.1–6.3 in the same way. We find that in every case we get an equation of the form

$$Ax^2 + Cy^2 + Dx + Ey + F = 0.$$

If we have a parabola, then either $A = 0$ or $C = 0$; if we have an ellipse, then A and C are both positive or both negative; if we have a hyperbola, then A and C have opposite signs.

It is convenient to express the above conditions on A and C in terms of the product AC. Either $A = 0$ or $C = 0$ is equivalent to $AC = 0$; A and C have the same signs if and only if AC is positive; they have opposite signs if and only if AC is negative. The above results are summarized in the following theorem.

THEOREM 6.4

Every conic with axis (or axes) parallel to or on a coordinate axis (or axes) may be represented by an equation of the form

$$Ax^2 + Cy^2 + Dx + Ey + F = 0,$$

where A and C are not both zero. Furthermore, $AC = 0$ if the conic is a parabola, $AC > 0$ if it is an ellipse, and $AC < 0$ if it is a hyperbola.

The equations of Theorems 6.1–6.3 are called the **standard forms for conics**, while the equation of Theorem 6.4 is called the **general form**. It is a simple matter to go from the standard form to the general form; one merely carries out the indicated multiplications. We go from the general form to the standard form by completing the square, just as we did for a circle. Once we have the conic in standard form, it is a simple matter to carry out a translation.

EXAMPLE 1   Sketch and discuss the parabola $9y^2 + 36x - 6y - 23 = 0$.

SOLUTION   First we put the equation into standard form by completing the square on the y terms. To do this we first isolate the y terms on one side of the equation. Then we make the coefficient on y^2 one by dividing by 9. Finally we divide the resulting coefficient of y by 2 and square; this number is then added to both sides, making one side a perfect square.

$$9y^2 - 6y = -36x + 23$$

$$y^2 - \frac{2}{3}y = -4x + \frac{23}{9}$$

$$y^2 - \frac{2}{3}y + \frac{1}{9} = -4x + \frac{24}{9}$$

$$\left(y - \frac{1}{3}\right)^2 = -4\left(x - \frac{2}{3}\right)$$

The equations of translation,

$$x' = x - \frac{2}{3} \quad \text{and} \quad y' = y - \frac{1}{3},$$

now give

$$y'^2 = -4x'.$$

In the $x'y'$ system, we have a parabola with vertex at the origin, axis the x' axis ($y' = 0$), focus at $(-1, 0)$, directrix $x' = 1$, and a latus rectum of length 4. By using the equations of translation, we express the vertex, axis, and so on, in the xy coordinate system. Thus we have a parabola with vertex at $(2/3, 1/3)$, axis $y = 1/3$, focus at $(-1/3, 1/3)$, directrix $x = 5/3$, and a latus rectum of length 4. The parabola, with both sets of coordinate axes, is shown in Figure 6.5.

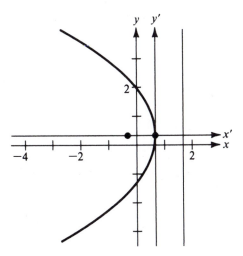

Figure 6.5

Note that the process of completing the square here is exactly the same one that we used in Section 4.1. If the coefficient on the square term is not one, we make it one by division. Then we complete the square by dividing the coefficient of the first-degree term by two and squaring. Since this method of completing the square works only when the coefficient on the square term is one, the first step is essential. This is easily carried out when the x^2 and y^2 terms both have the same coefficient or when one of them is zero. If they are nonzero and different, we cannot make them both one by dividing. In that case, instead of dividing, we factor. For example,

$$9x^2 + 18x = 9(x^2 + 2x).$$

Now we can complete the square inside the parentheses. This can be repeated on the y terms in the same way. This is illustrated in the following example.

EXAMPLE 2   Sketch and discuss the hyperbola

$$9x^2 - 4y^2 - 18x - 24y - 63 = 0.$$

SOLUTION   Here we must complete the square on both the x and the y terms. Thus we must make the coefficients of both x^2 and y^2 one. We do this by factoring.

$$9x^2 - 18x - 4y^2 - 24y = 63$$
$$9(x^2 - 2x) - 4(y^2 + 6y) = 63$$
$$9(x^2 - 2x + 1) - 4(y^2 + 6y + 9) = 63 + 9 \cdot 1 - 4 \cdot 9$$
$$9(x - 1)^2 - 4(y + 3)^2 = 36$$
$$\frac{(x - 1)^2}{4} - \frac{(y + 3)^2}{9} = 1$$

Now the equations of translation,

$$x' = x - 1$$

and

$$y' = y + 3,$$

give

$$\frac{x'^2}{4} - \frac{y'^2}{9} = 1.$$

In the $x'y'$ system, we have a hyperbola with center at $(0, 0)$, $a = 2$, $b = 3$, and $c^2 = a^2 + b^2 = 13$. Thus we have vertices $(\pm 2, 0)$, foci $(\pm\sqrt{13}, 0)$, asymptotes $y' = \pm 3x'/2$, and latera recta of length $2b^2/a = 2 \cdot 9/2 = 9$. Using the equations of translation, we see that in the xy system we have a hyperbola with center at $(1, -3)$, vertices $(3, -3)$ and $(-1, -3)$, foci $(1 \pm \sqrt{13}, -3)$, asymptotes $y + 3 = \pm 3(x - 1)/2$, or $3x - 2y - 9 = 0$ and $3x + 2y + 3 = 0$. The length of the latera recta is, of course, unchanged by the translation. The hyperbola with both coordinate systems is given in Figure 6.6.

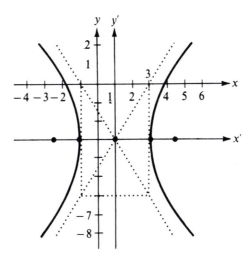

Figure 6.6

While Theorem 6.4 tells us that any conic with axis (or axes) on or parallel to the coordinate axes can be represented in the form

$$Ax^2 + Cy^2 + Dx + Ey + F = 0,$$

it does not follow that any equation in this form represents a parabola, ellipse, or hyperbola. This point is demonstrated by the following example.

EXAMPLE 3   Sketch and discuss $4x^2 + 3y^2 - 16x + 18y + 43 = 0$.

SOLUTION   Since $A = 4$ and $C = 3$, $AC = 12 > 0$. The equation appears to represent an ellipse. But completing the square gives

$$4(x^2 - 4x) + 3(y^2 + 6y) = -43$$
$$4(x^2 - 4x + 4) + 3(y^2 + 6y + 9)$$
$$= -43 + 4 \cdot 4 + 3 \cdot 9$$
$$4(x - 2)^2 + 3(y + 3)^2 = 0.$$

Since neither term on the left is negative, the sum can be zero only if both terms are zero. Thus $x = 2$ and $y = -3$. This equation is satisfied only by the point $(2, -3)$ as shown in Figure 6.7.

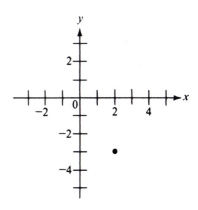

Figure 6.7

The foregoing example is called a **degenerate case of an ellipse**, since the equation has the form of an ellipse but does not actually represent an ellipse. It is comparable to the degenerate cases of a circle, which we saw in Chapter 4. The degenerate cases of the three conics are given in the table.

Conic	AC	Degenerate cases
Parabola	0	One line (two coincident lines)
		Two parallel lines*
		No graph*
Ellipse	+	Circle
		Point
		No graph*
Hyperbola	−	Two intersecting lines

*This is one of the cases mentioned in Section 5.1 in which an equation of the second degree does not represent a curve formed by the intersection of a plane with a right circular cone.

In the previous examples we have been working from the equation of a conic section to its graph. Let us now consider the reverse problem; that is, starting with a description of a conic section, we shall find an equation for it.

EXAMPLE 4

Find an equation(s) of the ellipse(s) with axes parallel to (or on) the coordinate axes and with vertex (3, 5) and covertex (1, 0).

SOLUTION

Since the axes are parallel to or on the coordinate axes and the vertex and covertex are the ends of the axes, we must have one of the situations shown in Figure 6.8. But (3, 5) and (1, 0) are given to be a vertex and covertex, respectively. This means that (3, 5) is an end of a major axis, and (1, 0) an end of a minor axis. Thus Figure 6.8a is the correct one. Now we can take the result directly from Figure 6.8a. Since the major axis is parallel to the y axis, we have the form

$$\frac{(y-k)^2}{a^2} + \frac{(x-h)^2}{b^2} = 1.$$

Furthermore, it is clear from the figure that the center is (3, 0), $a = 5$, and $b = 2$. Thus the desired equation is

$$\frac{(y-0)^2}{25} + \frac{(x-3)^2}{4} = 1$$

or

$$25x^2 + 4y^2 - 150x + 125 = 0.$$

Note that there is no need to actually carry out the translation in the above example; the standard form of Theorem 6.2 gives us all we need.

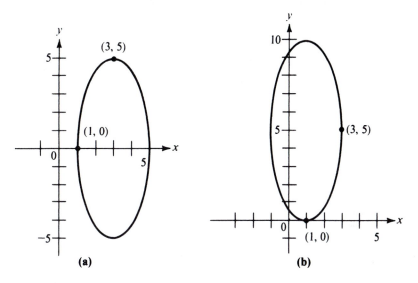

Figure 6.8

EXAMPLE 5 Find an equation(s) of the hyperbola(s) containing $(7, -2)$, $(-1, -2)$, $(8, 1)$, and $(-2, -5)$.

SOLUTION It is easier in this case to use the general form, $Ax^2 + Cy^2 + Dx + Ey + F = 0$, of Theorem 6.4 rather than one of the standard forms of Theorem 6.3. Since the four given points satisfy the equation of the conic section, we have the following four equations.

$$49A + 4C + 7D - 2E + F = 0$$
$$A + 4C - D - 2E + F = 0$$
$$64A + C + 8D + E + F = 0$$
$$4A + 25C - 2D - 5E + F = 0$$

Since there are four equations but five unknowns, we cannot solve for all of them; but let us solve for four of the unknowns in terms of the fifth. First let us subtract the second equation from the other three, eliminating F.

$$48A + 8D = 0$$
$$63A - 3C + 9D + 3E = 0$$
$$3A + 21C - D - 3E = 0$$

Adding the last pair, we get:

$$48A \qquad\quad + 8D = 0$$

$$66A + 18C + 8D = 0.$$

Finally we subtract the first from the second.

$$18A + 18C = 0$$

or

$$C = -A$$

Substituting back, we get:

$$D = -6A$$

$$E = -4A$$

$$F = -11A.$$

Thus the desired equation is

$$Ax^2 - Ay^2 - 6Ax - 4Ay - 11A = 0$$

or

$$x^2 - y^2 - 6x - 4y - 11 = 0.$$

Its graph is shown in Figure 6.9.

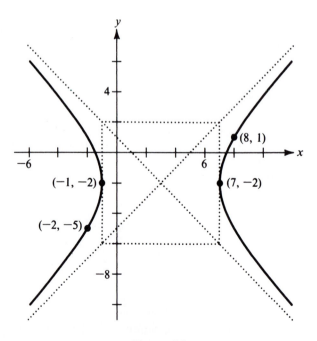

Figure 6.9

When using a graphing calculator, we again have a choice of our usual three methods of graphing; namely, use the quadratic formula to solve for y, complete the square and solve for y, or use the conic graphing program. Generally, either the second or the third method is the most reasonable choice.

If we want merely the graph of the equation without locating vertex, center, foci, directrices, and so forth, the graphing program is the most reasonable choice. The one additional thing the graphing program gives us is the graph of the asymptotes of the hyperbola. While certain other things, such as vertices and covertices, can be determined from the graph, a more in-depth discussion requires another method. In that case, completing the square is the method of choice. The calculations of the special characteristics of the conics, which we have already discussed in this section, give us this information.

The graphing program is based on the quadratic formula. Thus the use of this formula gives us nothing that we cannot get with the program. Moreover, using the quadratic formula reintroduces a tedious exercise that the program is designed to eliminate. So the use of the quadratic formula is not a good choice here when the equation is more complicated than the standard forms of Chapter 5.

EXAMPLE 6　　Use a graphing calculator to sketch and discuss the hyperbola $9x^2 - 4y^2 - 18x - 24y - 63 = 0$.

SOLUTION　　The simplest method of graphing is by the graphing program. In that case, we have

$$A = 9 \qquad\qquad B = 0 \qquad\qquad C = -4$$
$$D = -18 \qquad\qquad E = -24 \qquad\qquad F = -63,$$

giving the graph of Figure 6.10. This graph includes the asymptotes.

However, we cannot determine much about the hyperbola from this method of graphing. To determine other things, we must resort to completing the square. This was done in Example 2, giving

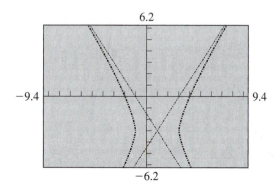

Figure 6.10

$$\frac{(x - 1)^2}{4} - \frac{(y + 3)^2}{9} = 1.$$

We can use the equations of translation to determine the center, vertices, foci, length of the latera recta, and the equations of the asymptotes, as was done in Example 2. Finally, to get the graph from this equation, we must first solve for y.

$$\frac{(y + 3)^2}{9} = \frac{(x - 1)^2}{4} - 1$$

$$(y + 3)^2 = 9\left[\frac{(x - 1)^2}{4} - 1\right]$$

$$y + 3 = \pm 3\sqrt{\frac{(x - 1)^2}{4} - 1}$$

$$y = -3 \pm 3\sqrt{\frac{(x - 1)^2}{4} - 1}$$

We now have the two equations to enter into the calculator. Of course the result is the same as that of Figure 6.10 except the asymptotes are not graphed unless they are entered separately.

PROBLEMS

In Problems 1–20, translate so that the center or vertex of the conic section is at the origin of the new coordinate system. Sketch the curve showing both the old and new axes.

A   **1.** $(x - 3)^2 = 8(y - 2)$ **2.** $16(x - 3)^2 + 9(y + 1)^2 = 144$

   3. $16(x + 3)^2 + 25y^2 = 400$ **4.** $(y - 4)^2 = 2x$

   5. $9(x + 2)^2 + 4(y + 1)^2 = 36$ **6.** $9(x + 2)^2 - 16(y - 1)^2 = -144$

B   **7.** $y^2 - 4x - 2y + 9 = 0$ **8.** $x^2 - 8x - 8y + 8 = 0$

   9. $4x^2 + y^2 + 24x - 2y + 21 = 0$ **10.** $x^2 + 9y^2 - 10x + 36y + 52 = 0$

   11. $9x^2 - 4y^2 + 90x + 32y + 125 = 0$ **12.** $9x^2 - 16y^2 + 72x + 96y + 144 = 0$

   13. $9x^2 + 4y^2 - 72x + 16y + 160 = 0$ **14.** $4x^2 - y^2 - 40x + 6y + 91 = 0$

   15. $4x^2 - 4x - 4y - 5 = 0$ **16.** $16x^2 + 36y^2 + 48x - 180y + 257 = 0$

   17. $4x^2 - 16y^2 + 12x + 16y + 69 = 0$ **18.** $y^2 - 5y + 6 = 0$

   19. $25x^2 + 4y^2 - 150x + 40y + 350 = 0$ **20.** $4x^2 - y^2 - 2x - y = 0$

In Problems 21–30, find an equation(s) of the conic section(s) described.

21. Parabola with focus $(3, 5)$ and directrix $x = -1$

22. Ellipse with vertices $(-1, 8)$ and $(-1, -2)$, containing $(1, 0)$

23. Hyperbola with vertices $(4, 1)$ and $(0, 1)$, and focus $(6, 1)$

24. Parabola with axis parallel to the y axis and containing $(0, 6)$, $(3, -6)$, and $(8, 14)$

25. Hyperbola with vertex $(6, -1)$ and asymptotes $3x - 2y - 6 = 0$ and $3x + 2y - 2 = 0$

26. Ellipse with covertices $(-5, 0)$ and $(1, 0)$ and having latera recta of length 9/2

27. Ellipse with axes parallel to the coordinate axes and containing $(6, -1)$, $(-4, -5)$, $(6, -5)$, and $(-12, -3)$

28. Hyperbola with axes parallel to the coordinate axes and containing $(2, -2)$, $(-3, 8)$, $(-1, -1)$, and $(2, 8)$

29. Hyperbola with foci $(4, 0)$ and $(-6, 0)$ and eccentricity 5/2

30. Ellipse with vertex $(8, -1)$, focus $(6, -1)$ and eccentricity 3/5

C 31. If a parabola with a vertical axis contains the points (x_0, y_0), (x_1, y_1), and (x_2, y_2), show that its equation can be put into the form

$$\begin{vmatrix} x^2 & x & y & 1 \\ x_0^2 & x_0 & y_0 & 1 \\ x_1^2 & x_1 & y_1 & 1 \\ x_2^2 & x_2 & y_2 & 1 \end{vmatrix} = 0.$$

32. What happens to the determinant of Problem 31 if the three given points are collinear? (*Hint*: See Problem 33, page 148.]

GRAPHING CALCULATOR

Graph the equations in Problems 33–38. Determine the following from your graph. Parabola: vertex; ellipse: center, vertices, covertices; hyperbola: center, vertices, asymptotes.

33. $(x + 1)^2 = 4(y - 3)$

34. $9(x + 2)^2 + 16(y - 1)^2 = 144$

35. $y^2 - 6x + 4y + 10 = 0$

36. $x^2 + 4y^2 - 4x + 24y + 24 = 0$

37. $2x^2 - y^2 + 8x + 8y - 8 = 0$

38. $16x^2 - 9y^2 - 64x - 54y + 127 = 0$

APPLICATIONS

39. A comet on a parabolic orbit comes to within 30 million miles of the sun. Give an equation of the comet's orbit when the axes are placed with the x axis on the axis of the parabola and the sun at the origin.

40. A ball is thrown upward and outward from the top edge of a 100-foot building. It reaches its highest point 20 feet out from and 30 feet above the building. Give an equation for its path with the x axis along the ground and the y axis along the side of the building.

41. A parabolic arch spans a width of 50 feet and is 70 feet high. Give an equation of the arch if the x axis joins the feet of the arch and the origin is halfway between them.

42. A suspension bridge has towers 600 feet apart and extending 120 feet above the road. The center of the main cable is 20 feet above the road. Find an equation for the cable

with the x axis along the road surface and the origin halfway between the towers. What is its equation if the origin is at the left tower?

43. Give an equation for the orbit of the planet Mercury (see the table on page 172), where the x axis is the major axis and the origin is the sun.

44. A bridge consists of vertical pillars with semiellipses between them. Each semiellipse spans a width of 50 feet, is 10 feet high at the pillars, and is 20 feet high at the center of the span. Given an equation for the semiellipse if the x axis is along the ground with the origin at the middle of the span.

45. A Cassegrain focus telescope (see page 180) has a parabolic mirror with focus 10 feet from the vertex and a hyperbolic mirror with 12 feet between foci; it is placed 9 feet from the vertex of the parabola. Give equations for both mirrors if the x axis is the axis of the parabola (and the transverse axis of the hyperbola) and the origin is the vertex of the parabola. Give their equations if the origin is midway between the foci of the hyperbola.

46. A comet approaches the sun on a hyperbolic orbit, starting from a path that is inclined at an angle of 30° with the transverse axis and approaching to within 40 million miles of the sun. Find an equation for its orbit with the x axis the transverse axis and the origin at the sun.

6.2

TRANSLATION OF GENERAL EQUATIONS

The method of completing the square is simple to use, but it is rather limited in scope. It can be used only on second-degree equations with no xy term. If there is an xy term or if the equation is not of the second degree, another method, illustrated by the following examples, can be used.

EXAMPLE 1   Translate axes so that the constant and the x term of $x^2 - 2xy + y^2 + 4x - 6y + 10 = 0$ are eliminated.

SOLUTION   Since we have an xy term, completing the square will never give us one of the forms of Theorems 6.1–6.3. Thus we cannot determine h and k before translating. Our only alternative is to use the equations of translation,

$$x = x' + h \qquad \text{and} \qquad y = y' + k,$$

and see what values of h and k are needed to eliminate the specified terms. Substituting the equations of translation into the given equation, we have

$$(x' + h)^2 - 2(x' + h)(y' + k) + (y' + k)^2 + 4(x' + h) - 6(y' + k) + 10 = 0$$
$$x'^2 + 2hx' + h^2 - 2x'y' - 2kx' - 2hy' - 2hk + y'^2 + 2ky' + k^2$$
$$+ 4x' + 4h - 6y' - 6k + 11 = 0$$
$$x'^2 - 2x'y' + y'^2 + (2h - 2k + 4)x' + (-2h + 2k - 6)y'$$
$$+ (h^2 - 2hk + k^2 + 4h - 6k + 10) = 0.$$

Since we want the constant term and the coefficient of x' to be zero, we must choose h and k so that

$$h^2 - 2hk + k^2 + 4h - 6k + 10 = 0$$
$$2h - 2k + 4 = 0.$$

Solving the second equation for h in terms of k and substituting into the first, we have

$$h = k - 2$$
$$(k - 2)^2 - 2(k - 2)k + k^2 + 4(k - 2) - 6k + 10 = 0$$
$$k^2 - 4k + 4 - 2k^2 + 4k + k^2 + 4k - 8 - 6k + 10 = 0$$
$$-2k + 6 = 0$$
$$k = 3$$
$$h = k - 2 = 1.$$

Thus the equations of translation are

$$x = x' + 1 \quad \text{and} \quad y = y' + 3,$$

and the result in the new coordinate system is

$$x'^2 - 2x'y' + y'^2 - 2y' = 0.$$

The graph of the given equation (or equivalently of the new equation) showing both sets of coordinate axes is given in Figure 6.11. The graph appears to

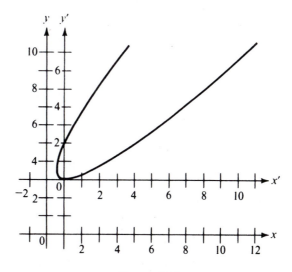

Figure 6.11

be that of a parabola. Is it really a parabola, or is it some other curve that is only similar to a parabola? As we shall see in Section 6.4, any second-degree equation represents a conic or degenerate conic. This is really a parabola. Later we shall see how to identify some of its special characteristics, such as the vertex, axis, focus, and so on.

EXAMPLE 2   Translate axes to eliminate the first-degree terms of $x^2 - 4xy + 3x - 2y + 4 = 0$.

SOLUTION   Substituting the equations of translation,

$$x = x' + h \quad \text{and} \quad y = y' + k,$$

into the given equation, we have

$$(x' + h)^2 - 4(x' + h)(y' + k) + 3(x' + h) - 2(y' + k) + 4 = 0$$
$$x'^2 + 2hx' + h^2 - 4x'y' - 4kx' - 4hy' - 4hk + 3x' + 3h - 2y' - 2k + 4 = 0$$
$$x'^2 - 4x'y' + (2h - 4k + 3)x' + (-4h - 2)y' + (h^2 - 4hk + 3h - 2k + 4) = 0.$$

Setting the coefficients of x' and y' equal to zero and solving, we have

$$2h - 4k + 3 = 0$$
$$-4h - 2 = 0$$

giving

$$h = -\frac{1}{2} \quad \text{and} \quad k = \frac{1}{2}.$$

Thus the equations of translation are

$$x = x' - \frac{1}{2} \quad \text{and} \quad y = y' + \frac{1}{2}$$

and the equation in the new coordinate system is

$$x'^2 - 4x'y' + \frac{11}{4} = 0$$

or

$$4x'^2 - 16x'y' + 11 = 0.$$

The result is shown graphically in Figure 6.12.

Both of these examples carried out translations on a general second-degree equation

$$Ax^2 + Bxy + Cy^2 + Dx + Ey + F = 0.$$

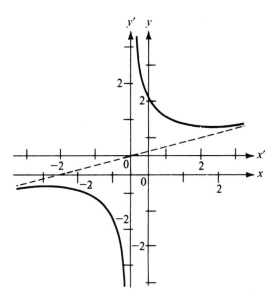

Figure 6.12

Several things might be noted here. Except for multiplying through by a constant, as we did in Example 2 to eliminate a fraction, the translations did not change the values of A, B, and C. A, B, and C are said to be *invariant* under a translation. Of the remaining three coefficients, we may dictate the values of only two of them. This does not mean that we may dictate the values of *any* two of the three; we could not have eliminated both first-degree terms in Example 1.

It might also be noted that, when $B \neq 0$, we may *not* use AC to determine the type of conic section we have. In Example 1 it was seen that the graph is a parabola although AC is positive (not zero). In Example 2 we had a hyperbola although $AC = 0$. Thus the table on page 194 is valid only when $B = 0$.

The method that we have used here is more generally applicable—its use is not restricted to second-degree equations.

EXAMPLE 3   Translate axes so that the constant and the x term of $y = x^3 - 5x^2 + 7x - 5$ are eliminated.

SOLUTION   Since we do not know what values of h and k to choose, we simply use the equations of translation,

$$x = x' + h \quad \text{and} \quad y = y' + k,$$

and see what values of h and k are needed to eliminate the terms specified.

$$y' + k = (x' + h)^3 - 5(x' + h)^2 + 7(x' + h) - 5$$
$$y' = x'^3 + (3h - 5)x'^2 + (3h^2 - 10h + 7)x' + (h^3 - 5h^2 + 7h - 5 - k)$$

Now we must choose h and k so that

$$3h^2 - 10h + 7 = 0$$
$$h^3 - 5h^2 + 7h - 5 - k = 0.$$

The first of these two equations gives

$$h = 1 \quad \text{or} \quad h = 7/3.$$

Substituting these values into the second, we have

$$k = -2 \quad \text{or} \quad k = -86/27.$$

Using $h = 1$ and $k = -2$, we get

$$y' = x'^3 - 2x'^2.$$

Using $h = 7/3$ and $k = -86/27$, we get

$$y' = x'^3 + 2x'^2.$$

The graphs of both the cases described in Example 3 are given in Figure 6.13. While there are two different translations giving two different equations, both

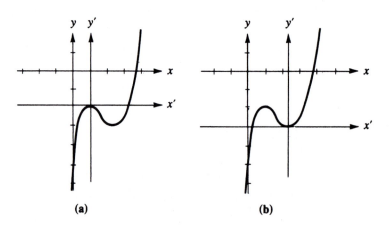

(a)   (b)

Figure 6.13

graphs are the same when referred to the original xy system. As an added bonus, this method has located the local maximum, $(1, -2)$, and minimum, $(7/3, -86/27)$. This method can also be used on second-degree equations with no xy term, but completing the square is so much simpler that most would prefer to use it.

Graphing calculators and graphing programs are very convenient for graphing equations of the type seen here without translation. Equations of the form given in Example 3 are very easily graphed, since they can be entered with no change. Only slightly more difficult is the equation of Example 2. Because it is linear in y, it is relatively easy to solve for y as a function of x:

$$x^2 - 4xy + 3x - 2y + 4 = 0$$
$$4xy + 2y = x^2 + 3x + 4$$
$$(4x + 2)y = x^2 + 3x + 4$$
$$y = \frac{x^2 + 3x + 4}{4x + 2}.$$

Equations of the type given in Example 1 give the most difficulty. We must look upon the equation as quadratic in y, using the quadratic formula to solve:

$$x^2 - 2xy + y^2 + 4x - 6y + 10 = 0$$
$$y^2 - (2x + 6)y + (x^2 + 4x + 10) = 0.$$

This is now in the form $ay^2 + by + c = 0$, with $a = 1$, $b = -(2x + 6)$, and $c = x^2 + 4x + 10$. By the quadratic formula, we have

$$y = \frac{2x + 6 \pm \sqrt{(2x + 6)^2 - 4(x^2 + 4x + 10)}}{2}.$$

There is no need to simplify this expression—the computer can handle it with no difficulty.

Since any one of the given equations in the problems can be graphed electronically, no separate graphing calculator problems are given here. Nevertheless, you are invited to graph any of these equations with a graphing calculator or program.

PROBLEMS

B *In Problems 1–14, translate to eliminate the terms indicated.*

1. $x^2 - 2xy + 4y^2 + 8x - 26y + 38 = 0$; first-degree terms
2. $2x^2 - xy - y^2 + 5x - 8y - 3 = 0$; first-degree terms
3. $x^2 + 4xy - y^2 - 2x - 14y - 3 = 0$; first-degree terms

4. $3x^2 + xy + y^2 - 16x - 10y + 30 = 0$; first-degree terms

5. $xy - 5x + 4y - 4 = 0$; first-degree terms

6. $x^2 + xy + 9x + 5y + 20 = 0$; first-degree terms

7. $y = x^3 - 6x^2 + 11x - 8$; constant, x^2 term

8. $y = x^3 - 3x + 6$; constant, x term

9. $y = x^3 - 3x^2 + 3x + 5$; constant, x term

10. $y = x^3 - 9x^2 + 24x + 3$; constant, x term

11. $y = x^4 - 8x^3 + 24x^2 - 28x + 7$; constant, x term

12. $y = x^4 - 10x^3 + 37x^2 - 120x + 138$; constant, x term

13. $x^2y - 2x^2 + 2xy + y - 4x - 6 = 0$; second-degree terms

14. $x^2y + x^2 + 2xy + 2x - 3y - 1 = 0$; second-degree terms

C **15.** Prove that any translation on

$$Ax^2 + Bxy + Cy^2 + Dx + Ey + F = 0$$

leaves A, B, and C invariant.

16. Prove that any translation on $y = P(x)$, where $P(x)$ is a polynomial in x, leaves invariant the coefficient of the highest degree term of $P(x)$.

6.3

ROTATION

The second transformation of the axes that we wish to consider is a rotation of the axes about the origin (see Figure 6.1b). If the axes are rotated through an angle θ, then every point of the plane has two representations: (x, y) in the original coordinate system and (x', y') in the new coordinate system. Alternatively, every vector $\mathbf{v}$ in the plane has two representations: $\mathbf{v} = x\mathbf{i} + y\mathbf{j}$ in the original coordinate system and $\mathbf{v} = x'\mathbf{i}' + y'\mathbf{j}'$ in the new coordinate system (see Figure 6.14). In order to find the relationships between the x and y of one coordinate system and the x' and y' of the other, let us consider the relationships of $\mathbf{i}$ and $\mathbf{j}$ with $\mathbf{i}'$ and $\mathbf{j}'$. Remembering that $\mathbf{i}$, $\mathbf{j}$, $\mathbf{i}'$, and $\mathbf{j}'$ are all unit vectors, we see from Figure 6.15 that

$$\mathbf{i}' = \cos\theta\,\mathbf{i} + \sin\theta\,\mathbf{j}$$
$$\mathbf{j}' = -\sin\theta\,\mathbf{i} + \cos\theta\,\mathbf{j}.$$

Thus

$$\mathbf{v} = x'\mathbf{i}' + y'\mathbf{j}'$$
$$= x'(\cos\theta\,\mathbf{i} + \sin\theta\,\mathbf{j}) + y'(-\sin\theta\,\mathbf{i} + \cos\theta\,\mathbf{j})$$
$$= (x'\cos\theta - y'\sin\theta)\mathbf{i} + (x'\sin\theta + y'\cos\theta)\mathbf{j}.$$

Since $\mathbf{v} = x\mathbf{i} + y\mathbf{j}$, we have

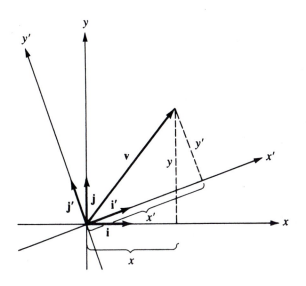

Figure 6.14

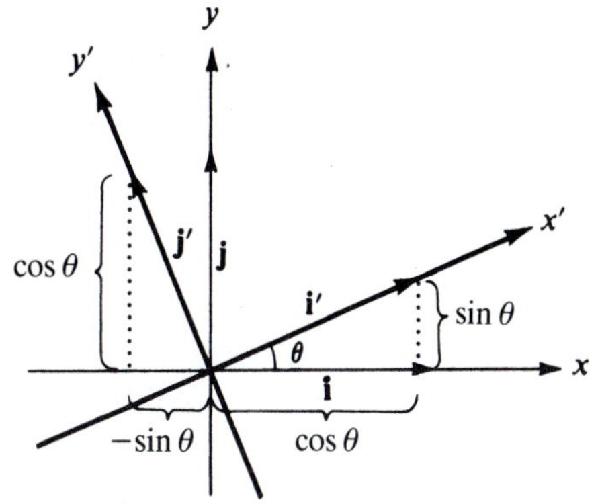

Figure 6.15

$$x = x'\cos\theta - y'\sin\theta$$

and

$$y = x'\sin\theta + y'\cos\theta,$$

called the **equations of rotation**.

An alternative method of finding the above equations of rotation without the use of vectors is the following. Recalling again that every point P of the plane has two representations, (x, y) in the original coordinate system and (x', y') in the new system, we see from Figure 6.16 that

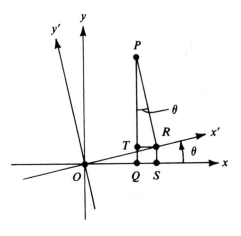

Figure 6.16

$$x = \overline{OQ}, \qquad x' = \overline{OR},$$
$$y = \overline{PQ}, \qquad y' = \overline{PR}.$$

Let us now consider the relations between x and y, and x' and y'. Noting first of all that

$$\sphericalangle ROQ = \sphericalangle RPQ = \theta,$$

we find from triangle ORS

$$\sin \theta = \frac{\overline{RS}}{\overline{OR}} \qquad \cos \theta = \frac{\overline{OS}}{\overline{OR}},$$
$$\overline{RS} = \overline{OR} \sin \theta \qquad \overline{OS} = \overline{OR} \cos \theta$$
$$= x' \sin \theta, \qquad = x' \cos \theta,$$

and from triangle PRT

$$\sin \theta = \frac{\overline{TR}}{\overline{PR}} \qquad \cos \theta = \frac{\overline{PT}}{\overline{PR}},$$
$$\overline{TR} = \overline{PR} \sin \theta \qquad \overline{PT} = \overline{PR} \cos \theta$$
$$= y' \sin \theta, \qquad = y' \cos \theta.$$

Now

$$\begin{aligned} x &= \overline{OQ} & y &= \overline{PQ} \\ &= \overline{OS} - \overline{QS} & &= \overline{TQ} + \overline{PT} \\ &= \overline{OS} - \overline{TR} & &= \overline{RS} + \overline{PT} \\ &= x' \cos \theta - y' \sin \theta; & &= x' \sin \theta + y' \cos \theta. \end{aligned}$$

Thus we have the equations of rotation

$$x = x' \cos \theta - y' \sin \theta,$$
$$y = x' \sin \theta + y' \cos \theta.$$

EXAMPLE 1 Find the new representation of

$$x^2 - xy + y^2 - 2 = 0$$

after rotating through an angle of 45°. Sketch the curve, showing both the old and new coordinate systems.

SOLUTION Since $\sin 45° = \cos 45° = 1/\sqrt{2}$, the equations of rotation are

$$x = \frac{x' - y'}{\sqrt{2}} \qquad \text{and} \qquad y = \frac{x' + y'}{\sqrt{2}}.$$

Substituting into the original equation, we have

$$\frac{(x' - y')^2}{2} - \frac{x' - y'}{\sqrt{2}} \cdot \frac{x' + y'}{\sqrt{2}} + \frac{(x' + y')^2}{2} - 2 = 0$$

$$\frac{x'^2 - 2x'y' + y'^2 - x'^2 + y'^2 + x'^2 + 2x'y' + y'^2}{2} = 2$$

$$x'^2 + 3y'^2 = 4.$$

Figure 6.17 shows the final result.

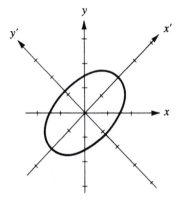

Figure 6.17

EXAMPLE 2   Find a new representation of $x^2 + 4xy - 2y^2 - 6 = 0$ after rotating through an angle θ = Arctan 1/2. Sketch the curve, showing both the old and new coordinate systems.

SOLUTION   Since θ = Arctan 1/2, θ is the first-quadrant angle whose tangent is 1/2. This results in the triangle of Figure 6.18. From that figure we can easily determine the sine and cosine of θ:

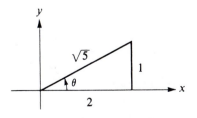

Figure 6.18

$$\sin \theta = \frac{1}{\sqrt{5}} \quad \text{and} \quad \cos \theta = \frac{2}{\sqrt{5}},$$

giving equations of rotation

$$x = \frac{2x' - y'}{\sqrt{5}} \quad \text{and} \quad y = \frac{x' + 2y'}{\sqrt{5}}.$$

Substituting into the original equation, we have

$$\frac{(2x' - y')^2}{5} + 4\frac{2x' - y'}{\sqrt{5}} \cdot \frac{x' + 2y'}{\sqrt{5}} - 2\frac{(x' + 2y')^2}{5} - 6 = 0,$$

$$\frac{4x'^2 - 4x'y' + y'^2 + 8x'^2 + 12x'y' - 8y'^2 - 2x'^2 - 8x'y' - 8y'^2}{5} = 6,$$

$$2x'^2 - 3y'^2 = 6.$$

Figure 6.19 shows the final results. Note that Figure 6.18 can be used to determine the position of the new coordinate axes; the x' axis contains the origin and the point $(2, 1)$.

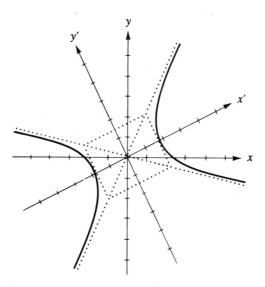

Figure 6.19

In both of these examples, we have seen that the given rotation has eliminated the xy term. Of course, not every rotation will do so—the angle of rotation must be specially chosen. We shall see in the next section how to choose θ to eliminate the xy term.

PROBLEMS

In Problems 1–10, find a new representation of the given equation after rotating through the given angle. Sketch the curve, showing both the old and new coordinate systems.

A   **1.** $2x + 3y = 6$; $\theta = $ Arctan $3/2$ **2.** $3x - y = 5$; $\theta = $ Arctan 3

       3. $xy = 4$; $\theta = 45°$ **4.** $2x^2 - xy + 2y^2 - 15 = 0$; $\theta = 45°$

B   **5.** $x^2 - 2xy + y^2 + x + y = 0$; $\theta = 45°$

       6. $31x^2 + 10\sqrt{3}xy + 21y^2 - 144 = 0$; $\theta = 30°$

       7. $x^2 + 2\sqrt{3}xy + 3y^2 + 8\sqrt{3}x - 8y = 0$; $\theta = 60°$

       8. $11x^2 - 50\sqrt{3}xy - 39y^2 + 576 = 0$; $\theta = 60°$

       9. $8x^2 + 5xy - 4y^2 - 4 = 0$; $\theta = $ Arctan $1/5$

      10. $6x^2 - 5xy - 6y^2 + 26 = 0$; $\theta = $ Arctan $(-1/5)$

In Problems 11–16, find a new representation of the given equation after rotating through the given angle.

      11. $3x^2 - 3xy - y^2 + 4 = 0$; $\theta = $ Arctan $(-1/2)$

      12. $4x^2 + 3xy - 5 = 0$; $\theta = $ Arctan $1/2$

      13. $4x^2 + 3xy - 5 = 0$; $\theta = 45°$

      14. $x^2 - 3xy + y^2 + 5 = 0$; $\theta = 30°$

      15. $3x^2 - 3xy - y^2 + 4 = 0$; $\theta = 60°$

      16. $x^2 - 5xy + 2 = 0$; $\theta = $ Arctan $1/3$.

C   **17.** Show that $x^2 + y^2 = 25$ is invariant under rotation through any angle.

      18. Show that a second form for the equations of rotation is

$$x' = x \cos \theta + y \sin \theta$$

and

$$y' = -x \sin\theta + y \cos \theta.$$

6.4

THE GENERAL EQUATION OF SECOND DEGREE

We have seen that any conic section with axes parallel to the coordinate axes can be represented by a second-degree equation with $B = 0$; furthermore, any second-degree equation with $B = 0$ represents a conic or degenerate conic with axes parallel to the coordinate axes. We now extend this concept to conic sections in any position. It is an easy matter to see that any conic can be represented by a second-degree equation, starting from our standard forms and translating and rotating.

Suppose, given a second-degree equation with $B \neq 0$, we rotate axes through an angle θ. If our assumption that this equation represents a conic or degenerate conic is correct, then a rotation of axes through some positive angle less than 90° should give us a conic with axis (or axes) on or parallel to the coordinate axes. Thus such a rotation should eliminate the xy term, and we shall assume throughout this discussion that $0° < \theta < 90°$ and

$$Ax^2 + Bxy + Cy^2 + Dx + Ey + F = 0.$$

Substituting the equations of rotation,

$$x = x' \cos \theta - y' \sin \theta,$$
$$y = x' \sin \theta + y' \cos \theta,$$

we have

$$A(x' \cos \theta - y' \sin \theta)^2 + B(x' \cos \theta - y' \sin \theta)(x' \sin \theta + y' \cos \theta)$$
$$+ C(x' \sin \theta + y' \cos \theta)^2 + D(x' \cos \theta - y' \sin \theta)$$
$$+ E(x' \sin \theta + y' \cos \theta) + F = 0.$$

After carrying out the multiplication and combining similar terms, we find that the coefficient of $x'y'$ is

$$(C - A)2 \sin \theta \cos \theta + B(\cos^2 \theta - \sin^2 \theta) = (C - A) \sin 2\theta + B \cos 2\theta.$$

We want this coefficient to be zero for the proper choice of θ. Let us set it equal to zero and see what θ should be.

$$(C - A) \sin 2\theta + B \cos 2\theta = 0$$

At this point we divide the argument into two cases.

Case I: If $A = C$, then

$$B \cos 2\theta = 0$$
$$\cos 2\theta = 0$$
$$2\theta = 90°$$
$$\theta = 45°.$$

Case II: If $A \neq C$, then

$$(A - C) \sin 2\theta = B \cos 2\theta$$
$$(A - C)\frac{\sin 2\theta}{\cos 2\theta} = B$$
$$(A - C) \tan 2\theta = B$$
$$(A - C)\frac{2 \tan \theta}{1 - \tan^2 \theta} = B \quad \text{(by a trigonometric identity)}$$
$$2(A - C) \tan \theta = B - B \tan^2 \theta$$
$$B \tan^2 \theta + 2(A - C) \tan \theta - B = 0$$

This is a quadratic equation in $\tan\theta$. Solving by the quadratic formula, we have

$$\tan\theta = \frac{2(C - A) \pm \sqrt{4(A - C)^2 + 4B^2}}{2B}$$

$$= \frac{(C - A) \pm \sqrt{(C - A)^2 + B^2}}{B}.$$

Now we have two values of $\tan\theta$. Which one do we want? It is not difficult to see (refer to Problem 17) that the values of θ that we get from them must differ by an odd multiple of 90° and that the two values of θ have opposite signs. Thus either value should eliminate the xy term. Since we are assuming that $0° < \theta < 90°$, we want the positive value of $\tan\theta$. Since

$$\sqrt{(C - A)^2 + B^2} > \sqrt{(C - A)^2} = |C - A|$$

(that is to say, the radical in the numerator is always numerically larger than $C - A$), it follows that the sign of the numerator corresponds to the sign on the radical. By taking the sign on the radical to agree with the sign of B, we can be sure that the result is always positive. Once we have $\tan\theta$ it is a simple matter to find $\sin\theta$ and $\cos\theta$ and substitute them into the equations of rotation. Thus we are always able to rotate axes to eliminate the xy term. The resulting equation must then represent a conic or degenerate conic.

THEOREM 6.5   *Any conic section can be represented by the second-degree equation*

$$Ax^2 + Bxy + Cy^2 + Dx + Ey + F = 0$$

where A, B, and C are not all zero. Any second-degree equation represents either a conic or a degenerate conic.

Let us sum up the results of the previous discussion. If $B \neq 0$, then the axes may be rotated to eliminate the xy term as follows:
If $A = C$, then $\theta = 45°$. If $A \neq C$, then

$$\tan\theta = \frac{(C - A) \pm \sqrt{(C - A)^2 + B^2}}{B},$$

where the sign on the radical is taken to agree with the sign of B.
    An alternative approach for Case II is the following:

$$(A - C)\sin 2\theta = B\cos 2\theta$$

$$\tan 2\theta = \frac{\sin 2\theta}{\cos 2\theta} = \frac{B}{A - C}.$$

From the identity

$$\tan^2 2\theta + 1 = \sec^2 2\theta = \frac{1}{\cos^2 2\theta}$$

we get

$$\cos 2\theta = \frac{\pm 1}{\sqrt{1 + \tan^2 2\theta}}$$

Because $0° < 2\theta < 180°$, the signs of $\cos 2\theta$ and $\tan 2\theta$ must agree. Finally, using the half angle formulas

$$\sin\frac{x}{2} = \pm\sqrt{\frac{1 - \cos x}{2}}$$

$$\cos\frac{x}{2} = \pm\sqrt{\frac{1 + \cos x}{2}}$$

with $x = 2\theta$, we get

$$\sin\theta = \sqrt{\frac{1 - \cos 2\theta}{2}}$$

$$\cos\theta = \sqrt{\frac{1 + \cos 2\theta}{2}},$$

which can be used in the equations of rotation.

EXAMPLE 1   Rotate axes to eliminate the xy term of $x^2 + 4xy - 2y^2 - 6 = 0$. Sketch, showing both sets of axes.

SOLUTION

$$\tan\theta = \frac{(C - A) \pm \sqrt{(C - A)^2 + B^2}}{B}$$

$$= \frac{(-2 - 1) + \sqrt{(-2 - 1)^2 + 4^2}}{4}$$

$$= \frac{-3 + \sqrt{9 + 16}}{4}$$

$$= \frac{1}{2}$$

Using the triangle of Figure 6.20, we have

$$\sin\theta = \frac{1}{\sqrt{5}} \quad \text{and} \quad \cos\theta = \frac{2}{\sqrt{5}}.$$

Thus the equations of rotation are

$$x = \frac{2x' - y'}{\sqrt{5}} \quad \text{and} \quad y = \frac{x' + 2y'}{\sqrt{5}}$$

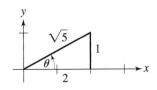

Figure 6.20

Substituting into the original equation (see Example 2 of the previous section), we have

$$2x'^2 - 3y'^2 = 6.$$

The sketch is given in Figure 6.19 on page 210.

EXAMPLE 2

Rotate axes to eliminate the xy term of

$$2x^2 - xy + 2y^2 - 2 = 0.$$

Sketch, showing both sets of axes.

SOLUTION

Since $A = C$, $\theta = 45°$ and the equations of rotation are

$$x = \frac{x' - y'}{\sqrt{2}} \quad \text{and} \quad y = \frac{x' + y'}{\sqrt{2}}.$$

Substituting these into the original equation, we have the following.

$$2\frac{(x' - y')^2}{2} - \frac{x' - y'}{\sqrt{2}}\frac{x' + y'}{\sqrt{2}} + 2\frac{(x' + y')^2}{2} - 2 = 0$$

$$\frac{2x'^2 - 4x'y' + 2y'^2 - x'^2 + y^2 + 2x'^2 + 4x'y' + 2y'^2}{2} = 2$$

$$3x'^2 + 5y'^2 = 4$$

$$\frac{x'^2}{4/3} + \frac{y'^2}{4/5} = 1$$

The sketch is given in Figure 6.21.

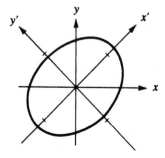

Figure 6.21

EXAMPLE 3

Rotate axes to eliminate the xy term of

$$21x^2 + 50xy - 99y^2 + 6\sqrt{26}x - 82\sqrt{26}y - 494 = 0.$$

Sketch, showing both sets of axes.

$$
\begin{aligned}
\tan \theta &= \frac{(C - A) \pm \sqrt{(C - A)^2 + B^2}}{B} \\[2mm]
&= \frac{(-99 - 21) + \sqrt{(-99 - 21)^2 + 50^2}}{50} \\[2mm]
&= \frac{-120 + \sqrt{120^2 + 50^2}}{50} \\[2mm]
&= \frac{-120 + 10\sqrt{12^2 + 5^2}}{50} \\[2mm]
&= \frac{-120 + 130}{50} \\[2mm]
&= \frac{1}{5}
\end{aligned}
$$

From Figure 6.22 we have

$$\sin \theta = \frac{1}{\sqrt{26}} \qquad \cos \theta = \frac{5}{\sqrt{26}},$$

giving the equations of rotation

$$x = \frac{5x' - y'}{\sqrt{26}} \qquad y = \frac{x' + 5y'}{\sqrt{26}}.$$

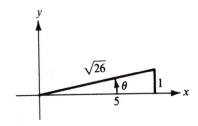

Figure 6.22

Substituting into our original equation, we have

$$21\frac{(5x' - y')^2}{26} + 50\frac{5x' - y'}{\sqrt{26}} \cdot \frac{x' + 5y'}{\sqrt{26}} - 99\frac{(x' + 5y')^2}{26}$$

$$+ 6\sqrt{26}\frac{5x' - y'}{\sqrt{26}} - 82\sqrt{26}\frac{x' + 5y'}{\sqrt{26}} - 494 = 0$$

$$\frac{21(25x'^2 - 10x'y' + y'^2) + 50(5x'^2 + 24x'y' - 5y'^2)}{26}$$

$$- \frac{99(x'^2 - 10x'y' + 25y'^2)}{26} + 6(5x' - y') - 82(x' + 5y') - 494 = 0$$

$$26x'^2 - 104y'^2 - 52x' - 416y' - 494 = 0$$

$$x'^2 - 4y'^2 - 2x' - 16y' - 19 = 0.$$

Now let us put this into standard form by completing the square on the x' and y' terms.

$$(x'^2 - 2x') - 4(y'^2 + 4y') = 19$$

$$(x'^2 - 2x' + 1) - 4(y'^2 + 4y' + 4) = 19 + 1 - 4 \cdot 4$$

$$(x' - 1)^2 - 4(y' + 2)^2 = 4$$

$$\frac{(x' - 1)^2}{4} - \frac{(y' + 2)^2}{1} = 1$$

The graph is shown in Figure 6.23.

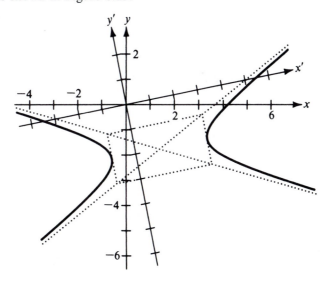

Figure 6.23

There is a method of determining which conic we have before we rotate the axes. It is based on the fact that certain expressions are invariant under rotation; that is, they have the same value before and after any rotation. Although there are several such expressions (see Problem 15), the one in which we are interested is $B^2 - 4AC$ for the equation

$$Ax^2 + Bxy + Cy^2 + Dx + Ey + F = 0.$$

If we substitute the equations of rotation,

$$x = x' \cos \theta - y' \sin \theta$$

and

$$y = x' \sin \theta + y' \cos \theta,$$

into this equation, we get a new second-degree equation,

$$A'x'^2 + B'x'y' + C'y'^2 + D'x' + E'y' + F' = 0,$$

with

$$A' = A \cos^2 \theta + B \sin \theta \cos \theta + C \sin^2 \theta,$$
$$B' = -2A \sin \theta \cos \theta + B(\cos^2 \theta - \sin^2 \theta) + 2C \sin \theta \cos \theta,$$

and

$$C' = A \sin^2 \theta - B \sin \theta \cos \theta + C \cos^2 \theta.$$

From this, we find that the expression for $B'^2 - 4A'C'$ then simplifies to $B^2 - 4AC$. This gives the following theorem.

THEOREM 6.6   *If the equation*

$$Ax^2 + Bxy + Cy^2 + Dx + Ey + F = 0$$

is transformed into the equation

$$A'x'^2 + B'x'y' + C'y'^2 + D'x' + E'y' + F' = 0$$

by rotating the axes, then

$$B^2 - 4AC = B'^2 - 4A'C'.$$

That is, $B^2 - 4AC$ is invariant under rotations.

If we choose the angle of rotation properly, $B' = 0$ and the type of conic can be determined by looking at A' and C' (see the table on page 194). Thus we have the following results.

THEOREM 6.7   *The equation*

$$Ax^2 + Bxy + Cy^2 + Dx + Ey + F = 0$$

represents a hyperbola, an ellipse, or a parabola (or a degenerate case of one of these) according to the value of $B^2 - 4AC$, as indicated in the following table.

Conic Section	$B^2 - 4AC$
Hyperbola	Positive
Parabola	Zero
Ellipse	Negative

Let us apply this to the foregoing examples. In Example 1, the equation

$$x^2 + 4xy - 2y^2 - 6 = 0$$

gives

$$B^2 - 4AC = 16 - 4 \cdot 1 \cdot (-2) = 24,$$

indicating that the conic is a hyperbola, which we have seen to be the case. After rotation we have

$$2x'^2 - 3y'^2 = 6,$$

giving

$$B'^2 - 4A'C' = 0 - 4 \cdot 2 \cdot (-3) = 24.$$

In Example 2, the equation

$$2x^2 - xy + 2y^2 - 2 = 0$$

gives

$$B^2 - 4AC = 1 - 4 \cdot 2 \cdot 2 = -15.$$

This shows the conic to be an ellipse, which again is what we have already found. It might be noted that, after rotation, we get the result

$$\frac{3}{2}x'^2 + \frac{5}{2}y'^2 = 2$$

provided we do not multiply both sides by some constant. This again gives

$$B'^2 - 4A'C' = 0 - 4 \cdot \frac{3}{2} \cdot \frac{5}{2} = -15.$$

For the result

$$3x'^2 + 5y'^2 = 4,$$

we get a different value of $B'^2 - 4A'C'$, *which is still negative.* If, at some stage, we multiply both sides by some nonzero number k, $B^2 - 4AC$ is then multiplied by k^2. Since k^2 must be positive, the sign of $B^2 - 4AC$ is not changed, no matter what number k is.

Finally, in Example 3,

$$B^2 - 4AC = 50^2 - 4 \cdot 21(-99) = 10{,}816.$$

Since this is positive, the conic is a hyperbola. It is really not necessary to carry out all of the arithmetic in this case. It is necessary only to observe that both terms are positive, giving a positive result. This is sufficient to tell us that the conic is a hyperbola.

This use of $B^2 - 4AC$ is helpful here and will be used again in Section 7.6.

As we can see from the above examples, rotating the axes is not a barrel of fun. It is long, tedious, and very error prone. Graphing calculators and programs are a welcome alternative. The only question is the method used to enter the equation. When there is an xy term, we cannot use the method of completing the square. Thus using the quadratic formula or using the graphing program are the only alternatives. Once again the program is much simpler to use than the quadratic formula; it is the method recommended.

PROBLEMS

In Problems 1–14, rotate axes to eliminate the xy term. Sketch, showing both sets of axes.

B
1. $x^2 + xy + y^2 + 4\sqrt{2}x - 4\sqrt{2}y = 0$ 2. $5x^2 + 6xy + 5y^2 - 8 = 0$
3. $7x^2 + 6xy - y^2 - 32 = 0$ 4. $4x^2 + 4xy + y^2 + 8\sqrt{5}x - 16\sqrt{5}y = 0$
5. $8x^2 - 12xy + 17y^2 = 20$ 6. $9x^2 + 8xy - 6y^2 = 70$
7. $5x^2 - 4xy + 8y^2 - 36 = 0$ 8. $x^2 + 12xy + 6y^2 = 30$
9. $4x^2 + 12xy + 9y^2 + 8\sqrt{13}x + 12\sqrt{13}y - 65 = 0$
10. $6x^2 + 12xy + 11y^2 = 240$
11. $9x^2 - 6xy + y^2 - 12\sqrt{10}x - 36\sqrt{10}y = 0$
12. $x^2 + 8xy + 7y^2 - 36 = 0$
13. $8x^2 + 12xy - 8y^2 - 40 = 0$
14. $5x^2 - 6xy + 5y^2 = 72$

C
15. Given the equation

$$Ax^2 + Bxy + Cy^2 + Dx + Ey + F = 0,$$

which yields

$$A'x'^2 + B'x'y' + C'y'^2 + D'x' + E'y' + F' = 0$$

after rotation through the angle θ, show that $A' + C' = A + C$ for any value of θ; that is, $A + C$ is invariant under rotation.

16. It can easily be seen graphically that two conic sections have at most four points in common. But

$$2x^2 + xy - y^2 + 3y - 2 = 0$$

and

$$2x^2 + 3xy + y^2 - 6x - 5y + 4 = 0$$

have the five points $(1, 0)$, $(-2, 3)$, $(5, -4)$, $(-6, 7)$, and $(10, -9)$ in common. Why?

17. Suppose

$$\tan \theta_2 = \frac{(C - A) - \sqrt{(C - A)^2 + B^2}}{B}$$

and

$$\tan \theta_1 = \frac{(C - A) + \sqrt{(C - A)^2 + B^2}}{B}$$

GRAPHING CALCULATOR

Show that $\tan \theta_1 \cdot \tan \theta_2 = -1$. Use this result to show that $\tan \theta_1$ and $\tan \theta_2$ have opposite signs and θ_1 and θ_2 differ by an odd multiple of 90°.

18.–23. Use the graphing program to graph the equations of Problems 1–6.

APPLICATIONS

24. Three loran stations, A, B, and C, at $(-500, 0)$, $(0, 0)$, and $(300, 500)$, respectively, send out simultaneous impulses. A ship receiving the impulses has determined that it is 450 miles closer to B than to A and 100 miles closer to C than to B. What is the ship's position?

REVIEW PROBLEMS

In Problems 1–9, sketch and discuss.

A   **1.** $4x^2 + y^2 - 8x + 6y + 9 = 0$ **2.** $y^2 - x + 2y + 4 = 0$

3. $y^2 - 8x + 4y + 28 = 0$ **4.** $9x^2 - 16y^2 + 36x - 128y - 364 = 0$

5. $9x^2 + 25y^2 + 18x + 100y - 116 = 0$ **6.** $9x^2 - 16y^2 + 18x - 16y - 139 = 0$

7. $x^2 - 4x - 4y = 0$ **8.** $3x^2 + 4y^2 + 30x - 16y + 91 = 0$

9. $4x^2 - 9y^2 - 16x - 54y - 65 = 0$

B   **10.** Find an equation(s) of the ellipse(s) with center $(-4, 1)$, axes parallel to the coordinate axes, and tangent to both coordinate axes.

11. Find an equation(s) of the parabola(s) with axis parallel to a coordinate axis, focus $(3, 5)$ and directrix $x = -1$.

12. Find an equation(s) of the hyperbola(s) with asymptotes $5x - 4y + 22 = 0$ and $5x + 4y - 18 = 0$ and containing the point $(32/5, -7)$.

13. Find an equation(s) of the parabola(s) with horizontal or vertical axis, vertex $(2, -3)$, and containing $(6, -1)$.

14. Find an equation(s) of the ellipse(s) with axes parallel to the coordinate axes, focus $(-1, -1)$, and covertex $(3, 2)$.

15. Find an equation(s) of the hyperbola(s) with vertices $(1, -1)$ and $(7, -1)$ and focus $(-1, -1)$.

16. Translate axes to eliminate the constant and second-degree terms of

$$y = x^3 + 6x^2 + 3x - 14.$$

17. Translate axes to eliminate the constant and third-degree terms of

$$y = x^4 - 16x^3 + 88x^2 - 192x + 140.$$

18. Rotate axes to eliminate the xy term of $3x^2 + 12xy - 2y^2 + 42 = 0$. Sketch, showing both sets of axes.

19. Rotate axes to eliminate the xy term of $2x^2 - \sqrt{3}xy + y^2 - 10 = 0$. Sketch, showing both sets of axes.

20. Rotate axes to eliminate the xy term of $4x^2 - 4xy + y^2 + \sqrt{5}x + 2\sqrt{5}y - 10 = 0$. Sketch, showing both sets of axes.

C **21.** Show that $2x^2 - 2xy + y^2 - 9 = 0$ is an equation of an ellipse. This equation is a quadratic equation in y. Rearranging the terms, we have $y^2 - 2xy + (2x^2 - 9) = 0$, which can be solved by the quadratic formula to give $y = x \pm \sqrt{9 - x^2}$. Sketch $y = x$ and $y = \pm\sqrt{9 - x^2}$ (square both sides first in the latter equation) on the same set of axes. For each value of x, add the y coordinates for these two curves to get points on the original curve. Use this method to find the graph of the given curve.

Chapter **7** Curve Sketching

7.1

SYMMETRY AND INTERCEPTS

In the first chapter, we sketched the graph of an equation by the tedious process of point-by-point plotting—a method that sometimes causes one to overlook some "interesting" portions of the graph or to sketch certain portions incorrectly. Suppose, for example, you are asked to sketch the graph of

$$y = \frac{10x(x + 8)}{(x + 10)^2}.$$

The methods of Chapter 1 might lead you to the graph of Figure 7.1. A better sketch of the graph is given in Figure 7.2. While the earlier method produced correct results for the portion we were sketching, it provided no means for determining which portions of the curve are most "interesting."

x	y
−5	−6.00
−4	−4.44
−3	−3.06
−2	−1.88
−1	−0.86
0	0.00
1	0.74
2	1.39
3	1.95
4	2.45
5	2.89

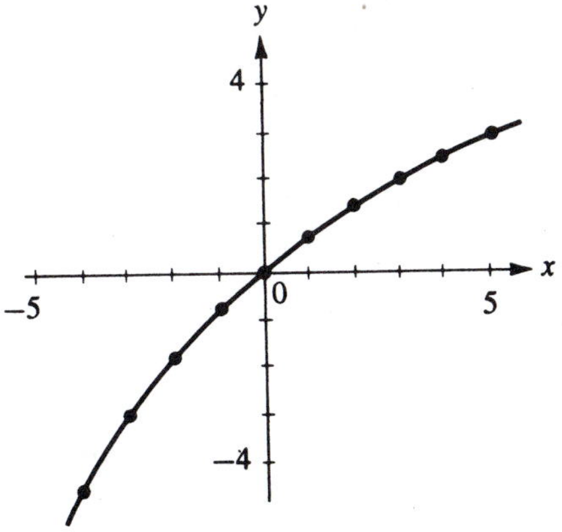

Figure 7.1

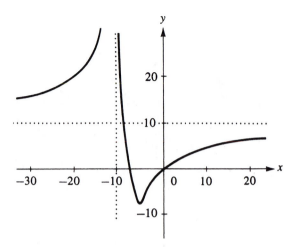

Figure 7.2

Let us consider one more example. Suppose we want to graph

$$y = \frac{2x(2x - 1)}{4x - 1}.$$

The methods of Chapter 1 lead us to the set of points shown in Figure 7.3. Now, what does the graph look like? How would you join the points? You might join them as indicated in Figure 7.4a. The correct graph is shown in Figure 7.4b. These examples demonstrate the need for better methods of sketching curves.

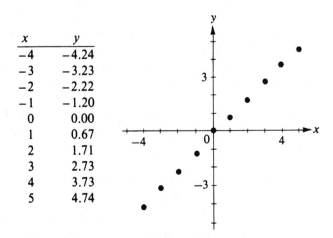

x	y
-4	-4.24
-3	-3.23
-2	-2.22
-1	-1.20
0	0.00
1	0.67
2	1.71
3	2.73
4	3.73
5	4.74

Figure 7.3

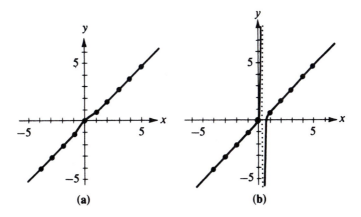

Figure 7.4

There are those who maintain that the use of graphing calculators makes it unnecessary to consider the methods of this chapter. There are, however, some problems in relying entirely upon graphing calculators. Even if you are sure that you will always have one available when you need it, you still have the same problems noted above. Because a graphing calculator is essentially a point-by-point plotter, it has no way of picking out the interesting parts of the graph—that is up to you. Nevertheless, a graphing calculator makes it easy for you to change the range of values used on the *x* and *y* axes and regraph. Because of this, the calculator makes it easier to search for the interesting features. On the other hand, why search by trial and error when learning a few simple rules makes that unnecessary?

A second problem with graphing calculators is that most of them have difficulty with asymptotes. Vertical asymptotes are generally not clearly indicated, and it is often difficult to guess just what nonvertical line is approached by a graph. If you have access to a graphing calculator, try using it with the above examples. Your success (or lack of it) is likely to depend upon the range of *x* and *y* values that you use.

Now let us consider some methods to help us sketch graphs quickly. An important concept that can be of some use in sketching graphs is that of **symmetry**. There are two types of symmetry: symmetry about a line and symmetry about a point. If a curve is symmetric about a line, then one-half of it is the mirror image of the other half, with the mirror as the line of symmetry. Let us state this more precisely.

DEFINITION   *A curve is symmetric about a line l if, for every point P of the curve not on l, there is a point P' on the curve such that PP' is perpendicular to, and bisected by, l.*

An example of this type of symmetry occurs with the graph of $y = 1/x^2$, in which the *y* axis is the line of symmetry (see Figure 7.5). It is easily seen that the right and left halves of this curve are mirror images of each other. We have also

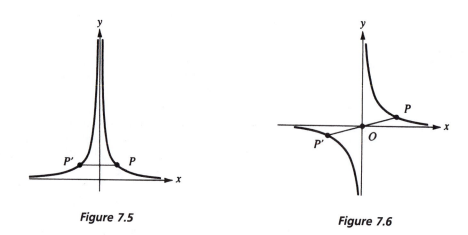

Figure 7.5   **Figure 7.6**

seen symmetry in connection with the conic sections. A parabola is symmetric about its axis, while an ellipse or hyperbola is symmetric about either of its axes. We see from this that a curve can have more than one line of symmetry. Furthermore, a curve can have infinitely many lines of symmetry. A circle is symmetric about any line through its center, and a line is symmetric about itself as well as about any line perpendicular to it.

Let us now consider the second type of symmetry—symmetry about a point.

DEFINITION   *A curve is symmetric about a point Q if, for every point P of the curve distinct from Q, there is a point P′ on the curve such that PP′ is bisected by Q.*

It is easily seen from Figure 7.6 that the origin is a point of symmetry of the graph of $y = 1/x$. A circle, ellipse, or hyperbola is symmetric about its center. Furthermore, a curve can have several points of symmetry; a line is symmetric about any of its points.

It is clear that knowing the symmetries of a curve can be an aid in sketching it; if we know one-half of the curve, we can get the other half by reflection. Now let us consider how we may determine these symmetries from the equation of the curve. While there is no restriction on the lines or points of symmetry that a curve might have, we shall consider only symmetry about the axes and about the origin because it is relatively easy to determine these symmetries from the equation. We begin by a consideration of symmetry about the y axis.

If a curve is symmetric about the y axis, then, corresponding to every point $P = (x, y)$ on the curve, there is a point $P′ = (-x, y)$ (see Figure 7.5) with the same y coordinate (and an x coordinate that is the negative of the x coordinate) of P. In this situation we get the same value for y whether we substitute a positive number x into the equation or its negative, $-x$. This results in the following theorem.

THEOREM 7.1   *If every x in an equation is replaced by −x and the resulting equation is equivalent to the original, then its graph is symmetric about the y axis.*

By a similar argument, we also have the following.

THEOREM 7.2   *If every y in an equation is replaced by −y and the resulting equation is equivalent to the original, then its graph is symmetric about the x axis.*

Before considering examples of these, let us consider the term "equivalent equations." Two equations are equivalent if every point that satisfies one of them satisfies the other. We have really been considering equivalent equations whenever we simplified or changed the form of equations. For example, the circle with center $(5, -2)$ and radius 3 has equation

$$(x - 5)^2 + (y + 2)^2 = 9.$$

By squaring, combining terms, and rearranging, we have the equivalent equation

$$x^2 + y^2 - 10x + 4y + 20 = 0.$$

The types of operations that we have used in the past to simplify or change the form of an equation will be used here to show equivalence.

EXAMPLE 1   Test $y = 1/x^2$ for symmetry about both axes.

SOLUTION   If we replace x by $-x$, we have

$$y = \frac{1}{(-x)^2} = \frac{1}{x^2}.$$

Since the substitution resulted in an equation that easily simplified to give the original, we have symmetry about the y axis.

On the other hand, the replacement of y by $-y$ gives

$$-y = \frac{1}{x^2},$$

which cannot be simplified to give our original equation. We do not have symmetry about the x axis.

EXAMPLE 2   Show that an ellipse with center at the origin is symmetric about both axes.

SOLUTION   The ellipse has equation

$$\frac{x^2}{a^2} + \frac{y^2}{b^2} = 1.$$

Replacing x by $-x$, we have

$$\frac{(-x)^2}{a^2} + \frac{y^2}{b^2} = 1.$$

Since $(-x)^2 = x^2$, this is equivalent to the original equation, showing that we have symmetry about the y axis.

Similarly, replacing y by $-y$ and noting that $(-y)^2 = y^2$ shows that we have symmetry about the x axis.

Finally, let us consider symmetry about the origin. If a curve is symmetric about the origin, then for every point $P = (x, y)$ on the curve, there is a point $P' = (-x, -y)$ on the curve (see Figure 7.6). This leads to the following theorem.

THEOREM 7.3   *If every x in an equation is replaced by $-x$ and every y by $-y$ and the resulting equation is equivalent to the original, then its graph is symmetric about the origin.*

EXAMPLE 3   Show that $y = 1/x$ is symmetric about the origin.

SOLUTION   Replacing x by $-x$ and y by $-y$, we have

$$-y = \frac{1}{-x}.$$

This is equivalent to the original equation, since we get the original by multiplying both sides by -1. Thus we have symmetry about the origin.

EXAMPLE 4   Show that an ellipse with center at the origin is symmetric about the origin.

SOLUTION   An equation of the ellipse is

$$\frac{x^2}{a^2} + \frac{y^2}{b^2} = 1.$$

Since $(-x)^2 = x^2$ and $(-y)^2 = y^2$, making both substitutions result in an equivalent equation, and we have symmetry about the origin.

It might be noted that Example 2 showed that any ellipse with center at the origin is symmetric about both axes. If this is the case, then symmetry about the origin follows from that. Why? See Problem 31.

Now let us turn our attention to intercepts. The **intercepts** of a curve are simply the points of the curve that lie on the coordinate axes; those on the x axis are the x intercepts, while those on the y axis are the y intercepts (the origin is both an x intercept and a y intercept). We determine the x intercepts (if any) by setting y equal to zero and solving for x; similarly, the y intercepts are found by setting x equal to zero and solving for y. While we want to avoid the method of point-by-point plotting of curves, intercepts are usually so easy to find that we shall want to use them in our curve sketching.

EXAMPLE 5   Find the intercepts of $\dfrac{x^2}{4} + \dfrac{y^2}{9} = 1$.

SOLUTION   When $y = 0$, we have

$$\frac{x^2}{4} = 1$$
$$x^2 = 4$$
$$x = \pm 2,$$

giving intercepts, $(2, 0)$ and $(-2, 0)$. Similarly, when $x = 0$,

$$\frac{y^2}{9} = 1$$
$$y^2 = 9$$
$$y = \pm 3,$$

giving $(0, 3)$ and $(0, -3)$.

A type of equation often encountered is one of the form $y = P(x)$ or $y = P(x)/Q(x)$, where $P(x)$ and $Q(x)$ are polynomials having no common factor. If $P(x)$ can be factored to give the form

$$P(x) = c(x - a_1)^{n_1}(x - a_2)^{n_2} \ldots (x - a_k)^{n_k},$$

where $c, a_1, \ldots, a_k$ are real numbers, then the x intercepts are

$$(a_1, 0), (a_2, 0), \ldots, (a_k, 0).$$

The y intercept (an equation in this form has at most one) is still found by setting x equal to zero.

EXAMPLE 6   Find the intercepts of $y = (x + 1)^2(x - 3)$.

SOLUTION   It is clear that $y = 0$ when either of the two factors on the right is 0. That is, $y = 0$ when $x + 1 = 0$ or $x = -1$ and when $x - 3 = 0$ or $x = 3$. Thus the x intercepts are $(-1, 0)$ and $(3, 0)$.

We find the y intercept simply by substituting $x = 0$. This gives

$$y = 1^2(-3) = -3$$

or y intercept $(0, -3)$.

EXAMPLE 7   Find the intercepts of $y = x^3 - 2x^2 + x - 2$.

SOLUTION   Let us begin by factoring the right-hand side.

$$\begin{aligned} x^3 - 2x^2 + x - 2 &= x^2(x - 2) + (x - 2) \\ &= (x - 2)(x^2 + 1) \end{aligned}$$

The second of these two factors cannot be factored any further without getting into complex numbers. Thus we have

$$y = (x - 2)(x^2 + 1).$$

Now $y = 0$ when $x - 2 = 0$, or $x^2 + 1 = 0$. The first gives $x = 2$, but the second gives

$$\begin{aligned} x^2 + 1 &= 0 \\ x^2 &= -1, \end{aligned}$$

which has no real solution. The result is the single x intercept $(2, 0)$.

Finally, substituting $x = 0$ into our original equation (or into the equivalent factored form), we have $y = -2$ or y intercept $(0, -2)$.

Suppose we are unable to carry out the factorization of the above example by the method we used there. We still have a way to factor by the following consideration. If an equation of the form

$$a_n x^n + a_{n-1} x^{n-1} + \cdots + a_1 x + a_0 = 0$$

with integer coefficients has a rational root, it must be of the form

$$x = \frac{a}{b},$$

where a is a divisor of a_0 and b is a divisor of a_n. Let us use this to solve Example 7.

In order to find the x intercept(s) we set $y = 0$ and solve for x. This gives

$$x^3 - 2x^2 + x - 2 = 0.$$

If we cannot factor this expression, we use the above method to find the roots. The coefficient of the leading term is 1, with divisors ± 1; the constant term is -2, with divisors ± 1 and ± 2. The only possible rational solutions are

$$x = \frac{\pm 1}{\pm 1} = \pm 1 \qquad \text{or} \qquad x = \frac{\pm 2}{\pm 1} = \pm 2.$$

Trying one after another, we have:

If $\quad x = 1, \qquad 1^3 - 2 \cdot 1^2 + 1 - 2 = -2 \neq 0.$

If $\quad x = -1, \qquad (-1)^3 - 2(-1)^2 - 1 - 2 = -6 \neq 0.$

If $\quad x = 2, \qquad 2^3 - 2 \cdot 2^2 + 2 - 2 = 0.$

If $\quad x = -2, \qquad (-2)^3 - 2(-2)^2 - 2 - 2 = -20 \neq 0.$

Thus the only rational root is $x = 2$.

We still do not know if there is an irrational root. Since $x = 2$ is a root, $x - 2$ must be a factor of our polynomial. Knowing that, we can divide $x^3 - 2x^2 + x - 2$ by $x - 2$ to get

$$
\begin{array}{r}
x^2 \qquad\quad + 1 \\
x - 2 \overline{)x^3 - 2x^2 + x - 2} \\
\underline{x^3 - 2x^2} \qquad\qquad \\
x - 2 \\
\underline{x - 2.}
\end{array}
$$

Since $x^2 + 1 = 0$ has no real root, we have only one x intercept, $(2, 0)$.

The y intercept $(0, -2)$ is still found as we did above—by substituting $x = 0$ to get $y = -2$.

The above can be quite tedious for third-degree equations and higher. If there are irrational roots, it is even more difficult. For those with graphing calculators, the process can be reversed. That is, instead of using the intercepts to help draw the graph, we can use the graph to find the intercepts. By magnifying portions of the graph, or, better, by using CALC, JUMP, or G SOLVE, we can determine the intercepts to any degree of accuracy that we want. This is perhaps one of the strongest points in favor of using a calculator.

EXAMPLE 8   Find the intercepts of

$$y = \frac{(x - 2)^2(x + 1)}{(x - 3)(x - 1)^2}.$$

SOLUTION   When dealing with a polynomial equation, we found the x intercepts by considering factors on the right-hand side; if any one of them is zero, then y is zero. In this case, we consider only the factors in the numerator; if the numerator is zero, then the quotient is zero. Thus we have the point $(2, 0)$ from the factor $(x - 2)^2$ and the point $(-1, 0)$ from the factor $(x + 1)$. Finally, by substituting $x = 0$, we get

$$y = \frac{(-2)^2(1)}{(-3)(-1)^2} = -\frac{4}{3}.$$

Therefore, the y intercept is $(0, -4/3)$.

Note that an intercept is a point. It is represented not by an equation, but by an ordered pair of numbers. For example, $x = 2$ is not an x intercept; it is an equation of a vertical line. If $y = 0$ when $x = 2$, then the point $(2, 0)$ is an x intercept.

PROBLEMS

A   *In Problems 1–10, check for symmetry about both axes and the origin.*

1. $y = x^4 - x^2$

2. $y = x^3 - x$

3. $y = x^3 - x^2$

4. $\dfrac{x^2}{4} + \dfrac{y^2}{9} = 1$

5. $y^2 = \dfrac{x + 1}{x}$

6. $y^3 = \dfrac{x + 1}{x}$

7. $xy = 1$

8. $x^2y^2 = 1$

9. $y = \dfrac{x}{x^2 + 1}$

10. $y = \dfrac{(x + 1)(x - 1)}{x^2}$

B   *In Problems 11–30, find all intercepts.*

11. $y = (x + 1)(x - 3)$

12. $y = (x + 2)(x - 1)^2$

13. $y = x^2 - 5x - 6$

14. $y = x^3 + x^2 - 2x$

15. $y = x^4 - x^2$

16. $y = x^3 - x$

17. $y = \dfrac{x + 1}{x}$

18. $y = \dfrac{x - 2}{x + 2}$

19. $y = \dfrac{(2x + 1)(x - 1)^2}{(x - 2)(x + 1)^2}$

20. $y = \dfrac{x - 3}{(x + 1)(x - 2)}$

21. $y = \dfrac{(x + 2)(x - 4)}{x - 1}$

22. $y = \dfrac{x - 1}{(x + 2)(x - 4)}$

23. $y = \dfrac{(x + 2)^2(x - 4)}{(x - 1)^2}$

24. $y = \dfrac{x}{x^2 + 1}$

25. $y = \dfrac{x^2 + 1}{x}$

26. $xy = 2x + 1$

27. $x^2y = 2x + 1$

28. $x^2y - y = x^2$

29. $x^2y - y = x^3$

30. $x^2y - y = x$

C **31.** Show that if a graph has any two of the three types of symmetry—about the x axis, about the y axis, about the origin—then it must have the third.

32. Give an example of a curve with exactly three lines of symmetry.

33. Can a graph have exactly two points of symmetry?

34. Give an example of a curve with infinitely many points of symmetry.

35. Show that if two perpendicular lines are lines of symmetry of a given curve, then their point of intersection is a point of symmetry.

36. An even function f is one for which $f(-x) = f(x)$; an odd function is one for which $f(-x) = -f(x)$. What can we say concerning the symmetry of the graph of an even function? Of an odd function?

37. A graph (with at least one point not on the x axis) is symmetric about the x axis. Can it be the graph of a function? Explain.

38. Show that if every x in an equation is replaced by $2k - x$ and the resulting equation is equivalent to the original, then its graph is symmetric about the line $x = k$.

GRAPHING CALCULATOR

In Problems 39–44, graph the equations to find the x intercepts.

39. $y = x^3 + 2x^2 - 5x + 1$

40. $y = x^3 + x^2 - 7x + 3$

41. $y = x^5 - 3x^3 + 2x$

42. $y = x^4 - 5x^3 + 2x^2 - x + 5$

43. $y = x^4 - 8x^3 + 2x - 7$

44. $y = x^6 - 20x^4 + 15x^2 - 40$

APPLICATIONS

45. A ball thrown upward from ground level at 80 feet per second will be at height $h = 80t - 16t^2$ after t seconds. When will it hit the ground?

46. An arrow shot upward from ground level at 176 feet per second will be at height $h = 176t - 16t^2$ after t seconds. When will it hit the ground?

47. A ball thrown upward and outward from a 100-foot building has trajectory $y = 100 + 8x - (x^2/5)$, where y is the height and x is the distance from the building. Where will it hit the ground?

48. It is estimated that the cost of producing n widgets is $C = 200n + 10,400$. The revenue (income) from selling those n widgets is $R = 800n - 4n^2$, giving a profit of

$$P = R - C = -4n^2 + 600n - 10,400.$$

For what value(s) of n does the manufacturer break even?

7.2

SKETCHING POLYNOMIAL EQUATIONS

Let us now use symmetry and intercepts to help us sketch polynomial equations. First, however, there are a few additional considerations that will aid us in the sketching. Since a polynomial equation has the form

$$y = a_nx^n + a_{n-1}x^{n-1} + \cdots + a_1x + a_0,$$

where n is a positive integer, it is easily seen that we have one and only one value of y for a given value of x. What are the things that would prevent us from getting a (real) value of y for a given x? There are just two things: an even root of a negative number and a zero in the denominator. These might occur if we have fractional or negative exponents, but the exponents in a polynomial must be nonnegative whole numbers. Thus we always get a value of y corresponding to a given value of x. Furthermore, there is no way that we can get two different values of y from a single value of x. Thus one x always results in one and only one y. Graphically, this says that any vertical line must contain one and only one point of the graph. Figures 7.7a and b are graphs of polynomials; Figures 7.7c, d, e, and f cannot be graphs of polynomials. It might be observed that, while a vertical line must contain one and only one point of the graph of a polynomial equation, a horizontal line may contain zero, one, or several points of the graph. This can be seen from Figure 7.7a and b. Finally, it might be noted that the single-valuedness of polynomials has one implication with regard to symmetry. The graph of a polynomial equation (except the trivial $y = 0$) cannot be symmetric about the x axis. Symmetry about the x axis implies two values of y for a single value of x: (x, y) and $(x, -y)$.

Now let us look into the behavior of a graph near an x intercept. We begin by considering $y = x^n$, which has the origin as an x intercept. Figure 7.8 on page 236 shows the graphs of $y = x^n$ for several values of n. Figure 7.8a shows graphs for odd values of n, while Figure 7.8b shows them for even values. The difference is clearly illustrated. When n is odd, the graph crosses the x axis; when n is even, the graph merely touches the x axis at a single point (the intercept), but remains above the x axis. The fact that our graphs go from below the x axis on the left to above the x axis on the right, or above the x axis both left and right of the intercept, depends upon the sign of the other factors. In this case, we may look

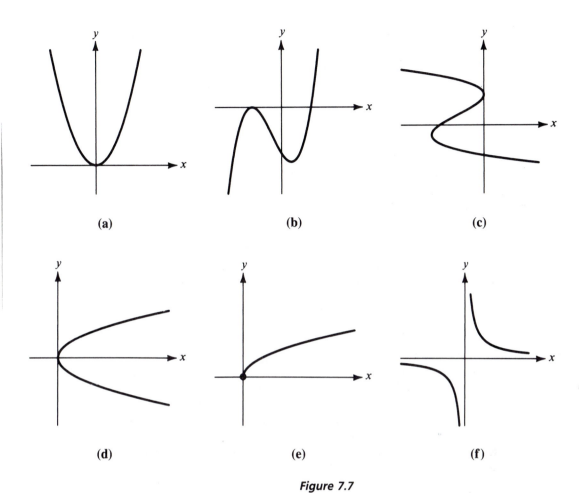

(a)   (b)   (c)

(d)   (e)   (f)

Figure 7.7

upon the other factor as $+1$. If we had $y = -x^n$, whose other factor is -1, or $y = x^n(x - 2)$, whose other factor, $(x - 2)$, is negative for x near zero, then the above situation is reversed; that is, the graph would cross the x axis by going from above it on the left to below it on the right, or it would remain below the x axis. Thus the important thing to consider here is that the graph crosses the x axis when the exponent is odd but merely touches the x axis without crossing it when the exponent is even.

A second thing that we might note from Figure 7.8a is that the graph crosses the x axis at an angle when the exponent is one, but levels off at $x = 0$ when the exponent is greater than one. This is true generally and is easily proved with the aid of calculus.

EXAMPLE 1   Describe the behavior of the graph of $y = (x - 1)^2(x - 2)(x + 1)^3$ at its x intercepts.

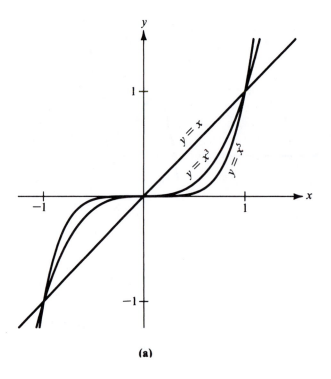

(a)

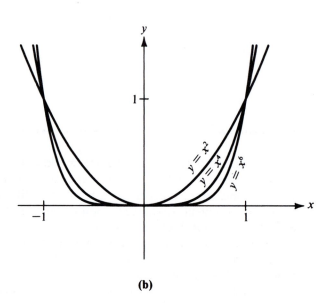

(b)

Figure 7.8

SOLUTION

The x intercepts are easily found to be $(1, 0)$, $(2, 0)$, and $(-1, 0)$. Each one of these came from a factor on the right-hand side of the equation. Let us record the exponent on each such factor.

From $(x - 1)^2$: $(1, 0)$, exponent even
From $(x - 2)$: $(2, 0)$, exponent odd (1)
From $(x + 1)^3$: $(-1, 0)$, exponent odd (>1)

We now see that the graph touches, but does not cross, the x axis at $(1, 0)$, it crosses at an angle at $(2, 0)$, and it levels off and crosses at $(-1, 0)$. The graph of this equation is shown in Figure 7.9.

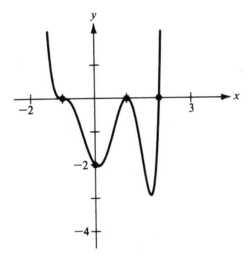

Figure 7.9

It might be observed that we are only considering x intercepts here. Because we have already noted that a polynomial must be single-valued, the graph cannot touch the y axis and turn back—it must cross. Furthermore, the y intercept does not come from a particular factor as does an x intercept; there is simply no way to consider y intercepts in the above way.

Finally, note that the even-or-odd considerations have nothing to do with the value of the intercept itself. It refers to the value of the exponent on the factor that determines the intercept.

A final consideration that we shall look into before sketching is the behavior of the graph at the ends. By "the ends" of the graph we mean the portion of the graph for numerically large (either positive or negative) values of x. It is not difficult to see that, for numerically large values of x, the value of y must also be numerically large. The only question here is whether y is positive or negative. If the equation is in factored form, we can determine the sign by the signs of the individual factors. If it is not in factored form, the term having x to the highest power will dominate the other terms; it can be used to determine the signs.

EXAMPLE 2   Describe the behavior of the graph of $y = (x - 1)^2(x - 2)(x + 1)^3$ at the ends.

SOLUTION   For large positive values of x, all factors are positive, giving a large positive product. For large negative values of x, we have (noting only the signs of the individual factors)

$$y = (-)^2(-)(-)^3 = (+)(-)(-) = +.$$

This may be expressed in the following shorthand:*

$$\text{As} \quad x \to +\infty, \quad y \to +\infty$$

$$\text{As} \quad x \to -\infty, \quad y \to +\infty$$

As we may observe in Figure 7.9, the graph goes up $(y \to +\infty)$ at both ends.

ALTERNATE SOLUTION   If the above equation had been given to us in the expanded, rather than the factored, form, it would have been

$$y = x^6 - x^5 - 4x^4 + 2x^3 + 5x^2 - x - 2.$$

We can easily determine the behavior at the ends without factoring merely by considering the term of highest degree (x^6). Since $(+)^6 = +$ and $(-)^6 = +$, we have

$$\text{As} \quad x \to +\infty, \quad y \to +\infty$$

$$\text{As} \quad x \to -\infty, \quad y \to +\infty$$

as before.

Now let us put all of this together to sketch the graphs of polynomial equations.

---

* This notation is taken from calculus. These expressions are another way of saying

$$\lim_{x \to +\infty} y = +\infty \quad \text{and} \quad \lim_{x \to -\infty} y = +\infty.$$

Do not look upon $+\infty$ and $-\infty$ as numbers; they only have meaning when taking limits.

EXAMPLE 3   Sketch $y = (x - 3)(x + 1)^2$.

SOLUTION   From the factors on the right, we get x intercepts $(3, 0)$ with an odd exponent and $(-1, 0)$ with an even exponent. If $x = 0$, then $y = -3$, which gives $(0, -3)$. As x gets large and positive, y gets large and positive; as x gets large and negative, y gets large and negative. It is easy to see that no symmetry exists about either axis or the origin. Summing up, we have:

Intercepts: $(3, 0)$ odd, $(-1, 0)$ even, $(0, -3)$

Ends: As $x \to +\infty$, $y \to +\infty$

As $x \to -\infty$, $y \to -\infty$

No symmetry

All of this is indicated in Figure 7.10. Let us sketch the graph, starting at the far left, and work to the right (this choice is quite arbitrary; we might just as well go from right to left or start in the middle and work outward). We keep in mind that the curve must go through all intercepts and that y is single-valued. Since $y \to -\infty$ as $x \to -\infty$, we start in the lower left-hand corner. Going to the right, we first reach the intercept $(-1, 0)$. Since it is an even intercept, the graph merely touches the x axis but does not cross it. Next, the graph goes through $(0, -3)$ and then turns back up in order to go through $(3, 0)$. Since $(3, 0)$ is an odd intercept, the graph crosses the x axis there and proceeds upward. The result is given in Figure 7.11.

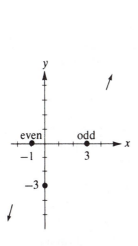

Figure 7.10

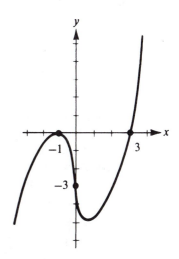

Figure 7.11

Note that we put the lowest point of the "dip" at approximately $x = 1$. How did we know to put it there? We didn't. We made no attempt to locate it—we simply guessed. Without further work, the best we can say is that it is between $x = -1$ and $x = 3$. Furthermore, how do we know that the graph does not have some extra "turns" and "wiggles" and perhaps look like Figure 7.12? Again, we don't. As a general rule, unless there is some special reason to put in some extra "turn" or "wiggle," we shall leave it out. This rule will not necessarily give us the correct graph every time, but there is no point in needlessly complicating the situation. These methods give only a general idea of the graph. If you take a course in calculus, you will see how derivatives may be used to answer the questions raised here.

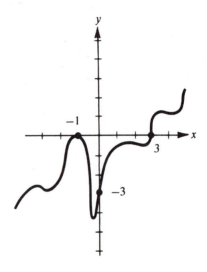

Figure 7.12

Note that, with the exception of the three intercepts, we have not plotted a single point! Yet we have some idea (within the restrictions noted above) of the main features of the curve. With a little practice, you should be able to sketch such curves quite quickly, and thus achieve the principal aim here.

EXAMPLE 4

Sketch $y = (x - 1)^2(x - 2)(x + 1)^3$.

SOLUTION

We have already considered the x intercepts and behavior at the ends in Examples 1 and 2. The y intercept is $(0, -2)$. Again it is easy to see that there is no symmetry about either axis or the origin. Summing up, we have:

Intercepts: (1, 0) even

(2, 0) odd (=1)

(−1, 0) odd (>1)

(0, −2)

Ends: As $x \to +\infty$, $y \to +\infty$

As $x \to -\infty$, $y \to +\infty$

No symmetry

This is indicated in graphical form in Figure 7.13. Again let us start at the far left and proceed to the right, hitting each intercept as we go. Starting from the upper left-hand corner, we first come to the intercept (−1, 0). As indicated, this is an odd intercept and the graph levels off here. Thus it levels off, crosses, and proceeds right through the y intercept (0, −2). From there we reach the intercept (1, 0). Since it is even, the graph merely touches the x axis but stays below it. The graph goes down from there but must eventually come back up to the intercept (2, 0). Finally, it crosses the x axis at an angle at (2, 0) and continues up to the upper right-hand corner. Figure 7.14 shows how you might have drawn the graph by the methods given above. Note that there are some differences between this graph and the more accurately drawn graph of Figure 7.9. Nevertheless, we do know the general shape within the limitations given above.

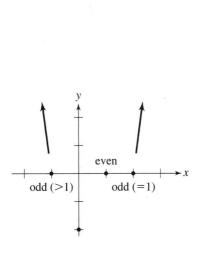

Figure 7.13

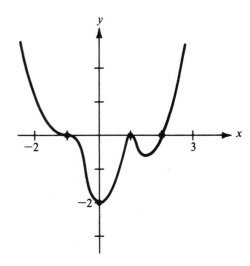

Figure 7.14

EXAMPLE 5

Sketch $y = x^4 - 1$.

SOLUTION

Let us first factor the right-hand side.

$$y = x^4 - 1 = (x^2 + 1)(x^2 - 1) = (x^2 + 1)(x + 1)(x - 1)$$

The factor $x^2 + 1$ yields no intercept because it can never be 0. The other two factors give

$$(-1, 0) \quad \text{odd} \ (= 1)$$
$$(1, 0) \quad \text{odd} \ (= 1).$$

Furthermore, $x = 0$ gives $y = -1$ or the y intercept $(0, -1)$. Since the dominant term is x^4, we have:

$$\text{As} \quad x \to +\infty, \quad y \to +\infty$$
$$\text{As} \quad x \to -\infty, \quad y \to +\infty.$$

Finally, since $(-x)^4 = x^4$, we have symmetry about the y axis.

These results are represented graphically in Figure 7.15. Again, starting at the far left, we cross the x axis at an angle at $(-1, 0)$ and continue down to $(0, -1)$. At this point we can now use symmetry to get the other half. The result is shown in Figure 7.16.

As mentioned earlier, the plotting of polynomial functions is what graphing

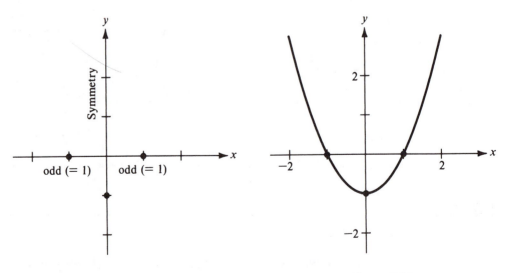

Figure 7.15   **Figure 7.16**

calculators and computer programs do best. Thus, it is assumed that those with electronic plotting capabilities will use them on most of the problems in this section, as well as in subsequent sections. The graphing calculator problems are only those that go beyond simple graph plotting.

EXAMPLE 6

Use a graphing calculator to graph $y = x^4 + 2x^3 - 13x^2 - 22x + 22$. Use the result to find the intercepts.

SOLUTION

Using the normal range of $x_{min} = -9.4$, $x_{max} = 9.4$, $y_{min} = -6.2$, $y_{max} = 6.2$, we get the graph shown in Figure 7.17. Clearly we need to change the range values. After some experimentation, we use $x_{min} = -4.7$, $x_{max} = 4.7$, $y_{min} = -50$, $y_{max} = 50$. Although this distorts the graph by compressing it vertically, it results in a much more comprehensible result, as shown in Figure 7.18. We now use CALC (or JUMP or G SOLVE) to find the x intercepts to be $(-3.317, 0)$, $(-2.732, 0)$, $(0.732, 0)$, and $(3.317, 0)$. The y intercept is easily found to be $(0, 22)$ by simply using $x = 0$ in the given equation.

This is an example of using the graph to help locate the intercepts rather than using the intercepts to help us determine the graph. The latter would require us to solve a fourth-degree equation—one that is not easily factorable.

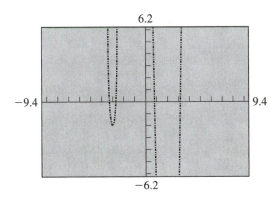

Figure 7.17

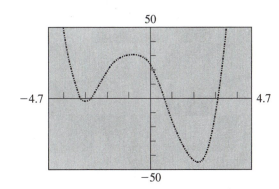

Figure 7.18

PROBLEMS

In Problems 1–20, sketch by the methods of this section. Do not plot the graph point by point.

A   **1.** $y = (x + 1)(x - 3)$ **2.** $y = (x + 2)(x - 1)^2$

        3. $y = x(x + 2)$ **4.** $y = (x - 1)^2(x + 2)^2$

5. $y = x^2 - 4$ **6.** $y = x^3 + 3x^2$

7. $y = x^2 + 2x + 1$ **8.** $y = (x - 1)^3$

B   **9.** $y = x^2(x - 1)(x + 2)^2$ **10.** $y = (x - 2)^2(x + 1)^3(x - 5)$

11. $y = x^3 - x$ **12.** $y = x^4 - x^2$

13. $y = x^5 - x^3$ **14.** $y = x^3 - 4x^2 + 3x$

15. $y = x^5 - x^4 - x^3 + x^2$ **16.** $y = x^3 - 3x + 2$

17. $y = (2x - 1)(x + 2)^2$ **18.** $y = (2x + 3)(x - 1)^3$

19. $y = 2x^2 - 5x - 3$ **20.** $y = 9x^4 - 12x^3 + 4x^2$

In Problems 21–26, shade the region bounded by the graphs of the given equations. Find the rightmost and leftmost points of the region.

21. $y = 4x - x^2, y = x$ **22.** $y = 2x - x^2, y = 2x - 4$

23. $y = x^2 - 5x + 4, y = -(x - 1)^2$ **24.** $y = x^2 - 1, y = 11 - 2x^2$

25. $y = x^3 - x, y = 3 - 3x^2$ **26.** $y = x(x - 2)^2, y = x$

27. Match the given equations with the *type* of curve illustrated in Figure 7.19. (There may be some distortion of the curves to bring out the main features.)

(**a**) $y = (x + 1)(x - 1)^2$ (**b**) $y = -(x + 1)(x - 1)$

(**c**) $y = -(x + 1)^3(x - 1)$ (**d**) $y = (x + 1)^2(x - 1)^2$

(**e**) $y = -(x + 1)^3(x - 1)^3$ (**f**) $y = (x + 1)^3(x - 1)^2$

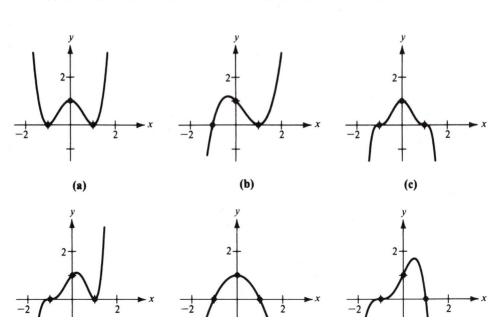

Figure 7.19

28. Match the given equations with the *type* of the curve illustrated in Figure 7.20. (There may be some distortion of the curve to bring out the main features.)

(a) $y = x^2(x + 2)^2$ (b) $y = x^3(x - 2)$

(c) $y = -x^2(x - 2)$ (d) $y = -x^2(x + 2)^3$

(e) $y = -x(x - 2)$ (f) $y = x^3(x + 2)$

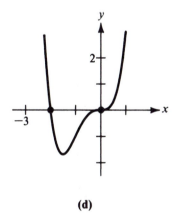

(a)

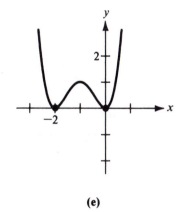

(b)

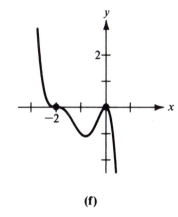

(c)

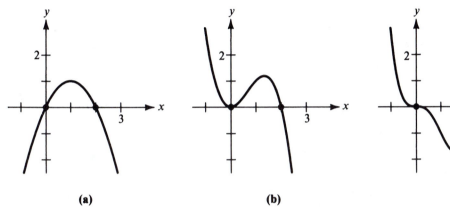

(d)

(e)

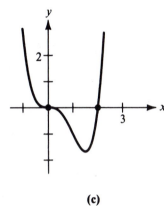

(f)

Figure 7.20

29. Which of the graphs illustrated in Figure 7.21 on page 246 cannot be the graph of a polynomial equation? Explain.

C   *In Problems 30–32, sketch using the methods of this section.*

30. $y = 3x^5 - 4x^4 + 3x^3 - 4x^2$

31. $y = 2x^4 - 9x^3 - 2x^2 + 39x - 18$

32. $y = 4x^4 + 8x^3 + 9x^2 + 5x + 1$

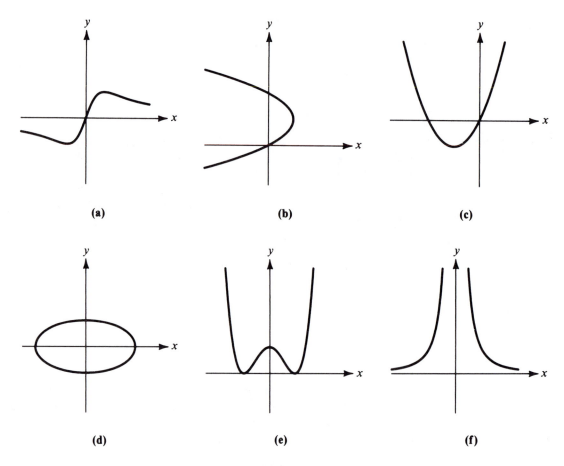

Figure 7.21

In Problems 33–36, use a graphing calculator to draw the graph and determine the intercepts.

33. $y = x^3 - 3.65x^2 - 0.84x + 8.94$

34. $y = x^3 - 1.46x^2 - 11.93x + 17.32$

35. $y = x^4 + 3x^3 - 14x^2 - 21x + 49$

36. $y = x^4 - 2x^3 - 16x^2 + 7x + 40$

37. Graph $y = (x - 2)(x - 1)^3$ and find the lowest point on it.

38. Graph $y = 18x - 2x^3$ and find the positive value of x that makes it a maximum. What is this maximum value?

39. Graph $L = 5x + (360{,}000/x)$ and find the positive value of x that makes L a minimum. What is this minimum value?

40. Graph $V = 9\pi r^2 - 3\pi r^3$ and find the positive value of r that makes V a maximum. What is this maximum value of V?

41. Graph $y_1 = x(x - 4)$ and $y_2 = 4 - x^2$. Shade the region above y_1 and below y_2.

42. Graph $y_1 = x^2$ and $y_2 = x + 2$. Shade the region above y_1 and below y_2.

43. Graph $y_1 = x^2(x - 4)$ and $y_2 = x - 4$. Shade the region between them from the leftmost to the rightmost point of intersection.

44. Graph $y_1 = x(x^2 - 4)$ and $y_2 = x$. Shade the region between them from the leftmost to the rightmost point of intersection.

APPLICATIONS

45. A farmer has 1000 yards of fencing material to fence three sides of a rectangular field, the fourth side being a straight river. If x is the length of a side perpendicular to the river, the area fenced in is $A = 1000x - 2x^2$. Sketch the graph of this equation. From what you know about this particular function, find the area of the largest such field.

46. Suppose we have a square piece of cardboard that is 12 inches on a side. We cut out equal squares from the four corners and turn up the sides to form a box. If x represents the sides of the squares cut out, give the volume of the box as a function of x. For what values of x does this function represent the situation described above? Graph this function. If you have a graphing calculator or program, find the maximum volume and the value of x that gives it.

7.3

ASYMPTOTES

Another feature of a curve that is of considerable aid in sketching is its asymptotes. We have already encountered asymptotes when we studied the hyperbola. As we indicated in Section 5.4, an **asymptote** is a line that the curve approaches more and more closely at its extremities. Curves other than hyperbolas can have asymptotes, as illustrated in Figure 7.2, in which $x = -10$ and $y = 10$ are asymptotes.

Let us see how we can determine the asymptotes from the equation. The easiest type to find is the vertical asymptote. Let us suppose that a curve has the vertical line $x = k$ as an asymptote (see Figure 7.22 on page 248). The curve must then approach the top end of the line as in Figure 7.22a or the bottom end as in Figure 7.22b, or the top on one side and the bottom on the other as in Figure 7.22c. Now let us imagine a point moving along the curve with its x coordinate approaching k. We see from Figure 7.22 that the y coordinate of the point must become numerically large (either positive or negative). Thus the problem of finding a vertical asymptote becomes one of finding a number k such that, for values of x near k, the corresponding values of y are numerically large.

Suppose we are dealing with an equation of the form

$$y = \frac{P(x)}{Q(x)},$$

where $P(x)$ and $Q(x)$ are polynomials. When dividing, the quotient is numerically large when the denominator is numerically small. Let us consider a particular example of this. Suppose we have

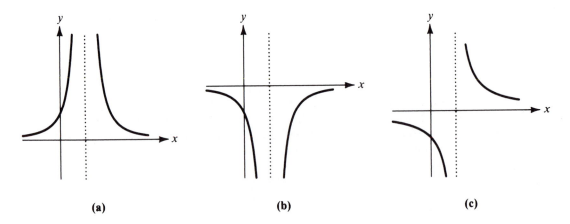

(a)   (b)   (c)

Figure 7.22

$$y = \frac{x + 1}{x - 2}.$$

The denominator is zero if $x = 2$, but this gives no value of y. Let us see what happens for values of x getting closer and closer to 2. We see from Table 7.1 that the value of y becomes numerically larger as x approaches 2; that is, $x = 2$ is an asymptote. Now, where do we get the number 2? It is simply a value of x that makes the denominator zero—the smallest numerical value we can have. This now provides us with a simple method of finding a vertical asymptote; we simply find those values of x that make the denominator zero (without making the numerator zero at the same time). As long as there are no common factors in the numerator and denominator of a rational fraction, they will not both be zero for the same value of x.

Table 7.1

x	y	x	y
1	-2	3	4
1.5	-5	2.5	7
1.9	-29	2.1	31
1.99	-299	2.01	301
1.999	-2999	2.001	3001

EXAMPLE 1   Determine the vertical asymptotes of

$$y = \frac{(x + 1)(x - 3)}{(2x - 1)(x + 2)^2}.$$

The denominator is zero when either one of its two factors is zero.

$$2x - 1 = 0 \qquad \text{gives} \qquad x = 1/2.$$
$$x + 2 = 0 \qquad \text{gives} \qquad x = -2.$$

Since neither value of x gives zero for the numerator, $x = 1/2$ and $x = -2$ are the vertical asymptotes. Figure 7.23 shows the graph with the vertical asymptotes. The method of sketching the graph is deferred until the next section.

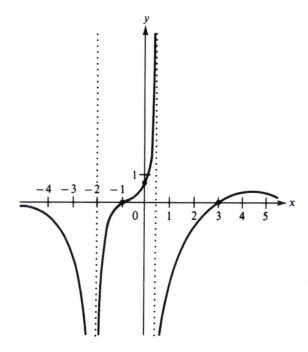

Figure 7.23

Let us now consider horizontal asymptotes. Of course, if the given equation is in the form

$$x = \frac{P(y)}{Q(y)},$$

or can easily be put into that form, we can simply use the methods given for vertical asymptotes. We merely reverse the roles of the x and y here. Unfortunately, it is often difficult or impossible to solve for x as function of y (consider the equation of Example 1), so another method must be found.

If $y = k$ is a horizontal asymptote for $y = f(x)$, then the distance between a point of the graph of $y = f(x)$ and the line $y = k$ must approach zero as x gets large in absolute value. Thus we must investigate the behavior of y as x gets large in one direction or the other; if y approaches the number k ($y \to k$) as $x \to \pm\infty$, then $y = k$ is a horizontal asymptote.

EXAMPLE 2   Determine the horizontal asymptote of

$$y = \frac{x^2 - 4}{x^2 + 3x}.$$

SOLUTION   As x gets large and positive, both the numerator and denominator are also getting large and positive. This fact alone tells us nothing about what number the quotient is approaching. Suppose we alter the equation by dividing both numerator and denominator by x^2. Then

$$y = \frac{x^2 - 4}{x^2 + 3x} = \frac{1 - 4/x^2}{1 + 3/x}.$$

Now, as x gets large,

$$\frac{4}{x^2} \to 0 \qquad \text{and} \qquad \frac{3}{x} \to 0.$$

Thus

$$y = \frac{1 - 4/x^2}{1 + 3/x} \to 1,$$

and $y = 1$ is the horizontal asymptote. Similarly, as x gets large and negative, $y \to 1$. Thus the asymptote $y = 1$ is approached by the curve in both directions. Note in Figure 7.24 that the graph crosses the horizontal asymptote at $(-4/3, 1)$.

In finding the asymptote in the preceding example, we first divided both numerator and denominator by the highest power of x (x^2 in this case) in the given expression. This trick often helps in finding asymptotes.

EXAMPLE 3   Determine the horizontal asymptote of

$$y = \frac{(x + 1)(x - 3)}{(2x - 1)(x + 2)^2}.$$

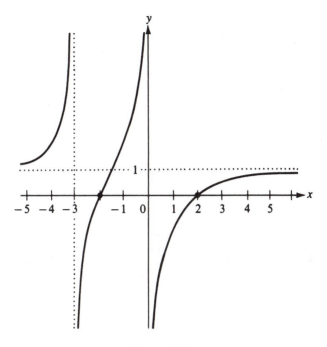

Figure 7.24

SOLUTION

If we multiplied out the numerator, the highest power of x would be x^2; in the denominator it would be x^3. Thus we shall divide the numerator and denominator by x^3 (do *not* divide the numerator by x^2 and the denominator by x^3; the result would *not* equal y).

$$y = \frac{(x+1)(x-3)}{(2x-1)(x+2)^2} = \frac{\left(\dfrac{x+1}{x}\right)\left(\dfrac{x-3}{x}\right)\left(\dfrac{1}{x}\right)}{\left(\dfrac{2x-1}{x}\right)\left(\dfrac{x+2}{x}\right)^2} = \frac{\left(1+\dfrac{1}{x}\right)\left(1-\dfrac{3}{x}\right)\left(\dfrac{1}{x}\right)}{\left(2-\dfrac{1}{x}\right)\left(1+\dfrac{2}{x}\right)^2}$$

As x gets large, all of the expressions with x in the denominator approach zero and

$$y \to \frac{(1+0)(1-0)(0)}{(2-0)(1+0)^2} = 0.$$

Thus $y = 0$ is the only horizontal asymptote. The graph was given in Figure 7.23. Again note that the graph crosses the horizontal asymptote at $(-1, 0)$ and $(3, 0)$.

EXAMPLE 4   Find the horizontal asymptote of

$$y = \frac{x(x+1)(x-2)}{(x-4)(x+2)}.$$

SOLUTION   The highest power of x in this expression is x^3. Dividing numerator and denominator by x^3, we have

$$y = \frac{x(x+1)(x-2)}{(x-4)(x+2)} = \frac{\left(\frac{x}{x}\right)\left(\frac{x+1}{x}\right)\left(\frac{x-2}{x}\right)}{\left(\frac{x-4}{x}\right)\left(\frac{x+2}{x}\right)\left(\frac{1}{x}\right)} = \frac{(1)\left(1+\frac{1}{x}\right)\left(1-\frac{2}{x}\right)}{\left(1-\frac{4}{x}\right)\left(1+\frac{2}{x}\right)\left(\frac{1}{x}\right)}.$$

As x gets large in absolute value, the numerator approaches 1 and the denominator, 0. Thus the fraction becomes arbitrarily large, rather than leveling off to some number k. There is, therefore, no horizontal asymptote. The graph is shown in Figure 7.25.

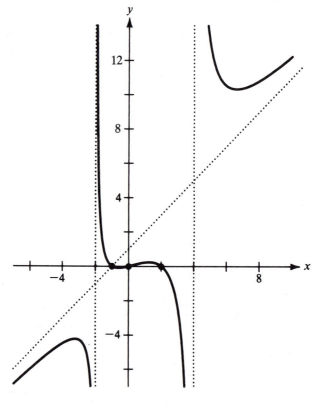

Figure 7.25

Finally, let us consider slant asymptotes, that is, asymptotes that are neither horizontal nor vertical. As in the previous cases, we consider here only equations of the form

$$y = \frac{P(x)}{Q(x)},$$

where $P(x)$ and $Q(x)$ are polynomials. Other types of equations will be considered later. For our present case, the curve will have a slant asymptote if and only if the degree of $P(x)$ is one more than that of $Q(x)$. The asymptote can be found by carrying out the division of $P(x)$ by $Q(x)$ and seeing what happens for numerically large values of x.

EXAMPLE 5   Find the slant asymptote of

$$y = \frac{x(x + 1)(x - 2)}{(x - 4)(x + 2)}.$$

SOLUTION   Let us first observe that the degree of the numerator is 3 and that of the denominator is 2. Since the degree of the numerator is one greater than that of the denominator, there is a slant asymptote. First we multiply in the numerator and denominator to get

$$y = \frac{x(x + 1)(x - 2)}{(x - 4)(x + 2)} = \frac{x^3 - x^2 - 2x}{x^2 - 2x - 8}.$$

Now we carry out the division.

$$
\begin{array}{r}
x + 1 \\
x^2 - 2x - 8 \overline{)\, x^3 - x^2 - 2x } \\
\underline{x^3 - 2x^2 - 8x} \\
x^2 + 6x \\
\underline{x^2 - 2x - 8} \\
8x + 8
\end{array}
$$

$$y = \frac{x(x + 1)(x - 2)}{(x - 4)(x + 2)} = \frac{x^3 - x^2 - 2x}{x^2 - 2x - 8} = x + 1 + \frac{8x + 8}{x^2 - 2x - 8}$$

We now observe that

$$\frac{8x + 8}{x^2 - 2x - 8} = \frac{\dfrac{8}{x} + \dfrac{8}{x^2}}{1 - \dfrac{2}{x} - \dfrac{8}{x^2}}.$$

For numerically large values of x, all of the terms with x in the denominator approach zero. Thus the whole expression approaches $0/1 = 0$. Going back to our original expression, we have

$$y = x + 1 + \frac{8x + 8}{x^2 - 2x - 8} \rightarrow x + 1$$

for numerically large values of x; that is, our original differs from $y = x + 1$ by an amount that approaches zero as x becomes numerically large. Thus

$$y = x + 1$$

is a slant asymptote of the curve. This is shown in Figure 7.25.

A graphing calculator is not an ideal instrument for determining asymptotes. Remember that when a graph is drawn using the connected mode, pixels with consecutive x values are joined by line segments (see page xi in the introduction to the graphing calculator). This sometimes gives the impression that vertical asymptotes are drawn. Suppose, for example, that $x = k$ is a vertical asymptote with the graph rising on the left and falling on the right (or vice versa) as it nears the asymptote. Then the last pixel to the left of the asymptote is joined to the first one on the right. Though not a vertical line, it is near vertical (it is only two pixels wide), giving an approximation of the asymptote.

However, if the graph rises (or falls) on both sides of the asymptote, then this joining of pixels gives nothing of any use to us. This is illustrated in Figure 7.26, showing all four of the above situations.

Horizontal and slant asymptotes are even more difficult to determine. They will never be shown on the graph. Since these asymptotes are determined by the behavior of the graph at the ends, we can get some idea of a horizontal asymptote by graphing for very large (or numerically large, but negative) values of x. For example, the graph of

$$y = \frac{x^2 - 4}{x^2 + 3x}$$

for x between 1000 and 1020 is a horizontal line with y coordinate 0.997; for x between -1020 and -1000, $y = 1.003$. It is reasonable to conclude that $y = 1$ is the horizontal asymptote. This will not work for slant asymptotes because of the problem of selecting the range of y values without prior knowledge of the existence or location of the asymptote.

The only real advantage gained is in the solution of the equation found by setting the denominator to 0. We have the same situation here that we had with x intercepts. That is, if the equation cannot be solved by algebraic means, we can always fall back on a graphical solution.

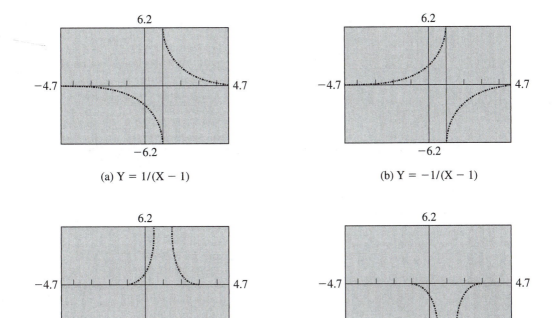

Figure 7.26

PROBLEMS

In Problems 1–20, find all asymptotes.

A **1.** $y = \dfrac{1}{x(x-1)}$ **2.** $y = \dfrac{1}{x-1}$

      3. $y = \dfrac{x^2}{(x+1)(x-2)}$ **4.** $y = \dfrac{4x-2}{x+1}$

      5. $y = \dfrac{x-1}{x+2}$ **6.** $y = \dfrac{x}{x+3}$

      7. $y = \dfrac{2x+1}{x-3}$ **8.** $y = \dfrac{x-3}{2x+1}$

      9. $y = \dfrac{x^2-1}{x^2+1}$ **10.** $y = \dfrac{x^2+1}{x^2-1}$

B   **11.** $y = \dfrac{x^3}{x - 1}$

12. $y = \dfrac{x(x^2 - 1)}{x - 2}$

13. $y = \dfrac{x(x - 1)}{x + 2}$

14. $y = \dfrac{(2x - 3)(x - 2)}{(x + 1)(x - 3)^2}$

15. $y = \dfrac{2x(x - 2)}{(x + 1)^2}$

16. $y = \dfrac{(x + 1)^2}{2x(x - 2)}$

17. $y = \dfrac{2x(x - 2)}{x + 1}$

18. $y = \dfrac{(4x - 7)(x - 1)^2}{(x + 1)(x + 2)(x + 3)}$

19. $y = \dfrac{(3x + 2)^3(x - 4)}{(2x + 3)^2(x + 1)^3}$

20. $y = \dfrac{(2x + 1)^2(x - 3)^3}{x(2x - 3)^2}$

C   **21.** If $y = \dfrac{a_nx^n + a_{n-1}x^{n-1} + \cdots + a_1x + a_0}{b_mx^m + b_{m-1}x^{m-1} + \cdots + b_1x + b_0}$, where $a_n \neq 0$ and $b_m \neq 0$, what can be said about horizontal asymptotes in case
(**a**) $n < m$?   (**b**) $n = m$?   (**c**) $n > m$?

GRAPHING CALCULATOR

In Problems 22 and 23, find the vertical asymptotes.

22. $y = \dfrac{x - 3.21}{x^2 - 1.23x + 10.25}$

23. $y = \dfrac{x^2 + 3x}{x^4 + 8x^2 - 209}$

APPLICATIONS

24. A certain chemical decomposition proceeds according to the formula

$$x = \frac{15}{15 + t} \quad (t \geq 0),$$

where t is the time in minutes and x is the proportion of undecomposed chemical; that is, x is the ratio of the weight of undecomposed chemical to the weight of the chemical when $t = 0$. Show that the chemical never decomposes entirely. How long does it take for 99% of the chemical to decompose?

25. Under certain conditions the size of a colony of bacteria is related to time by

$$x = 40{,}000\frac{2t + 1}{t + 1} \quad (t \geq 0),$$

where x is the number of bacteria and t is the time in hours. To what number does the size of the colony tend over a long period of time? In how many hours will the colony have achieved 95% of its ultimate size?

7.4

SKETCHING RATIONAL FUNCTIONS

Now let us use what we have learned to sketch rational functions, that is, equations of the form

$$y = \frac{P(x)}{Q(x)},$$

where $P(x)$ and $Q(x)$ are polynomials with no common factor. The things that we use to sketch the curve are: single-valuedness, symmetry, intercepts, asymptotes, and behavior at the ends. Let us review them.

Graphs of rational functions are single-valued, just as are graphs of polynomials. There is no way that we can get two values of y from a single value of x. However, there is one slight difference between rational functions and polynomials. Whereas graphs of polynomial equations must have a value of y corresponding to every value of x, this is not the case with rational functions. Since we now have a fraction, with a denominator, we get no value for y when this denominator is zero. We have seen in the previous section that this corresponds to a vertical asymptote. Thus we see that every vertical line except a vertical asymptote contains one and only one point of the graph; a vertical asymptote contains no point of the graph.

When sketching polynomial equations, we found that x intercepts came from factors on the right-hand side, and the exponents on these factors were significant. When dealing with rational functions, we get x intercepts from factors in the numerator of the fraction. Any value of x that makes such a factor zero makes the fraction—and therefore y—zero. The exponents on these factors have the same significance as in the past; that is, the graph crosses the x axis if the exponent is odd (crossing at an angle if it is one); it touches but remains on the same side of (above or below) the x axis if the exponent is even. A y intercept is found, as usual, by simply substituting $x = 0$ into the equation. There is no consideration of odd or even exponents on y intercepts—they do not come from identifiable factors.

We have seen that vertical asymptotes come from individual factors of the denominator. Once again, it is easily seen that the exponent on such a factor has some significance in terms of the appearance of the graph. The asymptote is approached by the graph on two sides—both the left and the right. If the factor giving that asymptote has an even exponent, the graph remains on the same side of (above or below) the x axis on the two sides of the asymptote, as in Figure 7.27 on page 258. On the other hand, if the exponent is odd, then the graph "crosses" the x axis; that is, the graph is above the x axis on one side of the asymptote and below it on the other side (see Figure 7.28 on page 258). It does not cross the x axis by running across as at an intercept; rather, it "hops" across the x axis. As we noted above, the vertical asymptote contains no point of the graph, either on or off the x axis.

Just as the odd-and-even considerations held for x intercepts but not y intercepts, the odd-and-even consideration on asymptotes holds only for vertical asymptotes—not for horizontal or slant asymptotes. Once again the vertical asymptote is determined by a single factor in the denominator; this is not the case with horizontal or slant asymptotes. Furthermore, while the graph cannot cross a vertical asymptote (after all, there can be no value of y corresponding to that value of x), there is no such restriction in the case of a horizontal or slant

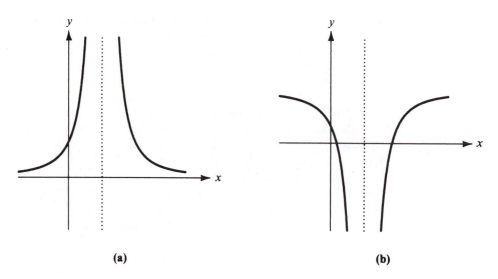

(a)

(b)

Figure 7.27

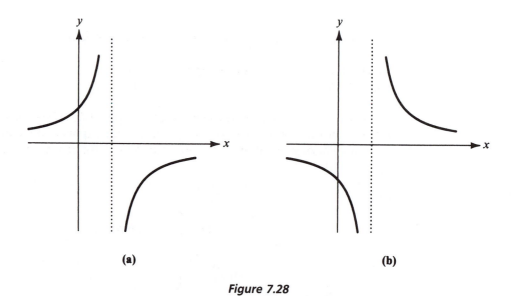

(a)

(b)

Figure 7.28

asymptote. Figures 7.24 and 7.25 give instances of graphs containing points of their horizontal or slant asymptotes.

Finally, let us note that if there are horizontal or slant asymptotes, the behavior of the graph at the ends is already determined—the graph approaches the asymptote at the far left and right. If there is no such asymptote, then the graph rises or falls rapidly at the ends. Once again we may test for large values of x to see if the y is positive or negative.

Now let us put this together to sketch the graph of a rational function.

EXAMPLE 1   Sketch $y = \dfrac{(2x - 1)(x + 2)^2}{(x + 1)^2(x - 3)}$.

SOLUTION   Intercepts: $(1/2, 0)$, odd $(=1)$; $(-2, 0)$, even; $(0, 4/3)$
Asymptotes: From the denominator: $x = -1$, even; $x = 3$, odd

$$y = \frac{(2x - 1)(x + 2)^2}{(x + 1)^2(x - 3)} = \frac{\left(\dfrac{2x - 1}{x}\right)\left(\dfrac{x + 2}{x}\right)^2}{\left(\dfrac{x + 1}{x}\right)^2\left(\dfrac{x - 3}{x}\right)}$$

$$= \frac{\left(2 - \dfrac{1}{x}\right)\left(1 + \dfrac{2}{x}\right)^2}{\left(1 + \dfrac{1}{x}\right)^2\left(1 - \dfrac{3}{x}\right)}$$

$$\text{As } x \to \pm\infty, \, y \to \frac{2 \cdot 1}{1 \cdot 1} = 2$$

Thus $y = 2$ is the horizontal asymptote.
No symmetry.

All of this information is summarized in Figure 7.29. If we begin sketching at one end or the other, we have the problem of not knowing whether the curve is approaching the asymptote from above or below. Similar problems exist at the

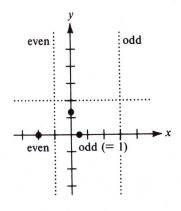

Figure 7.29

vertical asymptotes and x intercepts. Suppose, then, we start at (0, 4/3). Going to the right, we first come to (1/2, 0). Since it is an odd intercept (the exponent is one), the graph crosses the x axis at an angle there and then goes down to the vertical asymptote $x = 3$ (it cannot go up, since it cannot cross the x axis anywhere between $x = 1/2$ and $x = 3$). Since this asymptote is also odd, the graph now jumps above the x axis. Finally it comes down to the horizontal asymptote $y = 2$.

Going back to (0, 4/3) and proceeding to the left, we see that the graph must go up to the vertical asymptote $x = -1$ (remember there is nothing to prevent the graph from crossing a horizontal aymptote). Since $x = -1$ is an even asymptote, the curve stays above the x axis. It must then proceed down to the intercept $(-2, 0)$. This is also even, so the graph again remains above the x axis, finally going up to the horizontal asymptote. Thus, we have the graph indicated in Figure 7.30.

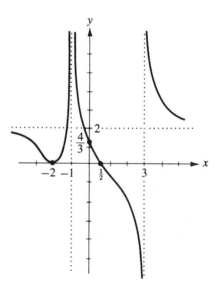

Figure 7.30

It must be admitted that we have the same problem with these sketches that we had in Section 7.2. That is, we do not know if the graph has some additional changes in direction that we have not detected. In particular, we have assumed that the graph approaches the horizontal asymptote $y = 2$ from above at the right end. Nevertheless, since the graph may cross a horizontal asymptote, it may dip down below the asymptote and come back up to approach from below as shown in Figure 7.31. Once again, unless we have some reason to believe that this is the case, we shall assume the simpler case of Figure 7.30. Of course, we can always

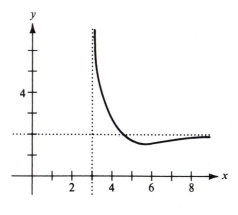

Figure 7.31

substitute values of x into the original equation to determine the graph at the ends, but that is what we are trying to avoid. Once again, the derivative can be used to settle such questions, but that is a subject of calculus.

EXAMPLE 2 Sketch $y = \dfrac{x^3(x - 2)}{(x + 1)(x - 4)}$.

SOLUTION Intercepts: $(0, 0)$, odd (>1); $(2, 0)$, odd $(=1)$
Asymptotes: $x = -1$, odd; $x = 4$, odd; no horizontal or slant asymptote
Ends: As $x \to +\infty$, $y \to +\infty$; as $x \to -\infty$, $y \to +\infty$
No symmetry

This information is summarized in Figure 7.32. Let us begin sketching at the left end. Since the graph increases as we go left, it must decrease as we come to

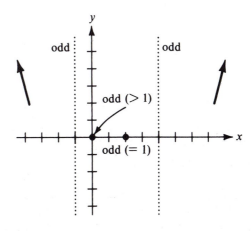

Figure 7.32

the right. But since it cannot cross the x axis to the left of $x = -1$, it must turn around and go back up to approach the asymptote $x = -1$. (We do not know the exact location of the low point of this section.) Since the asymptote is odd, the graph jumps from above the x axis to below it. Continuing to the right, we next encounter the intercept at $(0, 0)$. Because it is odd, the graph crosses the x axis; because it is greater than one, the graph levels off there. The graph must go up from there but come back down to the next intercept, $(2, 0)$. Again the graph crosses the x axis there, but at an angle because the exponent is one. The graph continues down to approach the vertical asymptote $x = 4$. Because this asymptote is odd, the graph jumps across the x axis. Finally, after descending from this vertical asymptote, the graph must turn around and go up at the right end. The result is shown in Figure 7.33.

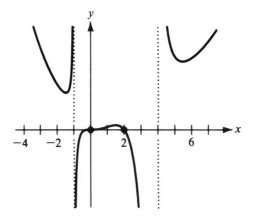

Figure 7.33

EXAMPLE 3   Sketch $y = \dfrac{x(x + 1)(x - 1)}{x^2 + 1}$.

SOLUTION   Intercepts: $(0, 0)$, odd $(=1)$; $(1, 0)$, odd $(=1)$; $(-1, 0)$, odd $(=1)$
Asymptotes: No vertical asymptote since $x^2 + 1$ cannot be zero. Slant asymptote $y = x$ since

$$y = \frac{x(x + 1)(x - 1)}{x^2 + 1} = \frac{x^3 - x}{x^2 + 1} = x - \frac{2x}{x^2 + 1} \to x$$

as $x \to \pm\infty$

Symmetry: About the origin, since

$$-y = \frac{-x(-x + 1)(-x - 1)}{(-x)^2 + 1} = -\frac{x(x - 1)(x + 1)}{x^2 + 1},$$

which is equivalent to the original equation.

This is summarized in Figure 7.34. Starting at the left end, the graph must increase from the asymptote to the intercept $(-1, 0)$. Since $(-1, 0)$ is above the asymptote, we assume that the graph approaches the asymptote from above. At $(-1, 0)$ the graph crosses the x axis at an angle. After increasing from the point $(-1, 0)$, it must decrease again to the next intercept, $(0, 0)$, where it again crosses at an angle. At this point, we may use the symmetry at the origin to get the remainder of the graph, as shown in Figure 7.35.

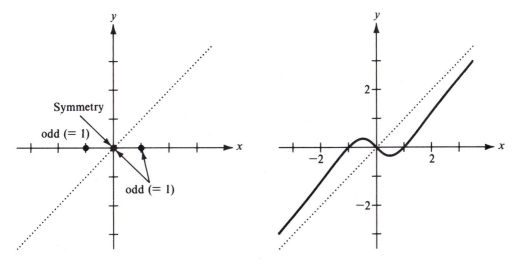

Figure 7.34   **Figure 7.35**

Electronic graphing of rational functions has certain drawbacks. In terms of eye appeal, vertical asymptotes are perhaps the least appealing. Many such programs tend to leave a gap at a vertical asymptote; some plot them as if they were intercepts with very large positive values on one side and large negative values on the other.

Nevertheless, electronic graphing can be quite helpful, especially when the denominator is not easily factorable. For example, the equation

$$y = \frac{x(x - 3)^2}{x^3 - 5x^2 + 3x + 5}$$

is difficult to sketch because it presents us with the problem of solving the cubic equation

$$x^3 - 5x^2 + 3x + 5 = 0.$$

Since this equation has no rational root, its solution poses a problem. However, electronic graphing shows that the given equation has three vertical asymptotes. Their equations are easily approximated from the graph. If more accuracy is needed, we can enlarge portions of the graph to determine them more precisely.

One way in which electronic graphing is lacking is in the determination of slant asymptotes. It is often not apparent from the graph that we have a slant asymptote. Furthermore, finding an equation of one from the graph is extremely difficult. Division is still the best way to find it.

PROBLEMS

In Problems 1–20, use the methods of this section to sketch the graph. Do not plot the graph point by point.

A   **1.** $y = \dfrac{x}{x^2 - 1}$ **2.** $y = \dfrac{x^2 - 1}{x}$

    3. $y = \dfrac{x + 1}{x}$ **4.** $y = \dfrac{x - 2}{x + 2}$

    5. $y = \dfrac{x^2 + 1}{x}$ **6.** $y = \dfrac{x}{x^2 + 1}$

    7. $x^2 y = 2x + 1$ **8.** $xy = 2x + 1$

    9. $x^2 y - y = x^3$ **10.** $x^2 y - y = x^2$

    11. $xy + y = 3x - 2$ **12.** $x^2 y - y = x$

B   **13.** $y = \dfrac{x(x + 1)}{(x - 2)^2}$ **14.** $y = \dfrac{(x - 2)^2}{x(x + 1)}$

    15. $y = \dfrac{(2x + 1)(x - 1)^2}{(x - 2)(x + 1)^2}$ **16.** $y = \dfrac{x - 3}{(x + 1)(x - 2)}$

    17. $y = \dfrac{(x + 2)(x - 4)}{x - 1}$ **18.** $y = \dfrac{x - 1}{(x + 2)(x - 4)}$

    19. $y = \dfrac{(x + 2)^2(x - 4)}{(x - 1)^2}$ **20.** $y = \dfrac{(x - 3)(2x - 5)}{x^2(x + 2)}$

In Problems 21–24, shade the region bounded by the graphs of the given equation. Find the rightmost and leftmost points of the region.

21. $x^2y = 4, 3x + y = 7$

22. $y = x, y = \dfrac{5x}{x^2 + 1}$

23. $y = \dfrac{(x + 2)(x - 2)^2}{x}, y = 4 - x^2$

24. $xy = 6 - 3x, y = 4 - x^2$

GRAPHING CALCULATOR

In Problems 25–26, verify that the given line is a slant asymptote of the graph of the other expression.

25. $y = x + 3; \quad y = \dfrac{x^3 + 3x^2 - 8x + 1}{x^2 + 1}$

26. $y = x - 6; \quad y = \dfrac{x^3 - 5x^2 + 2x + 3}{x^2 + x - 2}$

In Problems 27–30, graph the given equation and determine all intercepts and asymptotes.

27. $y = \dfrac{x^2 + 5x + 6}{x^3 + 4x^2 - 2x - 3}$

28. $y = \dfrac{x^3 - 2x^2 - 5x + 3}{x^3 - 6x^2 + 6x + 5}$

29. $y = \dfrac{x^3 - 3x^2 + 4x + 3}{x^4 - 3x^2 - 2x + 3}$

30. $y = \dfrac{6x^4 - 8x^3 - 7x^2}{3x^4 - x^3 + 2x - 8}$

APPLICATIONS

31. An unknown electrical resistance R is determined with a Wheatstone bridge by adjusting two things: a compensating resistance r read off a resistance box and the position x of a key along a slide wire of length l. The resistance R is then given by

$$R = r\frac{x}{l - x}.$$

Graph this equation for $r = 50$ ohms and $l = 100$ centimeters. What portion of the graph has physical significance?

32. The total cost function for a given commodity gives the cost C of producing x units of the commodity as a function of x. For example,

$$C = 0.01x^2 + 20x + 400.$$

The average cost function Q (or cost per unit) is simply C/x. Sketch the graph of Q for the given cost function. For what values of x does this function make sense? Estimate the smallest value of Q and the value of x at which it is reached.

7.5

RADICALS AND THE DOMAIN OF THE EQUATION

When we have y as a function of x, a single value of x gives at most one value for y. "At most one value" implies either one or none. When will we have none? We already have seen one such case—at a vertical asymptote. This corresponds to a zero in the denominator of the function.

What other situation might give no value for y? Since we are dealing with real numbers, we do not get a real result when we have a square root of a negative number. Since the square root of a number x is a number y such that $y^2 = x$, and y^2 can never be negative, we cannot take the square root of a negative number. This argument can also be used to show that we cannot take any even root (fourth root, sixth root, and so on) of a negative number. Thus, if we have a function with an even root, any value of x that makes the expression under the radical negative gives no value for y (there is one possible exception to this; it is illustrated in Example 1).

For example, if we have a function involving the expression $\sqrt{x - 4}$, then we have no value for that expression when $x - 4 < 0$ or when $x < 4$. Remember that inequalities can be handled as equations *except* that multiplying both sides of an inequality by a negative number reverses the inequality: $-x < 1$ implies that $x > -1$. Furthermore, $x^2 < 9$ implies that $-3 < x < 3$, and $x^2 > 9$ implies that either $x > 3$ or $x < -3$ ($|x| > 3$).

If the expression under the radical is more complicated, the best way to handle it is by factoring. For example, we write the inequality $x^2 - 3x < 0$ as $x(x - 3) < 0$. Now we know that the product of two numbers is negative if one is positive and the other negative. So let's see where each factor is positive and negative.

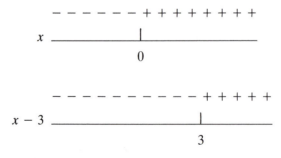

Now

Thus $x(x - 3) < 0$ implies that x is between 0 and 3, $0 < x < 3$.

Now let us consider functions with radicals.

EXAMPLE 1   Sketch $y = \dfrac{2x}{\sqrt{x^2 - 4}}$.

SOLUTION   Using the previous methods for determining intercepts, asymptotes, and symmetry, we have the following.

Intercepts: $(0, 0)$ odd
Asymptotes: $x = 2$, $x = -2$

    The radical is equivalent to the one-half power, which is neither odd nor even. We have a special problem in finding the horizontal asymptotes. The highest power of x in the numerator is clearly x. The highest power in the denominator appears to be x^2. But it is under the radical; so the highest power is really $(x^2)^{1/2} = x$. Thus we shall want to divide the numerator and denominator by x. But we shall want to put the x under the radical in the denominator, which leads to further complications. The symbol $\sqrt{}$ means the *nonnegative* square root. Thus $x = \sqrt{x^2}$ is true only when $x \geq 0$; when $x < 0$, $\sqrt{x^2} = -x$ (note that, since x itself is negative, $-x$ is positive), and we have two cases to consider.

$$\frac{2x}{\sqrt{x^2 - 4}} = \frac{\dfrac{2x}{x}}{\sqrt{\dfrac{x^2 - 4}{x^2}}} = \frac{2}{\sqrt{1 - \dfrac{4}{x^2}}} \qquad \text{(when } x > 0\text{)}$$

$$\frac{2x}{\sqrt{x^2 - 4}} = \frac{\dfrac{2x}{x}}{\sqrt{\dfrac{x^2 - 4}{x^2}}} = \frac{-2}{\sqrt{1 - \dfrac{4}{x^2}}} \qquad \text{(when } x < 0\text{)}$$

Thus,

$$\text{as } x \to +\infty, \ y = \frac{2x}{\sqrt{x^2 - 4}} = \frac{2}{\sqrt{1 - \dfrac{4}{x^2}}} = 2$$

and

$$\text{as } x \to -\infty, \ y = \frac{2x}{\sqrt{x^2 - 4}} = \frac{-2}{\sqrt{1 - \dfrac{4}{x^2}}} = -2$$

giving two horizontal asymptotes: $y = 2$, which is approached on the right, and $y = -2$, which is approached on the left. Replacing x by $-x$ and y by $-y$ gives

$$-y = \frac{2(-x)}{\sqrt{(-x)^2 - 4}} = \frac{-2x}{\sqrt{x^2 - 4}},$$

which is equivalent to the original equation. Thus we have symmetry about the origin.

    Finally, $\sqrt{x^2 - 4}$ represents a positive real number only when $x^2 - 4 > 0$, which gives

$$x^2 > 4 \qquad \text{or} \qquad \begin{cases} x > 2 \\ x < -2. \end{cases}$$

But *y* is real for one additional value of *x*, namely, $x = 0$. If $x = 0$, *y* equals zero divided by a complex number, which is still zero. Thus the domain is

$$\{\, x \mid x > 2 \quad \text{or} \quad x < -2 \quad \text{or} \quad x = 0\,\}.$$

We see here that $(0, 0)$ is an isolated point of the graph (see Note below).

All of this information is represented graphically in Figure 7.36, which shows the intercept as an isolated point; the fact that it is odd is of no use. Note one thing more: Since $\sqrt{x^2 - 4}$ is never negative, *y* is positive whenever *x* is positive and negative whenever *x* is negative. This additional information makes it easy for us to sketch the curve (see Figure 7.37).

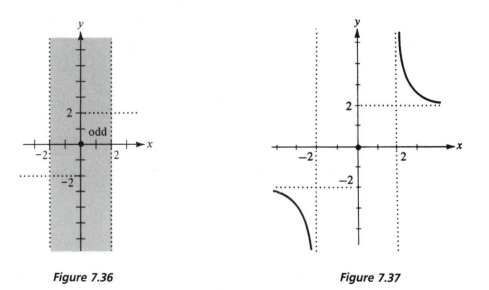

Figure 7.36   **Figure 7.37**

Note: The existence of an isolated point of this graph at the origin is open to controversy. On one hand are those who say that there is no point of the graph at $(0, 0)$. Their reasoning is as follows: Whenever we restrict ourselves to real functions of real variables, we say, in effect, that imaginary numbers do not exist. Thus, instead of having zero over an imaginary number, we have zero divided by no number at all, which yields no number. By this line of reasoning, every "part" of the equation must be real in order to yield a valid result. On the other hand are those who maintain that the result is independent of the means of obtaining it. The mere fact that we go from one real number to another by way of imaginary numbers, they say, does not invalidate the result. Exactly the same controversy arose when Cardan published his solution of a cubic equation in 1545. His rule for the solutions of $x^3 = 15x + 4$ leads to

$$x = \sqrt[3]{\sqrt{2} + \sqrt{-121}} + \sqrt[3]{\sqrt{2} - \sqrt{-121}},$$

which simplifies to $x = 4$, the only positive root.[*] It was decided then that the excursion into complex numbers did not invalidate the result. While we shall take the latter point of view throughout this text, it is good to bear in mind the controversial nature of the problem.

EXAMPLE 2 Sketch $y^2 = x^4 - x^2$.

SOLUTION To graph this equation, we use the following device: Since $y = \pm\sqrt{x^4 - x^2}$, we first graph $z = x^4 - x^2$ and then, from the values of z, get $y = \pm\sqrt{z}$. Graphing $z = x^4 - x^2 = x^2(x^2 - 1) = x^2(x + 1)(x - 1)$, we have the following.

Intercepts: $(0, 0)$, even; $(1, 0)$, odd; $(-1, 0)$, odd.

No asymptotes.

Symmetry about the z axis (see Figure 7.38a).

We see on this graph that, for each value of x, we have a value of $z = x^4 - x^2$. Now let us find the corresponding values for $y = \pm\sqrt{z}$. But first, we note the following points to keep in mind.

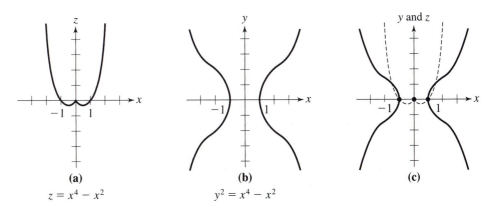

$$\begin{array}{ccc}
\textbf{(a)} & \textbf{(b)} & \textbf{(c)} \\
z = x^4 - x^2 & y^2 = x^4 - x^2 &
\end{array}$$

Figure 7.38

[*] In this connection, see Carl B. Boyer, *A History of Mathematics* (new York: John Wiley & Sons, Inc., 1968), pp. 310–316.

1. $\sqrt{z} = z$ if $z = 0$ or $z = 1$

2. $\sqrt{z} > z$ if $0 < z < 1$

3. $\sqrt{z} < z$ if $z > 1$

4. $\sqrt{z}$ is not real if $z < 0$

The final result is given by the graph of Figure 7.38b. The origin is again an isolated point of this graph. It is convenient to sketch both graphs on the same pair of axes. This is done in Figure 7.38c.

The same method could be used to sketch $y = \sqrt{x^4 - x^2}$. The only difference would be that we would have only the top half of the result in Figure 7.38b. We might also have used this method in Example 1, starting with

$$y^2 = \frac{4x^2}{x^2 - 4}.$$

In that case, we would have to be careful which branch we chose; we would have to choose the top portion when x is positive and the bottom portion when x is negative.

When there are radicals in the equation, we need new methods of finding asymptotes and intercepts. We have already seen this in Example 1. When we have a sum or difference of several terms, one of which is a radical, we can proceed as in the following example.

EXAMPLE 3   Sketch $y = x + \sqrt{x^2 + 1}$.

SOLUTION   When $x = 0$, $y = 1$, giving y intercept $(0, 1)$. We have an x intercept when

$$x + \sqrt{x^2 + 1} = 0$$
$$x = -\sqrt{x^2 + 1}$$
$$x^2 = x^2 + 1$$
$$0 = 1.$$

Since 0 is never equal to 1, there is no x intercept. Because there is no fraction, it would appear that there is no asymptote. But let us consider this more closely. For numerically large values of x, x^2 is considerably larger than 1. Thus $\sqrt{x^2 + 1}$ is very nearly $\sqrt{x^2} = |x|$. Then

$$y = x + \sqrt{x^2 + 1} \rightarrow x + |x| = \begin{cases} 2x & \text{if } x > 0, \\ 0 & \text{if } x < 0. \end{cases}$$

This implies that $y = 2x$ is a slant asymptote that is approached only at the right end and $y = 0$ is a horizontal asymptote approached only on the left. A second way of considering this is by rationalizing the numerator.

$$y = x + \sqrt{x^2 + 1} = \frac{(x + \sqrt{x^2 + 1})(x - \sqrt{x^2 + 1})}{x - \sqrt{x^2 + 1}}$$

$$= \frac{x^2 - (x^2 + 1)}{x - \sqrt{x^2 + 1}} = \frac{-1}{x - \sqrt{x^2 + 1}}$$

If x is numerically large and negative, the entire denominator is numerically large and negative, making the quotient nearly zero. Thus $y = 0$ is an asymptote that is approached only on the left. Unfortunately, this form is not so informative about the right end. For large positive values of x, the denominator is nearly zero and the quotient is large. This does not tell us that this large quotient approaches the line $y = 2x$.

Finally we see that the expression under the radical can never be negative; so there is no excluded region.

Putting this information together and sketching, we have the result shown in Figure 7.39. This curve is half of a hyperbola (see Problem 29).

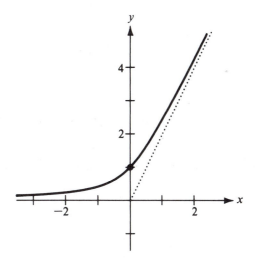

Figure 7.39

One final point. Let us recall that when we had an equation of the form

$$y = \frac{P(x)}{Q(x)},$$

HISTORICAL NOTE

The Witch of Agnesi

In Problem 10 of this section, we mention a curve called the **witch of Agnesi**. Its graph does not look even slightly like a witch. How did it acquire the name "witch," and who was Agnesi? Maria Gaetana Agnesi (pronounced ahn-YAY-see) was an eighteenth-century Italian mathematician at the University of Bologna. Among other accomplishments, she published (in 1784) one of the first comprehensive calculus textbooks. In it was a discussion of the so-called *versiera*, given by $xy^2 = a^2(a - x)$. In those days, many of the curves that they considered had special names. The name *versiera* had been given to the curve by Guido Grandi of Pisa and is derived from the Latin word *vertere*, meaning "to turn." It is also an abbreviation of the Italian word *avversiera*, which means "female devil" or "witch." Thus it was translated into English as "the witch," a name that has stuck.

we noted that x intercepts come from factors in the numerator and vertical asymptotes from factors in the denominator, *provided there is no value of x for which both numerator and denominator are zero.* In the examples we have been considering, this is equivalent to the provision that there is no factor common to both numerator and denominator. What happens if there *are* common factors? The answer is simple. You simply cancel the common factors and sketch the resulting equation. But remember that if you cancel the factor $x - a$, the original equation is not defined at $x = a$ (it gives 0/0) and there is no point on the graph with x coordinate a.

EXAMPLE 4   Sketch $y = (x^2 - 1)/(x - 1)$.

SOLUTION   Since the numerator and denominator have the common factor $x - 1$, we cancel them to get

$$y = x + 1,$$

which gives a straight line. But recall that the original equation gives no value of y when $x = 1$. Thus the point $(1, 2)$ should be deleted from the graph, as in Figure 7.40.

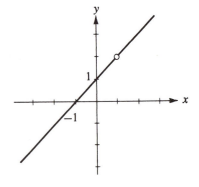

Figure 7.40

As we have already seen, in order to graph $y^2 = f(x)$ electronically, we graph the two functions $y = \sqrt{f(x)}$ and $y = -\sqrt{f(x)}$, since we must have functions rather than equations. In addition, the computer is very likely not to graph isolated points or single point deletions.

PROBLEMS

B *Sketch the graphs of the following equations.*

1. $y = x \sqrt{x^2 - 1}$

2. $y = \dfrac{x}{\sqrt{x - 1}}$

3. $y = \dfrac{-x}{\sqrt{x^2 - 1}}$

4. $y = \dfrac{x - 1}{\sqrt{x(x + 1)}}$

5. $y = \dfrac{x^2 - 4}{x - 2}$

6. $y = \dfrac{x^2 + x}{x}$

7. $y^2 = \dfrac{x}{x + 1}$

8. $y^2 = \dfrac{x^2}{(x + 1)(x - 2)}$

9. $y^2 = \dfrac{2x}{(x - 1)^2}$

10. $y^2 = \dfrac{1 - x}{x}$ (Witch of Agnesi)

11. $y^2 = \dfrac{x(x + 1)}{(x - 2)^2}$

12. $y^2 = \dfrac{x(x + 1)^2}{x - 2}$

13. $y^2 = \dfrac{x(x - 1)}{(x + 1)^2}$

14. $y^2 = \dfrac{x(1 - x)}{(x + 1)^2}$

15. $y^2 = \dfrac{(x^2 - 1)^2}{x - 2}$

16. $y^2 = (1 - x)(3 - x)^2$

17. $y^2 = (x - 1)(x - 3)^2$

18. $y^2 = -(x - 1)(x + 3)^2$

19. $y = x + \sqrt{x^2 - 1}$

20. $y = x - \sqrt{x^2 - 1}$

21. $y = \dfrac{x^2 + x}{x^2}$

22. $y = \dfrac{x^3 + x^2}{x}$

23. $y = \dfrac{x(x + 1)^2}{(x - 1)(x + 1)^3}$

24. $y = \dfrac{2x(x - 1)}{x(x + 1)}$

25. $y = \dfrac{1 - (1 + h)^2}{h}$

26. $y = \dfrac{-1 - [(1 + h)^2 - 2(1 + h)]}{h}$

[*Hint:* Simplify the numerator.]

27. $y = \dfrac{2 - \dfrac{2 + h}{1 + h}}{h}$

28. $y = \dfrac{1 - \sqrt{1 + h}}{h}$

C **29.** Show that the curve of Example 3 is a part of a hyperbola. [*Hint*: Eliminate the radical by squaring and use Theorem 6.7.]

30. What is the domain of $y = x + \sqrt{9 - x^2}$? Verify by graphing.

31. What is the domain of $y = x + \sqrt{x^2 - 9}$? Verify by graphing.

32. Find the distance from the point (x, y) on $y^2 = 4x$ to the point $(3, 0)$. Express this distance entirely in terms of x. Graph the resulting equation, and estimate the value of x that makes the distance a minimum.

33. Find the distance from the point (x, y) in the first quadrant on $xy = 3$ to the point $(-8, 0)$. Express entirely in terms of x and graph. Estimate the value of x that makes this distance a minimum.

34. A ship is anchored 4 miles off a straight shore. Opposite a point 9 miles down the coast, another ship is anchored 8 miles from the shore. A boat from the first ship is to land a passenger on the shore and then proceed to the other ship and pick up another passenger before returning. If the boat puts the passenger ashore x miles down the coast, find the total distance traveled as a function of x. Graph this function and estimate the value of x that makes it a minimum.

7.6

DIRECT SKETCHING OF CONICS

We have been able to sketch conics by putting them into a standard form in Chapters 5 and 6. But this was often quite tedious, especially when we had to rotate the axes. Let us see if we can determine some methods of sketching without going through the process of rotating axes.

Recall that any equation of second degree in x and y represents a conic or a degenerate conic. The type of conic can be determined by $B^2 - 4AC$, as indicated in Theorem 6.7 on page 218. Remember, however, that this test does not distinguish between the conics and their degenerate cases. The results are summarized in the following table.

Conic	$B^2 - 4AC$	Degenerate cases
Parabola	0	One line (two coincident lines) Two parallel lines No graph
Ellipse	−	Circle Point No graph
Hyperbola	+	Two intersecting lines

If we are dealing with a hyperbola, the greatest single aid in sketching the graph is determination of the asymptotes. If they are horizontal or vertical, the determination is relatively easy; so let us go to slant asymptotes. We shall consider two cases: the equation is linear in y ($C = 0$) and the equation is quadratic in y ($C \neq 0$). In either case, we first solve for y. Examples of each follow.

EXAMPLE 1   Sketch $x^2 - xy - 3y - 1 = 0$ without rotating axes.

SOLUTION   First of all, $B^2 - 4AC = (-1)^2 - 4(1)(0) = 1$, indicating that the conic is a hyperbola or a degenerate case of one. Solving for y, we have

$$y = \frac{x^2 - 1}{x + 3}.$$

The methods of this chapter give intercepts $(\pm 1, 0)$ and $(0, -1/3)$ and vertical asymptote $x = -3$. There is no horizontal asymptote, but we know that there must be a second asymptote. To find it, we carry out the division:

$$y = \frac{x^2 - 1}{x + 3} = x - 3 + \frac{8}{x + 3}.$$

We now see that, for numerically large values of x, $8/(x + 3)$ is almost zero and y is very near $x - 3$. Thus the slant asymptote is

$$y = x - 3.$$

With this, we can easily sketch the hyperbola (see Figure 7.41).

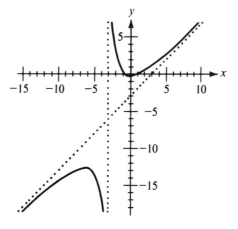

Figure 7.41

EXAMPLE 2   Sketch $7x^2 + 6xy - y^2 - 32 = 0$ without rotating axes.

SOLUTION   The equation is quadratic in y.

$$y^2 - 6xy + (32 - 7x^2) = 0.$$

Using the quadratic formula, we have

$$y = 3x \pm 4 \sqrt{x^2 - 2}.$$

Again, for large values of x, $\sqrt{x^2 - 2}$ is almost $\sqrt{x^2}$, and y is very near $3x \pm 4x$. Thus, the slant asymptotes are (see Figure 7.42)

$$y = 7x \quad \text{and} \quad y = -x.$$

This and the intercepts give us a good idea of the curve.

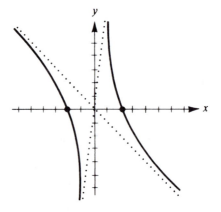

Figure 7.42

A method of sketching that is sometimes useful is that of addition of ordinates. If we have an equation of the form $y = y_1 + y_2$, where y_1 and y_2 are expressions in x, then we can sketch $y = y_1$ and $y = y_2$ on the same coordinate axes. Then, for each value of x, y is the sum of the ordinates (y coordinates) of these two. Let us consider some examples of this.

EXAMPLE 3   Sketch $2x^2 - 2xy + y^2 - 9 = 0$ without rotating axes.

SOLUTION   Since $B^2 - 4AC = -4$, the curve is an ellipse. Again, the equation is quadratic in y.

$$y^2 - 2xy + (2x^2 - 9) = 0$$

By the quadratic formula, we have

$$y = x \pm \sqrt{9 - x^2}.$$

Instead of trying to sketch this curve directly, let us sketch

$$y = x \qquad \text{and} \qquad y = \pm \sqrt{9 - x^2}.$$

By squaring both sides, we can put the second equation into the form

$$x^2 + y^2 = 9.$$

These two are easily sketched (see Figure 7.43).

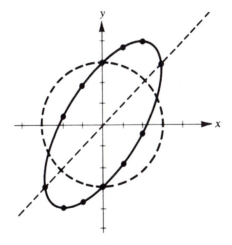

Figure 7.43

For each value of x in the interval $[-3, 3]$, there is an ordinate on the line and one (or two) on the circle. Adding them, we have the ellipse of Figure 7.43.

Since values of x outside the interval $[-3, 3]$ give complex values of y in the equation $y = x \pm \sqrt{9 - x^2}$, there is no graph to the right of $x = 3$ or to the left of $x = -3$.

EXAMPLE 4   Sketch $4x^2 - 4xy + y^2 - 5x + 2y + 1 = 0$ without rotating axes.

SOLUTION   Since $B^2 - 4AC = 0$, the curve is a parabola. Again solving for y, we have

$$y = 2x - 1 \pm \sqrt{x}.$$

This gives the two equations

$$y = 2x - 1 \qquad \text{and} \qquad y = \pm \sqrt{x},$$

where the latter can be written $y^2 = x$. Sketching these two and adding ordinates, we have the result given in Figure 7.44. The line $y = 2x - 1$ is *not* the axis of the parabola.

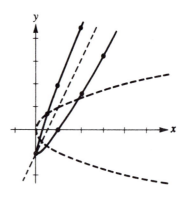

Figure 7.44

Addition of ordinates can be used for hyperbolas, as well as for ellipses and parabolas. One disadvantage of this method is that the two curves must be graphed relatively accurately, or the final result is likely to be extremely inaccurate.

Solving for y as a function (or pair of functions) of x is exactly what we have to do when graphing electronically. These methods are well suited to graphing by calculator. In fact, the conics graphing programs given in the introduction to the graphing calculator solve for y as a function or pair of functions of x by solving a linear or quadratic equation in y. Any of the first 20 problems below can be solved by the graphing program.

One final word; it is *not* maintained that the methods of this section will *always* provide the simplest method of sketching conics. They are alternate methods that are useful in many cases.

PROBLEMS

In Problems 1–20, sketch without rotating axes.

A
1. $xy - x + y + 3 = 0$
2. $2xy - x - y - 2 = 0$
3. $x^2 - xy - y - 4 = 0$
4. $x^2 - xy + x + 2y = 0$

B
5. $2x^2 - 2xy + y^2 - 1 = 0$
6. $5x^2 - 4xy + y^2 - 4 = 0$
7. $x^2 - 2xy + y^2 - x = 0$
8. $x^2 - 2xy + y^2 + x - 2y + 2 = 0$
9. $2xy - y^2 - 4 = 0$
10. $2x^2 - 2xy + y^2 + 4x - 4y - 5 = 0$
11. $2xy - y^2 + 6x - 6y - 18 = 0$
12. $3x^2 - 4xy + y^2 - 4x + 2y + 5 = 0$
13. $4x^2 + 4xy + y^2 - 3x + 2y + 1 = 0$
14. $x^2 - 2xy + y^2 - 12x + 8y + 24 = 0$
15. $3x^2 + 2xy - y^2 + 10x + 2y + 8 = 0$
16. $x^2 - 2xy + y^2 - 2x + 2y - 3 = 0$
17. $x^2 - xy - x - 2 = 0$
18. $xy - y^2 - y + 2 = 0$
19. $10x^2 - 6xy + y^2 + 12x - 4y + 4 = 0$
20. $4xy + y^2 - 4 = 0$

C
21. Sketch $\sqrt{x} + \sqrt{y} = \sqrt{a}$. Show that the graph is a portion of a parabola.

REVIEW PROBLEMS

In Problems 1–16, sketch the graphs of the equations as rapidly as possible without plotting points. Show all intercepts, asymptotes, and symmetry about either axis or the origin.

A
1. $y = (x + 1)^2(x - 2)$
2. $y = x^3(x - 3)^2$
3. $y = (x + 1)^2(x^2 + 1)$
4. $y = (x + 1)(x - 2)(x + 3)$
5. $y = \dfrac{x^2 - 1}{x^2 + 1}$
6. $y = \dfrac{x^2(x - 3)}{(x^2 - 4)^2}$
7. $y = \dfrac{x^2 - 9}{x^2}$
8. $y = \dfrac{x^3 - 1}{x^2 + 3x}$

B
9. $y = (x - 1)^2(x + 3)^3(2x - 3)$
10. $y = x + 1 - \dfrac{6}{x}$
11. $y = x - \sqrt{x^2 - 4}$
12. $y = \sqrt{\dfrac{(x - 1)(x + 2)}{x + 1}}$
13. $y^2 = \dfrac{x(x - 2)}{(x + 1)^2}$
14. $y^2 = \dfrac{2x - 1}{x + 2}$
15. $y = \dfrac{x^2 - 3x - 4}{x^2 + 4x + 3}$
16. $y = \dfrac{x^2}{x^2 - 2x}$

In Problems 17–20, sketch the graphs of the given conic sections without rotating axes.

17. $x^2 + xy - 2x + y = 0$

18. $x^2 - xy - y - 4 = 0$

19. $5x^2 - 4xy + y^2 - 4x + 2y - 8 = 0$

20. $4x^2 - 4xy + y^2 + 11x - 6y + 5 = 0$

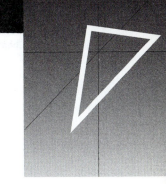

Chapter **8** *Polar Coordinates and Parametric Equations*

POLAR COORDINATES

Up to now, a point in the plane has been represented by a pair of numbers, (x, y), which represent (for perpendicular axes) the distances of the point from the y and x axes, respectively. Another way of representing points is by **polar coordinates**. In this case, we need only one axis (the **polar axis**) and a point on it (the **pole**). These correspond to the x axis and the origin of the rectangular coordinate system. Normally we shall include the y axis, even though it is not necessary to do so.

Before considering points in polar coordinates, let us recall that an angle in the standard position has its vertex at the origin (or pole) and its initial side on the positive end of the x axis (or polar axis). The terminal side is another ray (or half-line) with the origin as its endpoint. The ray with the same endpoint and on the same line as the terminal side is called the ray opposite the terminal side. For example, the terminal side of a 90° angle in standard position is the positive end of the y axis together with the origin; the ray opposite the terminal side is the negative end of the y axis together with the origin.

A point P is represented, in polar coordinates, by an ordered pair of numbers (r, θ) (see Figure 8.1). It is determined in the following way: first find the terminal side of the angle θ in standard position; if $r \geq 0$, then P is on this terminal side and at a distance r from the pole; if $r < 0$, then P is on the ray opposite the

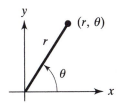

Figure 8.1

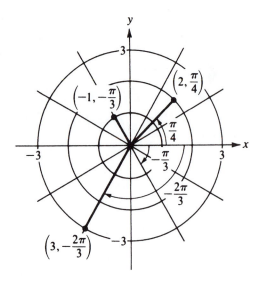

Figure 8.2

terminal side and at a distance $|r|$ from the pole. A few points are given with their polar coordinates in Figure 8.2.

It might be noted that while the terminal side of the angle $-\pi/3$ is in the fourth quadrant, $(-1, -\pi/3)$ is in the second quadrant. The quadrant that a point is in is *not* determined by the signs of the two polar coordinates, as it is with rectangular coordinates. It is determined by the size of θ and the sign of r. If r is positive, the point is in whatever quadrant θ is in; if r is negative, the point is in the opposite quadrant.

Polar coordinates present one problem we did not have with rectangular coordinates—a point has more than one representation. For example: $(2, \pi/2)$ and $(-2, -\pi/2)$ represent the same point. In fact, if (r, θ) is one representation of a point, then $(r, \theta + \pi n)$, where n is an even integer, and $(-r, \theta + \pi n)$, where n is an odd integer, are representations of the same point. Furthermore, $(0, \theta)$ is the pole for any choice of θ.

8.2

GRAPHS IN POLAR COORDINATES

Equations in polar coordinates can be graphed by point-by-point plotting, as we graphed rectangular coordinates.

EXAMPLE 1   Graph $r = \sin \theta$.

Note in Figure 8.3 that we have the entire graph for $0 \leq \theta < \pi$. The remaining values of θ simply repeat the graph a second time, since $(0, 0) = (0, \pi)$, $(0.5, \pi/6) = (-0.5, 7\pi/6)$, and so forth. Of course, values of θ outside the range $0 \leq \theta \leq 2\pi$ would give new points. As we shall see later, this is a circle.

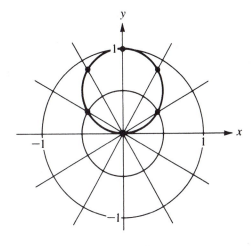

θ		r
$0 =$	$0°$	0.00
$\pi/6 =$	$30°$	0.50
$\pi/3 =$	$60°$	0.87
$\pi/2 =$	$90°$	1.00
$2\pi/3 =$	$120°$	0.87
$5\pi/6 =$	$150°$	0.50
$\pi =$	$180°$	0.00
$7\pi/6 =$	$210°$	-0.50
$4\pi/3 =$	$240°$	-0.87
$3\pi/2 =$	$270°$	-1.00
$5\pi/3 =$	$300°$	-0.87
$11\pi/6 =$	$330°$	-0.50
$2\pi =$	$360°$	0.00

Figure 8.3

This method of point-by-point plotting is quite cumbersome here, as it was in the case of rectangular coordinates. One way to simplify the proceedings is to represent the table of values of r and θ by means of a graph. This may sound as if we are going in circles—we can get the graph from a table of values of r and θ that is represented by a graph. Actually, this is not so bad as it sounds. We shall represent the table by a graph in **rectangular coordinates**.

For example, the table of values of r and θ used in Example 1 can be represented by the graph shown in Figure 8.4, which is the graph of $r = \sin \theta$ in rectangular coordinates.

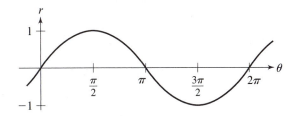

Figure 8.4

EXAMPLE 2   Graph $r = 1 + \cos \theta$.

SOLUTION   We can easily graph this equation in rectangular coordinates by using addition of ordinates. The result is given in Figure 8.5. Now we can read off values of r and θ just as we would from a table. As θ increases from 0 to $\pi/2$, r goes from 2 to 1. This gives the portion of the curve shown in Figure 8.6a. As θ goes from $\pi/2$ to π, r goes from 1 down to 0 (shown in Figure 8.6b). As θ goes from π to $3\pi/2$, r goes from 0 back up to 1 (as in Figure 8.6c); and finally, as θ goes from $3\pi/2$ to 2π, we see in Figure 8.6d that θ goes from 1 to 2. The same path is traced for values of θ

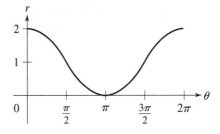

Figure 8.5

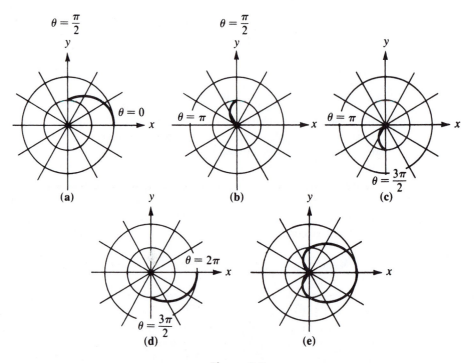

Figure 8.6

beyond 2π or less than 0. Putting all of this together, we have the desired graph, shown in Figure 8.6e. This curve is called a *cardioid*, which means "heart-shaped." It is a special case of a more general curve called a *limaçon* (French for snail), which has the form $r = a + b \sin \theta$ or $r = a + b \cos \theta$. If $|a| = |b|$, then we have a cardioid. A limaçon is easily identified by the form of its equation.

EXAMPLE 3   Graph $r = \sin 2\theta$.

SOLUTION   The graph is given in rectangular coordinates in Figure 8.7a. This is then put on the polar graph shown in Figure 8.7b. Note that for θ in the range $\pi/2 < \theta < \pi$, r is negative. Thus, instead of giving the loop in the second quadrant, it gives the one in the fourth quadrant. Similarly, for θ in the range $3\pi/2 < \theta < 2\pi$, r is negative. This gives the loop in the second quadrant. The resulting curve is called a four-leafed rose.

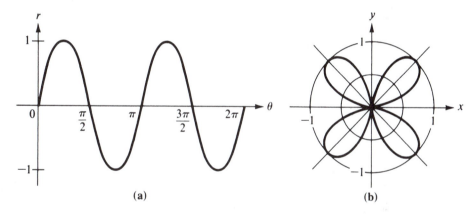

(a)   (b)

Figure 8.7

EXAMPLE 4   Graph $r^2 = \sin 2\theta$.

SOLUTION   Graphing in rectangular coordinates by the methods of Section 7.5, we have the result given in Figure 8.8a. There are a couple of things of interest here. First, $r^2 = \sin 2\theta$ has two values of r for each θ in the ranges $0 < \theta < \pi/2$ and $\pi < \theta < 3\pi/2$, while it has no value at all for $\pi/2 < \theta < \pi$ and $3\pi/2 < \theta < 2\pi$. Since it has two values in the range $0 < \theta < \pi/2$, we get both loops for $0 \leq \theta \leq \pi/2$, shown in Figure 8.8b. Similarly, we get both loops a second time for $\pi \leq \theta \leq 3\pi/2$. Because there is no value of r for $\pi/2 < \theta < \pi$ and $3\pi/2 < \theta < 2\pi$, there are no points of the graph in the second or fourth quadrants. This is called a lemniscate.

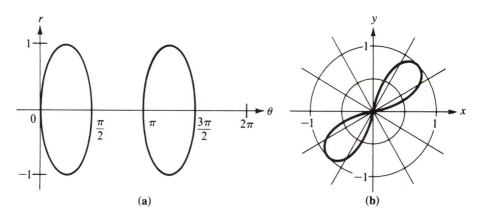

(a)   (b)

Figure 8.8

There are tests for symmetry in polar coordinates which are somewhat like those in rectangular coordinates. Suppose, for example, that for each point (r, θ) on a given curve there corresponds another point $(r, -\theta)$ on the same curve. Then (see Figure 8.9a) the curve is symmetric about the x axis. Thus if θ is replaced by $-\theta$ and the result is equivalent* to the original equation, then the graph is symmetric about the x axis. Figures 8.9b and 8.9c illustrate conditions leading to symmetry about the y axis and the pole, respectively. These tests are summarized in the following theorem.

THEOREM 8.1   (a) *If θ is replaced by $-\theta$ and the result is equivalent to the original equation, then the graph is symmetric about the x axis.*
(b) *If θ is replaced by $\pi - \theta$ and the result is equivalent to the original equation, then the graph is symmetric about the y axis.*
(c) *If r is replaced by $-r$ and the result is equivalent to the original equation, then the graph is symmetric about the pole.*

EXAMPLE 5   Test $r = 1 + \cos \theta$ for symmetry.

SOLUTION   The three tests are represented schematically below (the arrow is used to represent "is replaced by").

*Two equations are equivalent if any point that satisfies one of them also satisfies the other. To determine equivalence we use algebraic or trigonometric identities, add any expression to both sides of one equation, or multiply both sides of an equation by a nonzero constant in order to make that equation identical to the other.

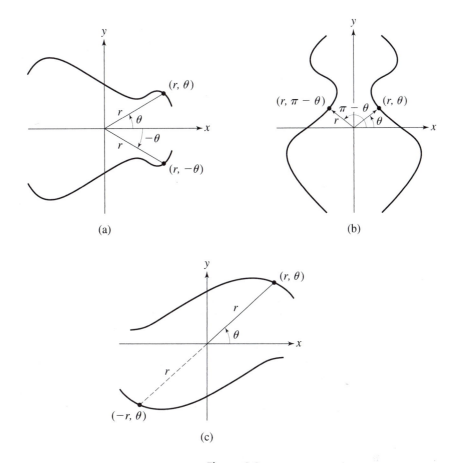

Figure 8.9

(a) $\begin{cases} r \to r \\ \theta \to -\theta \end{cases}$ (b) $\begin{cases} r \to r \\ \theta \to \pi - \theta \end{cases}$ (c) $\begin{cases} r \to -r \\ \theta \to \theta \end{cases}$

$\qquad r = 1 + \cos(-\theta)$ $\qquad r = 1 + \cos(\pi - \theta)$ $\qquad -r = 1 + \cos\theta$

$\qquad\quad = 1 + \cos\theta$ $\qquad\qquad = 1 - \cos\theta$

Since $r = 1 + \cos(-\theta)$ is equivalent to $r = 1 + \cos\theta$, we have symmetry about the x axis. The other two tests are negative; however, *this in itself is not enough to say that we do not have the other two types of symmetry* (we shall consider this in more detail in the next example). Nevertheless the graph of $r = 1 + \cos\theta$ (given in Figure 8.6e) indicates that we do have symmetry only about the x axis.

We see from this example that, while symmetry is sometimes an aid in graphing an equation, the graph can also be an aid in determining the presence or absence of symmetry. This is especially true in polar coordinates, because negative results in the tests of Theorem 8.1 do not necessarily imply a lack of symmetry, as the next example makes evident.

EXAMPLE 6   Test $r = \sin 2\theta$ for symmetry.

SOLUTION

(a) $\begin{cases} r \to r \\ \theta \to -\theta \end{cases}$

$r = \sin(-2\theta)$

$= -\sin 2\theta$

(b) $\begin{cases} r \to r \\ \theta \to \pi - \theta \end{cases}$

$r = \sin(2\pi - 2\theta)$

$= -\sin 2\theta$

(c) $\begin{cases} r \to -r \\ \theta \to \theta \end{cases}$

$-r = \sin 2\theta$

All three tests are negative. Yet we can see from Figure 8.7b that all three types of symmetry are present.

The reason for the rather strange behavior of the last example can be traced directly to the fact that one point has many different representations in polar coordinates. For example, $(r, -\theta) = (-r, \pi - \theta) = (r, 2\pi - \theta)$, and so forth. Thus there are many tests for symmetry about the x axis. The equalities above lead to the three tests:

$$\begin{cases} r \to r \\ \theta \to -\theta \end{cases} \qquad \begin{cases} r \to -r \\ \theta \to \pi - \theta \end{cases} \qquad \begin{cases} r \to r \\ \theta \to 2\pi - \theta. \end{cases}$$

If any one of these gives an equation that is equivalent to the original, there is symmetry about the x axis. Notice that while the first and third of these three tests give negative results in the equation of Example 6, the second gives a positive result. This is sufficient to assure us that there is symmetry about the x axis. If they all give negative results, nothing can be concluded, since there are still other possible tests. The student can easily devise other tests for all three types of symmetry. The result of all of this is that we must be content with tests for symmetry which do not guarantee a lack of symmetry when the test is negative.

Because of the multiplicity of the foregoing tests and the indecisiveness of negative results, you might prefer to rely upon the graphs of an equation. For example, Figure 8.7b suggests that we have all three types of symmetry for $r = \sin 2\theta$. Then the symmetry of Figure 8.7a assures us that we really do have the suspected symmetry. Note, however, that a particular type of symmetry in the rectangular coordinate graph does not necessarily imply the same type of symmetry in polar coordinates. Although the rectangular coordinate graph of $r^2 = \sin 2\theta$ (that is, of $y^2 = \sin 2x$) shows symmetry about the x axis, the polar graph does not (see Figure 8.8). Furthermore, the polar graph exhibits symmetry about the pole (or origin), while the rectangular coordinate graph does not. But the symmetry of the loops in rectangular coordinates does imply symmetry of the loops in polar coordinates and thus symmetry about the pole.

Graphs in polar coordinates are not so simple as in rectangular coordinates (in part because polar coordinates are less familiar). Thus, a graphing calculator or a computer program can be a great help. The most recent graphing calculators have the ability to graph polar functions expressed in the form $r = f(\theta)$. One merely sets the mode to polar. However, some of the older calculators do not have a polar mode. If you have such a calculator, you may graph $r = f(\theta)$ by setting the mode to parametric and entering $x = f(\theta)\cos\theta$, $y = f(\theta)\sin\theta$. The reason for this choice will be explained in Section 8.4.

In graphing a polar equation, you must select a range of values, not only for the x and y axes, but also for a range of θ values. This range is normally 0 to 2π ($= 6.2831853\ldots$) since θ is usually present only in trigonometric functions. When θ appears in nonperiodic functions, consider using a wider range of θ values.

PROBLEMS

A

1. Plot the following points: $(1, \pi/3)$, $(2, 45°)$, $(0, 30°)$, $(-2, 90°)$, $(-1, 3\pi/4)$, $(2, 300°)$.

*2. Give an alternate polar representation with $0° \leq \theta < 180°$: $(4, 330°)$, $(-2, 420°)$, $(1, 210°)$, $(0, 283°)$, $(-3, 270°)$, $(2, 240°)$.

*3. Give an alternate polar representation with $r \geq 0$ and $0 \leq \theta < 2\pi$: $(-4, 2\pi/3)$, $(3, -\pi/3)$, $(0, 7\pi/2)$, $(-1, 11\pi/6)$, $(-2, 13\pi/6)$, $(-2, 3\pi/4)$.

4. Give an alternate polar representation: $(1, 30°)$, $(-2, \pi)$, $(4, 210°)$, $(0, \pi/3)$, $(-1, 30°)$, $(2, \pi/2)$.

B

In Problems 5–20, sketch the graph of the given equation and name the curve.

5. $r = \cos\theta$ **6.** $r = 2\sin\theta$

7. $r = 1 - \cos\theta$ **8.** $r = 1 + \sin\theta$

9. $r = 1 - \sin\theta$ **10.** $r = \sin\theta - 1$

11. $r = \cos 2\theta$ **12.** $r = \sin 4\theta$

13. $r = \sin 3\theta$ **14.** $r = \cos 3\theta$

15. $r = \cos 5\theta$ **16.** $r = \sin 6\theta$

17. $r = 1 + 2\sin\theta$ **18.** $r = 1 - 2\cos\theta$

19. $r = 2 + \cos\theta$ **20.** $r = 2 + 3\sin\theta$

In Problems 21–34, sketch the graph.

21. $r = \tan\theta$ **22.** $r = \sec\theta$

23. $r^2 = \sin\theta$ **24.** $r^2 = \cos 3\theta$

* Answers are given for the first three parts of Problems 2 and 3 rather than for none of 2 and all of 3.

25. $r^2 = \cos 4\theta$

26. $r^2 = \sin^2 \theta$

27. $r^2 = 1 + \cos \theta$

28. $r^2 = 1 - \sin \theta$

29. $r = \theta$ (spiral of Archimedes)

30. $r = 1/\theta$ (hyperbolic spiral)

31. $r^2 = \theta^2$

32. $r = \dfrac{2}{1 - \cos \theta}$ (parabola)

33. $r = \dfrac{2}{1 - 2\cos \theta}$ (hyperbola)

34. $r = \dfrac{2}{2 - \cos \theta}$ (ellipse)

C   **35.** Find two tests for symmetry about the y axis that are different from the one given in this section.

36. Find two tests for symmetry about the pole that are different from the one given in this section.

37. Show that the distance between (r_1, θ_1) and (r_2, θ_2) is

$$d = \sqrt{r_1^2 + r_2^2 - 2r_1 r_2 \cos(\theta_1 - \theta_2)}.$$

[*Hint*: Use the law of cosines.]

GRAPHING
CALCULATOR

38. Graph $r = \sin n\theta$ for $n = 1, 2, 3, \ldots$ How many loops are there when n is even? When n is odd?

39. Graph $r = 1 + \sin \theta$, $r = 1 + \cos \theta$, $r = 1 - \sin \theta$, and $r = 1 - \cos \theta$. What is the relationship between them?

8.3

POINTS OF INTERSECTION

We can find points of intersection of a pair of curves in polar coordinates by solving their equations simultaneously, just as we did in finding points of intersection in rectangular coordinates. However, there is a special problem with polar coordinates that is caused, once again, by the fact that a point has many different polar representations. Solving the equations simultaneously is only a starting point in determining points of intersection of pairs of polar curves. We must then graph both equations to see if there are other points of intersection with different representations on the two curves. Let us consider some examples.

EXAMPLE 1   Find the points of intersection of $r = 1$ and $r = 2 \sin \theta$.

SOLUTION   Eliminating r from this pair of equations, we get

$$\sin \theta = \frac{1}{2}, \quad \text{or} \quad \theta = \frac{\pi}{6}, \frac{5\pi}{6},$$

giving the points $(1, \pi/6)$ and $(1, 5\pi/6)$. The graphs of these two curves are given in Figure 8.10. There are clearly only two points of intersection—points we have already found. Nothing else needs to be done.

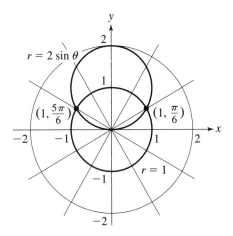

Figure 8.10

EXAMPLE 2

Find the points of intersection of $r = \sin\theta$ and $r = \cos\theta$.

SOLUTION

Eliminating r between the two equations, we have

$$\sin\theta = \cos\theta.$$

If we divide by $\cos\theta$, then

$$\tan\theta = 1$$

and

$$\theta = \frac{\pi}{4} + \pi \cdot n.$$

In the range $0 \le \theta < 2\pi$, we have $(1/\sqrt{2}, \pi/4)$ and $(-1/\sqrt{2}, 5\pi/4)$. But these are different representations for the same point. Thus we have found only one point of intersection. As we can see from Figure 8.11, there are really two points of intersection—the one we found and the pole.

The pole has many different representations. On the curve $r = \sin\theta$, it is represented by $(0, \pi n)$; on $r = \cos\theta$, it is represented by $(0, \pi/2 + \pi n)$. Thus, while the pole is common to both curves, it does not have a common representation that satisfies both equations. So we cannot find this point of intersection by finding simultaneous solutions of the two equations. This point has been identified in Figure 8.11 by both $(0, 0)$ and $(0, \pi/2)$ to show that there is no one representation satisfying both equations.

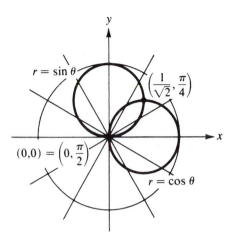

Figure 8.11

One convenient way to think of the phenomenon of this last example is to imagine the two curves as paths traced by points as θ increases uniformly with time. From this point of view, the curves both go through the origin; but they do so at different times. It is very much like two cars going through an intersection in different directions at different times. Clearly their paths intersect but, because they do so at different times, the cars do not collide.

A graphing calculator illustrates this idea very well. When a polar equation is graphed, the value of θ varies from 0 to 2π (or whatever range is set for it); and the path of the curve is traced over a time interval of a few seconds. When two or more equations are graphed, they are normally graphed one after the other. However, if the graphing mode is changed from sequential to simultaneous, points of intersection can easily be identified as having the same or different representations. In this case, θ literally corresponds to time.

This phenomenon of two curves crossing the same point at different times is easily seen with the two equations of Example 2. The path of $r = \sin\theta$ begins at the pole and proceeds to the right and upward, following a counterclockwise path around the circle. The path of $r = \cos\theta$, on the other hand, begins at the point $(1, 0)$, proceeding upward and to the left, again tracing a counterclockwise path around another circle. When the paths are graphed simultaneously, they both reach the point $(1/\sqrt{2}, \pi/4)$ at the same time. On the other hand, whereas the first path starts at the pole, the second does not reach that point until it has traced out a semicircle. By that time, a point on the first circle has reached the top of its circular path. Thus, one point of intersection has a common representation on the two curves, whereas the other point of intersection does not.

Once we have determined that a particular point of intersection has a common representation or different representations on the two curves, how can we find that (or those) representations? We use the TRACE command. We trace a path along one of the curves until the point of intersection is reached. The representation will be displayed on the screen. We can then switch the trace from one curve to the other. If the point has a common representation on the two curves, the position of the point is unchanged or moves only slightly. If the point has different representations on the two curves, switching to the second curve will result in the point jumping away from the point of intersection (to a point on the second curve with the same value of θ as the point of intersection on the first curve). By tracing the point along the second curve to the point of intersection, we can read the representation on the second curve from the screen.

EXAMPLE 3   Find all points of intersection of $r = \cos 2\theta$ and $r = \sin \theta$.

SOLUTION

$$\cos 2\theta = \sin \theta$$
$$1 - 2\sin^2\theta = \sin \theta$$
$$2\sin^2\theta + \sin \theta - 1 = 0$$
$$(2\sin \theta - 1)(\sin \theta + 1) = 0$$
$$\sin \theta = \frac{1}{2}, \quad \sin \theta = -1$$
$$\theta = \frac{\pi}{6}, \frac{5\pi}{6}, \frac{3\pi}{2}$$

Thus, we have the points $(1/2, \pi/6)$, $(1/2, 5\pi/6)$, and $(-1, 3\pi/2)$. In addition, we can see from Figure 8.12 that the pole is a point of intersection; it may be represented by $(0, \pi/4)$ satisfying the equation $r = \cos 2\theta$ or by $(0, 0)$ satisfying $r = \sin \theta$. It might also be noted that the point $(-1, 3\pi/2)$ can also be written $(1, \pi/2)$, but this form satisfies only $r = \sin \theta$.

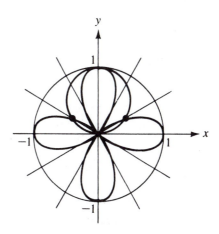

Figure 8.12

PROBLEMS

Find all points of intersection of the given curves.

A   **1.** $r = \sqrt{2}, r = 2\cos\theta$ **2.** $r = \sqrt{3}, r = 2\sin\theta$

   3. $r = 2, r = \sin\theta + 2$ **4.** $r = 1, r = 2\cos 2\theta$

   5. $r = \cos\theta, r = 1 - \cos\theta$

B   **6.** $r = \cos\theta, r = 1 + \sin\theta$

   7. $r = \sin 2\theta, r = \sin\theta$ **8.** $r = \sin 2\theta, r = \sqrt{2}\cos\theta$

   9. $r = \sec\theta, r = \csc\theta$ **10.** $r = \sec\theta, r = \tan\theta$

   11. $r = 3\cos\theta + 4, r = 3$ **12.** $r = \sin 2\theta, r = \cos 2\theta$

   13. $r = 2(1 + \cos\theta), r(1 - \cos\theta) = 1$ **14.** $r = 1 - \sin\theta, r(1 - \sin\theta) = 1$

   15. $r = 1 - \sin\theta, r = 1 - \cos\theta$ **16.** $r^2 = \sin\theta, r^2 = \cos\theta$

   17. $r^2 = \cos\theta, r^2 = \sec\theta$ **18.** $r = 2\cos\theta + 1, r = 2\cos\theta - 1$

   19. $r^2 = \sin\theta, r = \sin\theta$ **20.** $r^2 = \sin\theta, r = \cos\theta$

   21. $r = \sec\theta, r = 2\cos\theta$ **22.** $r = \csc\theta, r = 2\sin\theta$

8.4

RELATIONSHIPS BETWEEN RECTANGULAR AND POLAR COORDINATES

There are some simple relationships between rectangular and polar coordinates. These can be found easily by a consideration of Figure 8.13. They are given in the following theorem.

THEOREM 8.2   *If (x, y) and (r, θ) represent the same point in rectangular and polar coordinates, respectively, then*

$$x = r\cos\theta$$
$$y = r\sin\theta$$
$$r^2 = x^2 + y^2$$

$$\tan\theta = \frac{y}{x}$$

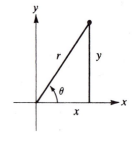

Figure 8.13

The last two, which may be solved for r and θ, would give us expressions involving $\pm$ and arctan. Thus, we prefer to leave them in their present form.

With these, we can now change from one coordinate system to the other.

EXAMPLE 1   Give a rectangular coordinate representation for the point having polar co-ordinates $(2, \pi/6)$.

SOLUTION

$$
\begin{aligned}
x &= r\cos\theta & y &= r\sin\theta \\
&= 2\cos\pi/6 & &= 2\sin\pi/6 \\
&= 2\cdot\frac{\sqrt{3}}{2} & &= 2\cdot\frac{1}{2} \\
&= \sqrt{3} & &= 1
\end{aligned}
$$

Thus the point $(2, \pi/6)$ in polar coordinates is the point $(\sqrt{3}, 1)$ in rectangular coordinates (see Figure 8.14).

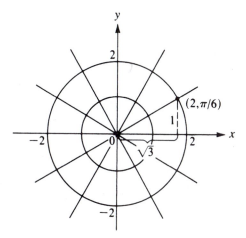

Figure 8.14

EXAMPLE 2   Express in polar coordinates the point having rectangular coordinates $(4, -4)$.

SOLUTION

$$
\begin{aligned}
r^2 &= x^2 + y^2 & \tan\theta &= \frac{y}{x} \\
&= 16 + 16 & &= \frac{-4}{4} \\
&= 32 & &= -1 \\
r &= \pm 4\sqrt{2} & \theta &= \frac{3\pi}{4} + \pi n
\end{aligned}
$$

We have a choice for both r and θ. The values of r and θ cannot be selected independently; the value we choose for one will limit the available choices for the other. In this case, the point $(4, -4)$ is in the fourth quadrant. Thus, we may choose either a fourth-quadrant angle and a positive r or a second-quadrant angle and a negative r. Thus (see Figure 8.15) the point $(4, -4)$ has any one of the following polar coordinate representations:

$$(4\sqrt{2}, 7\pi/4) = (-4\sqrt{2}, 3\pi/4) = (4\sqrt{2}, -\pi/4), \text{ and so forth.}$$

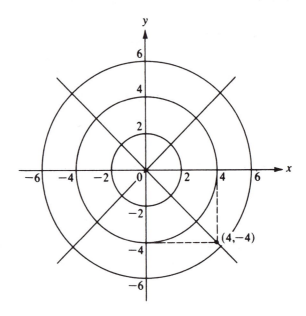

Figure 8.15

Of course we can use the equations of Theorem 8.2 to find a polar equation corresponding to one in rectangular coordinates, and vice versa.

EXAMPLE 3   Express $y = x^2$ in polar coordinates.

SOLUTION

$$y = x^2$$
$$r\sin\theta = r^2\cos^2\theta$$
$$\sin\theta = r\cos^2\theta \qquad \text{or} \qquad r = 0$$
$$r = \frac{\sin\theta}{\cos^2\theta}$$
$$r = \sec\theta\tan\theta$$

Since $r = 0$ represents only the pole and it is included in $r = \sec\theta\tan\theta$, we may drop $r = 0$. The result is

$$r = \sec\theta\tan\theta.$$

EXAMPLE 4   Express $r = 1 - \cos\theta$ in rectangular coordinates.

SOLUTION   First, multiply through by r.

$$r^2 = r - r\cos\theta$$

At this point, we could make the substitutions $r^2 = x^2 + y^2$, $r\cos\theta = x$, and $r = \pm\sqrt{x^2 + y^2}$. The last is rather bothersome, since it involves a $\pm$. In order to avoid this, let us isolate r on one side of the equation and square.

$$r = r^2 + r\cos\theta$$
$$r^2 = (r^2 + r\cos\theta)^2$$
$$x^2 + y^2 = (x^2 + y^2 + x)^2$$

We have done two things that might introduce extraneous roots: (1) Multiplying by r may introduce only a single point, the pole, to the graph. Since the pole is already a point of the graph of $r = 1 - \cos\theta$, no new point is introduced here. (2) Squaring may introduce several new points. The equation

$$r^2 = (r^2 + r\cos\theta)^2$$

is equivalent to

$$r = \pm(r^2 + r\cos\theta).$$

Now $r = r^2 + r\cos\theta$ is equivalent to our original equation, $r = 1 - \cos\theta$, while $r = -(r^2 + r\cos\theta)$ is equivalent to $r = -1 - \cos\theta$. Thus

$$x^2 + y^2 = (x^2 + y^2 + x)^2$$

is equivalent to $r = 1 - \cos\theta$ together with $r = -1 - \cos\theta$. But $r = 1 - \cos\theta$ and $r = -1 - \cos\theta$ have the same graph. Thus we have introduced no new points by squaring.

In Section 8.2 we indicated that some older graphing calculators and computer programs do not allow us to graph equations in polar coordinates, but they do allow the graphing of parametric equations. Thus, we can graph the polar equation

$r = f(\theta)$ by converting to the parametric form

$$x = f(\theta) \cos \theta, \qquad y = f(\theta) \sin \theta.$$

This comes directly from the relationships noted here. Since $x = r \cos \theta$ and $y = r \sin \theta$, a simple substitution of our polar equation, $r = f(\theta)$, gives this result.

While many calculators (even those that do not have graphing capabilities) can make single-point conversions from polar to rectangular coordinates and rectangular to polar coordinates, very few can convert a polar equation to its equivalent rectangular form or convert a Cartesian coordinate equation to polar coordinates. Thus, we have little electronic help in making these conversions.

PROBLEMS

A

1. The following points are given in polar coordinates. Give the rectangular coordinate representation of each: $(1, \pi)$, $(\sqrt{3}, \pi/3)$, $(-1, 3\pi)$, $(\sqrt{2}, 3\pi/4)$, $(2\sqrt{3}, 5\pi/3)$, $(-3, 7\pi/6)$, $(0, 5\pi/4)$, $(4, 0)$, $(-2, 7\pi/4)$.

2. The following points are given in rectangular coordinates. Give a polar coordinate representation of each: $(\sqrt{2}, -\sqrt{2})$, $(-1, \sqrt{3})$, $(4, 0)$, $(-1, -1)$, $(0, -2)$, $(0, 0)$, $(-2\sqrt{3}, 2)$, $(-3, 1)$, $(4, 3)$, $(-2, 4)$.

In Problems 3–12, express the given equation in polar coordinates.

3. $x = 2$	**4.** $y = 5$
5. $x^2 + y^2 = 1$	**6.** $x^2 - y^2 = 4$
7. $x = y^2$	**8.** $y = x^3$
9. $x + 2y - 4 = 0$	**10.** $x = y$
11. $y = 3x$	**12.** $y^2 = x^3$

In Problems 13–18, express the given equation in rectangular coordinates.

13. $r = a$	**14.** $\theta = \pi/4$
15. $\theta = \pi/3$	**16.** $r = 2 \sin \theta$
17. $r = 4 \cos \theta$	**18.** $r = \sin 2\theta$

B   *In Problems 19–24, express the given equation in polar coordinates.*

19. $(x + y)^2 = x - y$	**20.** $x^2 + y^2 - 2x = 0$
21. $x^2 + y^2 - 2x - 2y + 1 = 0$	**22.** $x^2 + 9y^2 = 9$
23. $xy = 1$	**24.** $y = \dfrac{x}{x + 1}$

In Problems 25–34, express the given equation in rectangular coordinates.

25. $r = \cos 2\theta$	**26.** $r = 1 - \cos \theta$

27. $r = 3 + 2 \sin \theta$ **28.** $r^2 = \sin \theta$

29. $r^2 = 1 + \sin \theta$ **30.** $r^2 = \sin 2\theta$

31. $r = \dfrac{1}{1 - \cos \theta}$ **32.** $r = \dfrac{1}{1 + \sin \theta}$

33. $r = 2 \sin \theta + 3 \cos \theta$ **34.** $r = \sec \theta$

In Problems 35–38, describe the graph of the given equation.

35. $r = 2h \cos \theta$ **36.** $r = 4c \sec \theta \tan \theta$

37. $r^2 = \dfrac{a^2 b^2}{b^2 \cos^2 \theta + a^2 \sin^2 \theta}$ **38.** $r^2 = \dfrac{a^2 b^2}{b^2 \cos^2 \theta - a^2 \sin^2 \theta}$

GRAPHING CALCULATOR

39. It was indicated in Example 4 that, because of squaring, the resulting rectangular coordinate equation is equivalent to a combination of $r = 1 - \cos \theta$ and $r = -1 - \cos \theta$. Use a graphing calculator to show the equations to be equivalent, first by graphing each individually and noting their similarity, and then by graphing them on the same coordinate axes to show them to be identical.

8.5

CONICS IN POLAR COORDINATES

We found earlier that the equations of conic sections (in rectangular coordinates) have very simple forms if the center or vertex is at the origin and the axes are the coordinate axes. There are, however, three different forms corresponding to the three different types of conics. We find that conics can be easily represented in polar coordinates if a focus is at the origin and one axis is a coordinate axis. Furthermore, the same type of equation represents all three types of conics if we use the unifying concept of eccentricity.

Recall that any conic can be determined by a single focus, the corresponding directrix, and the eccentricity. If P is a point on the conic, then the distance from P to the focus divided by the distance from P to the directrix equals the eccentricity. The particular conic we get depends upon the eccentricity; the eccentricity is a positive number and

> if $e < 1$, the conic is an ellipse,
>
> if $e = 1$, the conic is a parabola,
>
> if $e > 1$, the conic is a hyperbola.

This is illustrated in Figure 8.16 on page 300, where we have an ellipse, a parabola, and a hyperbola, all having the same focus and directrix. In this case the ellipse has eccentricity 1/2, and the hyperbola has eccentricity 2.

If $P = (r, \theta)$ is a point on a conic with focus 0, directrix $x = p$ (p positive), and eccentricity e, then (see Figure 8.17 on page 300).

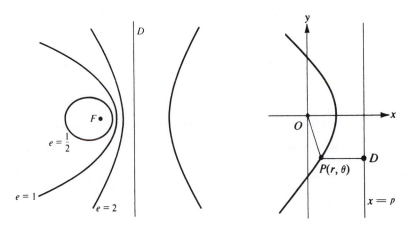

Figure 8.16   **Figure 8.17**

$$\frac{\overline{OP}}{\overline{PD}} = e, \quad \text{or} \quad \frac{|r|}{|p - r\cos\theta|} = e.$$

There are now two cases to consider.

$$\frac{r}{p - r\cos\theta} = e \quad \text{and} \quad \frac{r}{p - r\cos\theta} = -e$$

Either of these yields an equation of the desired conic (see Problems 23 and 24); however, the first yields the commonly used form. Solving for r in this equation, we have

$$r = \frac{ep}{1 + e\cos\theta}.$$

If the directrix is $x = -p$ (p positive), then the equation is

$$r = \frac{ep}{1 - e\cos\theta}.$$

If the directrix is $y = \pm p$ (p positive), then the equation is

$$r = \frac{ep}{1 \pm e\sin\theta}.$$

THEOREM 8.3   *The conic section with focus at the origin, directrix $x = \pm p$ (p positive), and eccentricity e has polar equation*

$$r = \frac{ep}{1 \pm e\cos\theta};$$

if the directrix is y = ±p (p positive), it has equation

$$r = \frac{ep}{1 \pm e\sin\theta}.$$

EXAMPLE 1   Describe $r = 6/(4 + 3\cos\theta)$.

SOLUTION   Dividing numerator and denominator by 4, we have

$$r = \frac{\dfrac{3}{2}}{1 + \dfrac{3}{4}\cos\theta} = \frac{\dfrac{3}{4}\cdot 2}{1 + \dfrac{3}{4}\cos\theta}.$$

Thus the eccentricity is 3/4 and the directrix is $x = 2$. The conic is an ellipse with focus at the origin, directrix $x = 2$, and eccentricity 3/4.

EXAMPLE 2   Sketch $r = 15/(2 - 3\cos\theta)$.

SOLUTION   Dividing by 2, we have

$$r = \frac{\dfrac{15}{2}}{1 - \dfrac{3}{2}\cos\theta} = \frac{\dfrac{3}{2}\cdot 5}{1 - \dfrac{3}{2}\cos\theta}.$$

Thus we have a hyperbola with focus at the origin, eccentricity 3/2, and directrix $x = -5$. The vertices are on the x axis, one between the focus and directrix and the other to the left of the directrix. When $\theta = 0$, $r = -15$; when $\theta = \pi$, $r = 3$. Thus the vertices are $(-15, 0)$ and $(3, \pi)$. When $\theta = \pi/2$ or $3\pi/2$, $r = 15/2$. Thus, the ends of one of the latera recta are $(15/2, \pi/2)$ and $(15/2, 3\pi/2)$. This information is enough to give a reasonably accurate picture of the hyperbola. If the asymptotes are desired, they can best be found by considering some of the above points in rectangular coordinates. Thus the vertices are $(-3, 0)$ and $(-15, 0)$, and the center is $(-9, 0)$, giving $a = 6$ and $c = 9$. We can now use the equation

$$b^2 = c^2 - a^2$$

to find $b^2 = 45$ or $b = 3\sqrt{5}$. Once we have this, the asymptotes are easily found (see Figure 8.18).

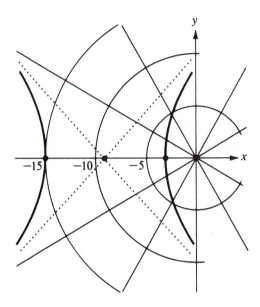

Figure 8.18

EXAMPLE 3   Find a polar equation of the parabola with focus at the origin and directrix $y = -4$.

SOLUTION   The equation is in the form

$$r = \frac{ep}{1 - e\sin\theta},$$

since the directrix is a horizontal line below the focus. Furthermore, $e = 1$, since the conic is a parabola, and directrix $y = -4$ gives $p = 4$. Thus the equation is

$$r = \frac{4}{1 - \sin\theta}.$$

Polar coordinates are used extensively in astronomy. When describing the orbit of a heavenly body, it is much more convenient to place the axes with the origin at the focus of an ellipse than to have it at the center. This is especially so when applying Kepler's second law of planetary motion, which says that a line from the sun to a planet sweeps out equal areas in equal times.

PROBLEMS

A In Problems 1–8, state the type of conic and give a focus and its corresponding directrix and the eccentricity.

1. $r = \dfrac{4}{1 + 2\cos\theta}$

2. $r = \dfrac{12}{1 - 3\sin\theta}$

3. $r = \dfrac{4}{3 + 2\sin\theta}$

4. $r = \dfrac{5}{4 - 4\cos\theta}$

5. $r = \dfrac{3}{1 + \sin\theta}$

6. $r = \dfrac{10}{5 - 2\cos\theta}$

7. $r(3 + 2\sin\theta) = 6$

8. $r(2 - 4\cos\theta) = 5$

In Problems 9–14, find a polar equation of the conic with focus at the origin and the given eccentricity and directrix.

9. Directrix: $x = 5$; $e = 2/3$

10. Directrix: $y = -3$; $e = 2$

11. Directrix: $y = 2$; $e = 1$

12. Directrix: $x = -4$; $e = 1$

13. Directrix: $x = 5$; $e = 5/4$

14. Directrix: $y = 3$; $e = 3/4$

B In Problems 15–22, sketch the given conic.

15. $r = \dfrac{2}{1 + \cos\theta}$

16. $r = \dfrac{16}{5 - 3\cos\theta}$

17. $r = \dfrac{16}{4 - 5\sin\theta}$

18. $r(3 - 5\cos\theta) = 9$

19. $r(13 + 12\sin\theta) = 25$

20. $r(3 + 3\sin\theta) = 4$

21. $r(2 - \cos\theta) = 5$

22. $r(1 - 3\sin\theta) = 12$

23. Sketch $r = \dfrac{-2}{1 - \cos\theta}$. Compare it with the conic of Problem 15 (see the following problem).

24. Show that the conic section with focus at the origin, directrix $x = p$ (p positive), and eccentricity e has polar equation

$$r = \frac{-ep}{1 - e\cos\theta}.$$

25. Suppose, in the equation $r = \dfrac{ep}{1 + e\cos\theta}$, $e \to 0$ and $p \to +\infty$ in such a way that ep remains constant. What happens to the shape of the conic? What happens to the equation of the conic?

C 26. Find a polar equation of a circle with center (k, α) and radius a by using the law of cosines (see Figure 8.19). What is the result when $k = 0$?

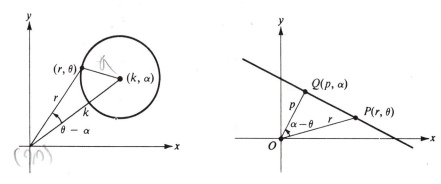

Figure 8.19   **Figure 8.20**

27. By using the trigonometry of right triangles, show that the line PQ (Figure 8.20) can be represented by the equation

$$x\cos\alpha + y\sin\alpha - p = 0.$$

This is called the **normal form** of the line, since it is expressed in terms of the polar coordinates of the point Q, which is the intersection of the original line and another perpendicular (or normal) to it and through the origin (see Problem 37, Section 3.3).

28. By using the identity

$$\sin^2\alpha + \cos^2\alpha = 1,$$

show that $Ax + By + C = 0$ can be put into the normal form by dividing through by $\pm\sqrt{A^2 + B^2}$ (see Problem 27 for the normal form).

29. Show that the distance from the point (x_1, y_1) to the line $Ax + By + C = 0$ is

$$d = \frac{|Ax_1 + By_1 + C|}{\sqrt{A^2 + B^2}}.$$

This result was found without the use of polar coordinates in Theorem 3.6, page 106. [*Hint:* Put the original line and the one parallel to it and through (x_1, y_1) into the normal form (see Problems 27 and 28).]

APPLICATIONS

30. Use the data in the table of Problem 35 on page 172 to find a polar equation for the earth's orbit. Assume that the sun is at one focus of the ellipse and the other focus is on the positive end of the *x* axis.

31. Use the data of Problem 34 on page 172 to find a polar equation of the moon's orbit about the earth. Place the axes as in Problem 30, using the earth instead of the sun.

32. A comet has a parabolic orbit with the sun at the focus. When the comet is 100 million miles from the sun, the line joining the sun and the comet makes an angle of 60° with the axis of the parabola (see Figure 8.21). How close to the sun will the comet get?

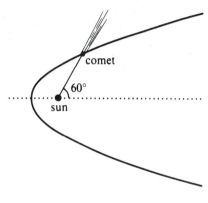

Figure 8.21

33. A satellite has an elliptic orbit with the earth at one focus. At its closest point it is 100 miles above the surface of the earth; at its farthest point, 500 miles. Find a polar equation of its path. (Take the radius of the earth to be 4000 miles.)

34. The distance of Halley's comet from the sun varies from a minimum of 50 million miles to a maximum of 3250 million miles. At **perihelion** (its closest approach to the sun), the comet's speed is 125,000 miles per hour. Estimate its speed at **aphelion** (its greatest distance from the sun). [*Hint:* Use Kepler's second law, which says that a line from the sun to an orbiting body sweeps out equal areas in equal times. Approximate the areas by using circular sectors.]

8.6

PARAMETRIC EQUATIONS

Up to now all of the equations we have dealt with have been in the form

$$y = f(x) \qquad \text{or} \qquad F(x, y) = 0.$$

In either case, a direct relationship between x and y is given. Sometimes, however, it is more convenient to express both x and y in terms of a third variable, called a **parameter**. That is,

$$x = f(t) \quad \text{and} \quad y = g(t).$$

Each value of the parameter t gives a value of x and a value of y.

t	x	y
0°	0.00	1.00
30°	0.50	0.87
60°	0.87	0.50
90°	1.00	0.00
120°	0.87	−0.50
150°	0.50	−0.87
180°	0.00	−1.00
210°	−0.50	−0.87
240°	−0.87	−0.50
270°	−1.00	0.00
300°	−0.87	0.50
330°	−0.50	0.87
360°	0.00	1.00

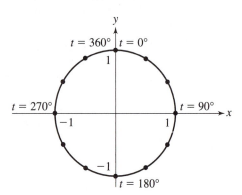

Figure 8.22

For instance, if $t = 0°$ in the parametric equations

$$x = \sin t \quad \text{and} \quad y = \cos t,$$

we see that $x = 0$ and $y = 1$. Thus the point $(0, 1)$ is a point of the graph. Note that we still have just the x and y axes; t does not appear on the graph. Continuing with this process gives the results shown in Figure 8.22. Of course, we could continue with values of t beyond 360°, but we would simply go over the same points again. Although the value of t need not appear anywhere on the graph, we have labeled several points with their corresponding values of t. Once the points are plotted, they are joined in order of increasing (or decreasing) values of t.

The result seems to resemble a circle. How can we be *sure* it is a circle? If we had a single equation in x and y, we could easily see by the form of the equation whether or not we have a circle. We can easily eliminate the parameter t between the equations $x = \sin t$ and $y = \cos t$.

$$\sin^2 t + \cos^2 t = 1$$
$$x^2 + y^2 = 1$$

We now see that we have a circle with center at the origin and radius 1.

Elimination of the parameter not only assures us that this particular curve is a circle, it also gives us a basis for sketching more rapidly than can be done by point-by-point plotting. However, we must be careful with the domain of the resulting equation. Let us illustrate this with some examples and see how the domain of $F(x, y) = 0$ plays an important role in sketching the graph.

EXAMPLE 1

Graph the following two pairs of parametric equations by eliminating the parameter.

$$x = t, y = t \quad \text{and} \quad x = t^2, y = t^2$$

SOLUTION

Elimination of the parameter gives $y = x$ in both cases. But the graphs are not the same, as the domains in the two cases will show. The domain can be determined from the first of the two parametric equations in each case. In the first case, $x = t$ and, since there is no restriction on t, there is none on x; the domain is the set of all real numbers. In the second case, $x = t^2$. The domain of $y = x$ is the range of $x = t^2$, which is $\{x \mid x \geq 0\}$. Thus we have a restricted domain here that we did not have in the first case. The graphs are given in Figure 8.23.

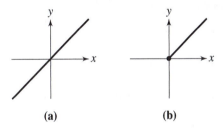

(a)   (b)

Figure 8.23

EXAMPLE 2

Eliminate the parameter between $x = t + 1$ and $y = t^2 + 3t + 2$ and sketch the graph.

SOLUTION

Solving $x = t + 1$ for t, we have

$$t = x - 1.$$

If this is substituted into $y = t^2 + 3t + 2$, then

$$y = (x - 1)^2 + 3(x - 1) + 2$$
$$= x^2 + x.$$

Note that there is no restriction on x; the domain of $y = x^2 + x$ is the set of all real numbers. It is now a simple matter to sketch the curve; it is given in Figure 8.24 on page 308.

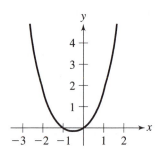

Figure 8.24

Eliminating the parameter is not always so simple as in these examples. Occasionally it is difficult or impossible. In such cases, the curve must be plotted point by point as we did with $x = \sin t$, $y = \cos t$. Closely related to parametric equations are vector-valued functions. They are represented in the following way.

$$\mathbf{f}(t) = f_1(t)\mathbf{i} + f_2(t)\mathbf{j}$$

Thus, when a value of t is substituted into the equation, the function takes on a vector value. For example, if

$$\mathbf{f}(t) = t\mathbf{i} + t^2\mathbf{j},$$

then

$$\mathbf{f}(1) = \mathbf{i} + \mathbf{j}$$

and

$$\mathbf{f}(2) = 2\mathbf{i} + 4\mathbf{j}.$$

Recall that, in graphing vectors, we graph only representatives. Thus, in graphing vector functions, let us graph representatives of the vectors, each having its tail at the origin. Thus,

$$\mathbf{f}(t) = t\mathbf{i} + t^2\mathbf{j}$$

has the graphical representation shown in Figure 8.25a. Normally we shall omit the directed line segments and show only their heads, as in Figure 8.25b. Thus, the result is equivalent to graphing the curve represented parametrically by

$$x = t \quad \text{and} \quad y = t^2.$$

EXAMPLE 3   Sketch the curve $\mathbf{f}(t) = (t + 2)\mathbf{i} + (t^2 + 7t + 12)\mathbf{j}$.

SOLUTION   This is equivalent to the parametric equations

$$x = t + 2 \quad \text{and} \quad y = t^2 + 7t + 12.$$

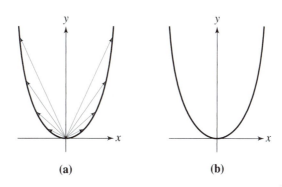

(a)   **(b)**

Figure 8.25

Eliminating the parameter, we have

$$y = x^2 + 3x + 2.$$

Its graph is given in Figure 8.26.

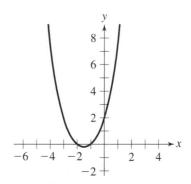

Figure 8.26

Let us consider a line determined by two of its points, $A = (x_1, y_1)$ and $B = (x_2, y_2)$. The directed line segments $\overrightarrow{OA}$ (where O is the origin) and $\overrightarrow{AB}$ represent vectors **u** and **v**, respectively (see Figure 8.27 on page 310). Now suppose $P = (x, y)$ is any point on the line, and **w** is the vector represented by $\overrightarrow{OP}$. Since $\overrightarrow{AB}$ and $\overrightarrow{AP}$ have the same directions (or opposite directions if P is above A),

$$\overrightarrow{AP} = r\overrightarrow{AB}$$

for some number r. (Note that r is negative if $\overrightarrow{AB}$ and $\overrightarrow{AP}$ have opposite

directions.) Then

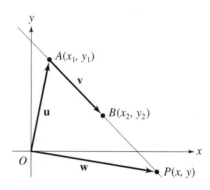

Figure 8.27

$$\overrightarrow{OP} = \overrightarrow{OA} + r\overrightarrow{AB}.$$

$$\mathbf{w} = \mathbf{u} + r\mathbf{v}.$$

But

$$\mathbf{w} = x\mathbf{i} + y\mathbf{j},$$

$$\mathbf{u} = x_1\mathbf{i} + y_1\mathbf{j},$$

$$\mathbf{v} = (x_2 - x_1)\mathbf{i} + (y_2 - y_1)\mathbf{j},$$

which gives

$$x\mathbf{i} + y\mathbf{j} = x_1\mathbf{i} + y_1\mathbf{j} + r[(x_2 - x_1)\mathbf{i} + (y_2 - y_1)\mathbf{j}]$$
$$= [x_1 + r(x_2 - x_1)]\mathbf{i} + [y_1 + r(y_2 - y_1)]\mathbf{j}.$$

Thus we have a vector-valued function of r which represents points on the given line. By tracing the argument backward, we see that for each value of r there is a vector which represents a point on the line. Thus the above vector-valued function represents the given line. The corresponding parametric form for the line is

$$x = x_1 + r(x_2 - x_1)$$
$$y = y_1 + r(y_2 - y_1).$$

These are the familiar point-of-division formulas that we saw on page 12. This use of the point-of-division formulas as a parametric representation for a line is a convenient one; it is one of the principal ways we shall have for representing lines in three-dimensional space (see page 335).

EXAMPLE 4   Find a parametric representation for the line through $(1, 5)$ and $(-2, 3)$.

SOLUTION   Letting $(1, 5)$ and $(-2, 3)$ be the first and second points, respectively, of

$$x = x_1 + r(x_2 - x_1)$$

and

$$y = y_1 + r(y_2 - y_1),$$

we then have

$$x = 1 - 3r$$

and

$$y = 5 - 2r.$$

The choice of (1, 5) and (−2, 3) as the first and second points in the preceding example is quite arbitrary; we could have reversed the designation. In that case the parametric representation would be

$$x = -2 + 3s$$

and

$$y = 3 + 2s.$$

While the two representations appear to have little in common, it is easily seen that they represent the same line. We get point (1, 5) when $r = 0$ or when $s = 1$; we get (−2, 3) when $r = 1$ or $s = 0$. In fact, whatever point we get for a given value of r, we get the same point for $s = 1 - r$. By using other points on the line, we can get still more parametric representations. Thus a line does not have a unique parametric representation.

We have already noted in Section 1.7 that we much prefer functions, which are single-valued, to equations, which may be many-valued when solved for y. Unfortunately, not all graphs are so obliging. That is where parametric equations can help. Parametric equations allow us to deal with very complicated graphs with loops and whorls by means of two functions. Thus, their use can considerably simplify certain problems.

Nevertheless, the graphing of parametric equations can be extremely difficult at times. Fortunately, most graphing calculators and computer graphing programs are able to handle parametric equations with no difficulty. This has already been noted as a way of graphing equations in polar coordinates.

PROBLEMS

A *In Problems 1–6, eliminate the parameter and sketch the curve.*

1. $x = t^2 + 1$, $y = t + 1$

2. $x = t^2 + t - 2$, $y = t + 2$

3. $x = t - 1$, $y = t^2 - 2t$

4. $x = 2t^2 + t - 3$, $y = t - 1$

5. $\mathbf{f}(t) = (t - 1)\mathbf{i} + t^2\mathbf{j}$

6. $\mathbf{f}(t) = (t^2 + 1)\mathbf{i} + t^2\mathbf{j}$

HISTORICAL NOTE

SPECIAL CURVES AND THE CLASSICAL GEOMETRIC CONSTRUCTIONS

Three classical Greek problems had defied all efforts at solution. All three were geometric constructions that required the exclusive use of a straightedge and compass. They were:

1. Trisect an arbitrary angle.
2. Double the cube; that is, given an arbitrary cube, find another having twice the volume.
3. Square the circle; that is, given an arbitrary circle, find a square having the same area.

It has subsequently been proved that these constructions are impossible using only a straightedge and compass. Nevertheless, the attempts to solve them led to a great deal of mathematical research. It has already been mentioned that the conic sections were first investigated in an attempt to solve these problems. Many other special curves were also invented for the same purpose. These include the **conchoid of Nicomedes**, the **cissoid of Diocles**, and the **quadratrix of Hippias**. Let us consider one of them—the cissoid.

Diocles used a somewhat different definition from that given in Problem 31 on page 314. The following is his construction. If *AB* is a horizontal line through the center of a circle (see Figure 8.28) and *XY* is a vertical line intersecting it, we begin by finding the point *X'* such that $\overline{OX} = \overline{OX'}$. *Y'* is then a point of intersection of the circle with the vertical line *X'Y'*. Finally, the point of intersection *P* of *XY* and *BY'* is a point of the cissoid. In this way, we can find the point of the cissoid corresponding to any vertical line within the circle.

Diocles showed that $\overline{XY}$ and $\overline{BX}$ are two geometric means between $\overline{AX}$ and $\overline{XP}$; that is,

$$\frac{\overline{AX}}{\overline{XY}} = \frac{\overline{XY}}{\overline{BX}} = \frac{\overline{BX}}{\overline{XP}}.$$

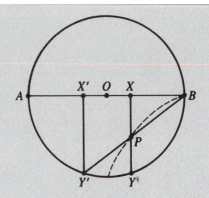

Figure 8.28

Let us see why. Since $\overline{BX'} = \overline{AX}$ and $\overline{X'Y'} = \overline{XY}$, it follows that

$$\frac{\overline{BX'}}{\overline{X'Y'}} = \frac{\overline{AX}}{\overline{XY}}.$$

By the similarity of triangles *AXY* and *YXB*,

$$\frac{\overline{AX}}{\overline{XY}} = \frac{\overline{XY}}{\overline{BX}} = \frac{\overline{BX'}}{\overline{X'Y'}},$$

and by the similarity of triangles *BX'Y'* and *BXP*,

$$\frac{\overline{BX'}}{\overline{X'Y'}} = \frac{\overline{BX}}{\overline{XP}}.$$

From these, we see that

$$\frac{\overline{AX}}{\overline{XY}} = \frac{\overline{XY}}{\overline{BX}} = \frac{\overline{BX}}{\overline{XP}}.$$

Now let us see how the cissoid can be used to double a cube. Suppose we have the cissoid *CB* (see Figure 8.29) and a cube with edge of length *a*. We find the point *Q* on *OC* such that

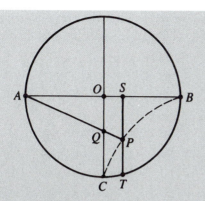

Figure 8.29

$$\frac{\overline{AO}}{\overline{OQ}} = \frac{\sqrt{2}a}{a}.$$

AQ is then extended to the point P on the cissoid, and ST is the vertical line through P. By the similarity of triangles AOQ and ASP, it follows that

$$\frac{\overline{AS}}{\overline{SP}} = \frac{\overline{AO}}{\overline{OQ}} = \frac{\sqrt{2}a}{a}.$$

By the double mean property of the cissoid, $\overline{ST}$ and $\overline{SB}$ are two geometric means between $\overline{AS}$ and $\overline{SP}$. That is,

$$\frac{\overline{AS}}{\overline{ST}} = \frac{\overline{ST}}{\overline{SB}} = \frac{\overline{SB}}{\overline{SP}}.$$

Since

$$\frac{\overline{AS}}{\overline{SP}} = \frac{\sqrt{2}a}{a},$$

there is a number such that

$$k \cdot \overline{AS} = \sqrt{2}a \quad \text{and} \quad k \cdot \overline{SP} = a.$$

Thus,

$$\frac{k \cdot \overline{AS}}{k \cdot \overline{ST}} = \frac{k \cdot \overline{ST}}{k \cdot \overline{SB}} = \frac{k \cdot \overline{SB}}{k \cdot \overline{SP}}.$$

Letting $x = k \cdot \overline{ST}$ and $y = k \cdot \overline{SB}$, we have

$$\frac{\sqrt{2}a}{x} = \frac{x}{y} = \frac{y}{a}.$$

Thus,

$$x^2 = \sqrt{2}ay \quad \text{and} \quad y^2 = ax$$
$$x^4 = 2a^2y^2 = 2a^2 \cdot ax = 2a^3x$$
$$x^3 = 2a^3.$$

This x is the edge of a cube having twice the volume of the original cube.

Note that the construction of a cissoid allows us to find any point on it intersecting an arbitrary vertical line. However, this construction does not give us the exact location of the point P that is on OQ, since we do not know what vertical line the point P will be on. For this reason, this does *not* allow us to double the cube using only a straightedge and compass.

Finally, we might note the origin of the name *cissoid*, which means "ivy-like." Diocles considered only the portion of the curve inside the circle that generates it. Since the interior of the circle above the curve has the shape of an ivy leaf (see Figure 8.30), it was named "cissoid."

Figure 8.30

In Problems 7–12, give equations in parametric form for the line through the given pair of points.

7. $(1, 5), (3, 1)$

8. $(4, 2), (-1, 3)$

9. $(2, 5), (-1, 2)$

10. $(4, 1), (-8, 3)$

11. $(2, 3), (5, 3)$

12. $(-3, 2), (-3, 5)$

B *In Problems 13–24, eliminate the parameter and sketch the curve.*

13. $x = t^2 + t, y = t^2 - t$

14. $x = t^2 + 1, y = t^2 - 1$

15. $x = t^3, y = t^2$

16. $x = a\cos\theta, y = b\sin\theta$

17. $x = 2 + \cos\theta, y = -1 + \sin\theta$

18. $x = 3 - \cos\theta, y = 2 + 4\sin\theta$

19. $x = 3 + 2\cos\theta, y = 2 + 3\sin\theta$

20. $x = 4 + 2\cos\theta, y = 1 - 4\sin\theta$

21. $\mathbf{f}(t) = t^2\mathbf{i} + t^3\mathbf{j}$

22. $\mathbf{f}(t) = (3t + 1)\mathbf{i} + (t^3 - 1)\mathbf{j}$

23. $\mathbf{f}(t) = \cos t\mathbf{i} + \sin t\mathbf{j}$

24. $\mathbf{f}(t) = t^2\mathbf{i} + e^t\mathbf{j}$

In Problems 25–31, sketch the curve.

25. $x = e^t, y = \sin t$

26. $x = \theta - \sin\theta, y = 1 - \cos\theta$

27. $x = \cos\theta + \theta\sin\theta, y = \sin\theta - \theta\cos\theta; \theta \geq 0$

28. $x = a\cos^3\theta, y = a\sin^3\theta$

29. $x = 2a\cos\theta - a\cos 2\theta, y = 2a\sin\theta - a\sin 2\theta$

30. $x = 2a\cot\theta, y = 2a\sin^2\theta$ (witch of Agnesi)

31. $x = 2a(\cot\theta - \sin\theta\cos\theta), y = 2a(1 - \sin^2\theta)$ (cissoid of Diocles)

C **32.** Show that the parametric representation

$$x = x_1 + r(x_2 - x_1) \quad \text{and} \quad y = y_1 + r(y_2 - y_1)$$

of the line through (x_1, y_1) and (x_2, y_2) is equivalent to the two-point form of the line given on page 88.

33. Sketch each of the following parametric equations and note the similarities and differences.

(a) $x = t, y = t^2$

(b) $x = \sqrt{t}, y = t$

(c) $x = e^t, y = e^{2t}$

(d) $x = \sin t, y = 1 - \cos^2 t$

34. Sketch each of the following parametric equations and note the similarities and differences.

(a) $x = t, y = \sqrt{t^2 - 1}$

(b) $x = \sqrt{t}, y = \sqrt{t - 1}$

(c) $x = \sec t, y = \tan t$

GRAPHING CALCULATOR

35.–41. The parametric equations of Problems 25–31 are difficult to sketch because of the difficulty of eliminating the parameter. This is where the graphing calculator is especially useful. Use it to sketch the graphs of Problems 25–31.

REVIEW PROBLEMS

A *In Problems 1–6, sketch the graph of the given equation and indicate any symmetry about either axis or the pole.*

1. $r = 1 + 3\cos\theta$ **2.** $r = \cos 4\theta$

3. $r = 1 + \sin 2\theta$ **4.** $r = e^\theta$

5. $r^2 = \sin 2\theta$ **6.** $r^2 = 1 + \cos\theta$

In Problems 7–12, find all points of intersection.

7. $r = 1,\ r = 1 + \cos\theta$ **8.** $r = 1 + \sin\theta,\ r = 1 - \cos\theta$

9. $r = \sin\theta,\ r = 1 + 2\sin\theta$ **10.** $r = 2\sin 2\theta,\ r = 1$

11. $r = \cos\theta,\ r = \cos 3\theta$ **12.** $r = 1 + \sin\theta,\ r = \cos\theta - 1$

13. **(a)** The following points are given in polar coordinates; give the rectangular coordinate representation of each: $(-1, \pi/2)$, $(2\sqrt{2}, 3\pi/4)$, $(6, 7\pi/6)$, $(-2\sqrt{3}, 2\pi/3)$.

   (b) The following points are given in rectangular coordinates; give a polar coordinate representation of each: $(-2, 2)$, $(-5, 0)$, $(1, -\sqrt{3})$, $(10, 4)$.

14. Express $x^2 + y^2 + 3y = 0$ in polar coordinates.

15. Give a polar equation for the ellipse with focus at the pole, the corresponding directrix $y = 3$, and eccentricity $2/3$.

16. Give a polar equation for the parabola with focus at the pole and directrix $x = -6$.

In Problems 17–20, eliminate the parameter and sketch the curve.

17. $x = t + 1,\ y = t^2 + 4t - 2$ **18.** $x = 1 - 3\cos\theta,\ y = 2 + 2\sin\theta$

19. $\mathbf{f}(t) = (t + 1)^2\mathbf{i} + t^2\mathbf{j}$ **20.** $\mathbf{f}(t) = e^t\mathbf{i} + \ln t\,\mathbf{j}$

21. Give a parametric representation of the line containing $(2, 3)$ and $(4, -5)$.

22. Give a parametric representation of the line containing $(-1, 4)$ and $(3, -3)$.

23. Find the point of intersection, if any, of $x = 1 - 3t,\ y = -3 + 7t$ and $x = -1 + 5s$, $y = s$.

24. Find the point of intersection, if any, of the segment from $(-3, 1)$ to $(2, -2)$ with the segment from $(1, -3)$ to $(7, 2)$.

B **25.** Express $r = 1 + \sin\theta$ in rectangular coordinates.

26. Express $r(3\cos\theta - 5) = 16$ in rectangular coordinates.

27. Sketch the conic $r = 2/(1 - \sin\theta)$. Identify the focus, directrix, and eccentricity.

28. Sketch the conic $r = 2/(1 - 2\cos\theta)$. Identify a focus, the corresponding directrix, and the eccentricity.

29. Eliminate the parameter and sketch $x = \sin\theta,\ y = \sin 2\theta$.

30. Sketch $\mathbf{f}(\theta) = \cos\theta(1 + \cos\theta)\mathbf{i} + \sin\theta(1 + \cos\theta)\mathbf{j}$.

31. Find the point(s) of intersection of $x = 5t,\ y = t + 3$ and $x = s^2 + s - 6,\ y = s + 1$.

Chapter **9** *Solid Analytic Geometry*

9.1

INTRODUCTION: THE DISTANCE AND POINT-OF-DIVISION FORMULAS

So far we have dealt almost exclusively with plane figures. Let us now consider the geometry of solid figures. Forming the bridge between algebra and geometry is the assignment of numbers to points in space, similar to the assignment we made to points in a plane; a point in space, however, is represented by a set of three numbers rather than two. We begin with a set of three lines, called **axes**, concurrent at a point (the origin). The only requirement is that these three lines not be coplanar—that is, that they not all lie in the same plane. However, we shall consider only the case in which the axes are mutually perpendicular. The three axes, labeled x, y, and z, with a scale on each, determine a set of three numbers, called **coordinates**, associated uniquely with any point in space. Since any pair of intersecting lines determines a plane, the three pairs of axes determine three **coordinate planes**, which we shall call the **xy plane**, the **xz plane**, and the **yz plane** (see Figure 9.1a). The x *coordinate* of a point P in space is the number associated with the point on the x axis that is the intersection of the x axis and the plane through P parallel to the yz plane. The y and z *coordinates* of P are defined in a similar fashion by considering the points of intersection of the y and z axes and with planes through P parallel to the xz and xy coordinate planes, respectively (see Figure 9.1b).

The coordinate planes separate space into eight **octants**. Although we shall not number all of them, the one in which all three coordinates are positive is called the **first octant**. Note that points of the xy plane have z coordinate 0, points of the xz plane have y coordinate 0, and points of the yz plane have x coordinate 0. Similarly, points of the x axis have y and z coordinates 0, and so on. Of course the origin has all of these coordinates 0.

The two basic geometric representatives of the axes are given in Figure 9.2; 9.2a shows a **right-hand system**, while 9.2b shows a **left-hand system**. Graphs of equations in the two systems are mirror images of each other. Since we shall

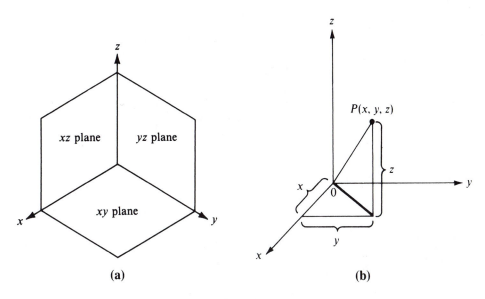

Figure 9.1

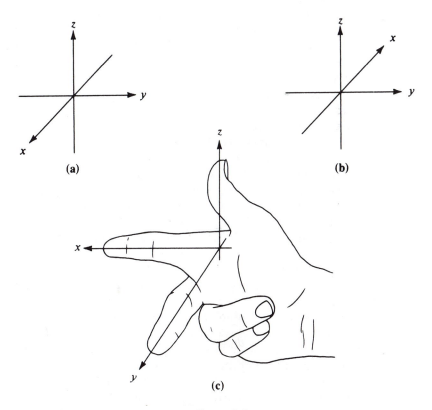

Figure 9.2

normally represent space by a right-hand system, the axes will usually appear in the positions indicated in Figure 9.2a. This is sometimes represented by the right-hand rule illustrated in Figure 9.2c. If the index and second fingers of the right hand point in the direction of the x and y axes, respectively, then the thumb points in the direction of the z axis.

Many of the formulas of solid analytic geometry are simple extensions of plane analytic geometry. The one that follows is an example.

THEOREM 9.1   *The distance between two points (x_1, y_1, z_1) and (x_2, y_2, z_2) is*

$$d = \sqrt{(x_1 - x_2)^2 + (y_1 - y_2)^2 + (z_1 - z_2)^2}.$$

PROOF   Suppose we project the points P_1 and P_2 onto the xy plane, giving Q_1 and Q_2 (see Figure 9.3). Since Q_1 and Q_2 are in the xy plane, we can use our distance formula for the plane to get

$$d_1 = \overline{Q_1 Q_2} = \sqrt{(x_2 - x_1)^2 + (y_2 - y_1)^2}.$$

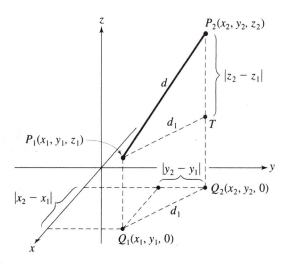

Figure 9.3

We now use the Pythagorean Theorem on triangle $P_1 T P_2$ with right angle at T.

$$d^2 = \overline{P_1 P_2}^2 = \overline{P_1 T}^2 + \overline{P_2 T}^2$$
$$= d_1^2 + |z_2 - z_1|^2$$
$$= (x_2 - x_1)^2 + (y_2 - y_1)^2 + (z_2 - z_1)^2$$
$$d = \sqrt{(x_2 - x_1)^2 + (y_2 - y_1)^2 + (z_2 - z_1)^2} \quad \blacksquare$$

If the line joining the two points is on or parallel to one of the coordinate planes, at least one of the three terms of this formula is zero, and it reduces to the plane case. Similarly, if the line joining the points is on or parallel to one of the axes, at least two terms are zero and the distance is the absolute value of the difference between the coordinates of the remaining pair.

EXAMPLE 1   Find the distance between $(1, -2, 5)$ and $(-3, 6, 4)$ (see Figure 9.4).

SOLUTION
$$d = \sqrt{(x_1 - x_2)^2 + (y_1 - y_2)^2 + (z_1 - z_2)^2}$$
$$= \sqrt{(1 + 3)^2 + (-2 - 6)^2 + (5 - 4)^2}$$
$$= \sqrt{16 + 64 + 1}$$
$$= \sqrt{81}$$
$$= 9$$

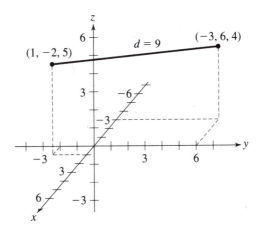

Figure 9.4

Another easy extension from two dimensions is the point-of-division formula.

THEOREM 9.2   *If $P_1 = (x_1, y_1, z_1)$, $P_2 = (x_2, y_2, z_2)$, and P is a point such that $r = \overline{P_1P}/\overline{P_1P_2}$, then the coordinates of P are*

$$x = x_1 + r(x_2 - x_1),$$
$$y = y_1 + r(y_2 - y_1),$$

and

$$z = z_1 + r(z_2 - z_1).$$

$(x, y, z) = (x_1, y_1, z_1) + r(x_2 - x_1, y_2 - y_1, z_2 - z_1)$

The proof is similar to the one for the two-dimensional case, and it is left to the student.

EXAMPLE 2   Find the point 1/3 of the way from $(-2, 4, 1)$ to $(4, 1, 7)$.

SOLUTION

$$x = x_1 + r(x_2 - x_1) = -2 + \frac{1}{3}(4 + 2) = 0$$

$$y = y_1 + r(y_2 - y_1) = 4 + \frac{1}{3}(1 - 4) = 3$$

$$z = z_1 + r(z_2 - z_1) = 1 + \frac{1}{3}(7 - 1) = 3$$

The desired point is $(0, 3, 3)$ (see Figure 9.5).

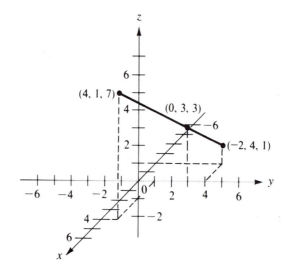

Figure 9.5

The following theorem is a direct corollary of the point-of-division formulas.

THEOREM 9.3   *If $P_1 = (x_1, y_1, z_1)$ and $P_2 = (x_2, y_2, z_2)$, then the coordinates of the midpoint of the segment P_1P_2 are*

$$x = \frac{x_1 + x_2}{2}, \qquad y = \frac{y_1 + y_2}{2}, \qquad z = \frac{z_1 + z_2}{2}.$$

EXAMPLE 3 Find the midpoint of the segment with ends $(4, -3, 1)$ and $(-2, 5, 3)$.

SOLUTION
$$x = \frac{x_1 + x_2}{2} = \frac{4 - 2}{2} = 1$$

$$y = \frac{y_1 + y_2}{2} = \frac{-3 + 5}{2} = 1$$

$$z = \frac{z_1 + z_2}{2} = \frac{1 + 3}{2} = 2$$

The point is $(1, 1, 2)$.

PROBLEMS

A *In Problems 1–10, plot both points and find the distance between them.*

1. $(2, 5, 0), (-3, 1, 3)$ **2.** $(4, -2, 1), (2, 2, -3)$

3. $(5, 4, -1), (2, 0, -1)$ **4.** $(2, 5, 3), (-2, 4, -1)$

5. $(-5, 0, 2), (4, 1, -5)$ **6.** $(-2, 5, 1), (-2, 8, 4)$

7. $(3, -1, 4), (3, 4, 4)$ **8.** $(2, 5, 0), (5, 5, 0)$

9. $(4, 7, -1), (3, -1, 3)$ **10.** $(5, 2, 3), (4, 5, -1)$

In Problems 11–16, find the point P such that $\overline{AP}/\overline{AB} = r$.

11. $A = (4, 3, -2), B = (-5, 0, 4), r = 2/3$

12. $A = (5, 2, 3), B = (-5, 7, -2), r = 2/5$

13. $A = (-2, 0, 1), B = (10, 8, 5), r = 1/4$

14. $A = (5, 5, 3), B = (2, -4, 0), r = 1/3$

15. $A = (3, 1, 5), B = (-3, 4, 2), r = 2$

16. $A = (-2, 5, 1), B = (4, -1, 2), r = 3/2$

In Problems 17–20, find the midpoint of segment AB.

17. $A = (5, -2, 3), B = (-3, 4, 7)$

18. $A = (4, 3, 5), B = (-2, -1, 2)$

19. $A = (-3, 2, 0), B = (5, 4, 3)$

20. $A = (4, 3, -1), B = (4, 8, -3)$

B **21.** Given $A = (5, -2, 3), P = (6, 0, 0)$, and $\overline{AP}/\overline{AB} = 1/3$, find B.

22. Given $B = (-4, 14, 4), P = (-1, 8, -4)$, and $\overline{AP}/\overline{AB} = 2/5$, find A.

23. Given $B = (6, 0, 9)$, $P = (4, 1, 6)$, and $\overline{AP}/\overline{AB} = 3/4$, find A.

24. Given $A = (5, 3, -2)$, $P = (1, 5, 2)$, and $\overline{AP}/\overline{AB} = 2/3$, find B.

In Problems 25–30, find the unknown quantity.

25. $A = (5, 1, 0)$, $B = (1, y, 2)$, $\overline{AB} = 6$

26. $A = (-2, 4, 3)$, $B = (x, -4, 2)$, $\overline{AB} = 9$

27. $A = (x, 4, -2)$, $B = (-x, -6, 3)$, $\overline{AB} = 15$

28. $A = (x, x, 5)$, $B = (-1, -2, 0)$, $\overline{AB} = 5\sqrt{2}$

29. $A = (x, x, x)$, $B = (1, -3, -1)$, $\overline{AB} = 2\sqrt{5}$

30. $A = (x, 2x, 3)$, $B = (1, 1, 2)$, $\overline{AB} = \sqrt{2}$

31. The point $(-1, 5, 2)$ is a distance 6 from the midpoint of the segment joining $(1, 3, 2)$ and $(x, -1, 6)$. Find x.

32. The point $(1, -2, 9)$ is a distance $5\sqrt{5}$ from the midpoint of the segment joining $(1, y, 2)$ and $(5, -1, 6)$. Find y.

C **33.** Prove Theorem 9.2.   **34.** Prove Theorem 9.3.

9.2

VECTORS IN SPACE

Vectors in three-dimensional space may be handled in much the same way as vectors in the plane. Vectors themselves, the sum and difference of two vectors, the absolute value of a vector, and the scalar multiple of a vector are defined in the same way as they were in Chapter 2; and Theorem 2.2 (see page 53) holds for vectors in space as well as for vectors in the plane. The following theorems and definitions are the three-dimensional analogs of theorems and definitions of Chapter 2. The proofs of the theorems are simple extensions of the corresponding arguments in two dimensions.

DEFINITION *If $O = (0, 0, 0)$, $X = (1, 0, 0)$, $Y = (0, 1, 0)$, and $Z = (0, 0, 1)$, then the vectors represented by $\overrightarrow{OX}$, $\overrightarrow{OY}$, and $\overrightarrow{OZ}$ are **i**, **j**, and **k**, respectively, and are called **basis vectors**.*

THEOREM 9.4 *Every vector in space can be written in the form*

$$a\mathbf{i} + b\mathbf{j} + c\mathbf{k}$$

in one and only one way. The numbers a, b, and c are called the first, second, and third components, respectively, of the vector.

THEOREM 9.5 If $\overrightarrow{AB}$, where $A = (x_1, y_1, z_1)$ and $B = (x_2, y_2, z_2)$, represents a vector $\mathbf{v}$ in space, then

$$\mathbf{v} = (x_2 - x_1)\mathbf{i} + (y_2 - y_1)\mathbf{j} + (z_2 - z_1)\mathbf{k}.$$

THEOREM 9.6
$$(a_1\mathbf{i} + b_1\mathbf{j} + c_1\mathbf{k}) + (a_2\mathbf{i} + b_2\mathbf{j} + c_2\mathbf{k}) = (a_1 + a_2)\mathbf{i} + (b_1 + b_2)\mathbf{j} + (c_1 + c_2)\mathbf{k}$$

$$(a_1\mathbf{i} + b_1\mathbf{j} + c_1\mathbf{k}) - (a_2\mathbf{i} + b_2\mathbf{j} + c_2\mathbf{k}) = (a_1 - a_2)\mathbf{i} + (b_1 - b_2)\mathbf{j} + (c_1 - c_2)\mathbf{k}$$

$$d(a\mathbf{i} + b\mathbf{j} + c\mathbf{k}) = da\mathbf{i} + db\mathbf{j} + dc\mathbf{k}$$

$$|a\mathbf{i} + b\mathbf{j} + c\mathbf{k}| = \sqrt{a^2 + b^2 + c^2}$$

DEFINITION If $\mathbf{u} = a_1\mathbf{i} + b_1\mathbf{j} + c_1\mathbf{k}$ and $\mathbf{v} = a_2\mathbf{i} + b_2\mathbf{j} + c_2\mathbf{k}$, then the **dot product** (scalar product, inner product) of $\mathbf{u}$ and $\mathbf{v}$ is

$$\mathbf{u} \cdot \mathbf{v} = a_1a_2 + b_1b_2 + c_1c_2.$$

Theorems 2.6–2.9 (pages 62–68) still hold for three-dimensional vectors. They are restated here for convenience.

THEOREM 9.7 If $\mathbf{u}$, $\mathbf{v}$, and $\mathbf{w}$ are vectors and k is a scalar, then

(a) $\mathbf{u} \cdot \mathbf{v} = \mathbf{v} \cdot \mathbf{u}$

(b) $(\mathbf{u} + \mathbf{v}) \cdot \mathbf{w} = \mathbf{u} \cdot \mathbf{w} + \mathbf{v} \cdot \mathbf{w}$

(c) $k(\mathbf{u} \cdot \mathbf{v}) = (k\mathbf{u}) \cdot \mathbf{v} = \mathbf{u} \cdot (k\mathbf{v})$

(d) $\mathbf{0} \cdot \mathbf{u} = 0$

(e) $\mathbf{u} \cdot \mathbf{u} = |\mathbf{u}|^2.$

DEFINITION The angle between two nonzero vectors $\mathbf{u}$ and $\mathbf{v}$ is the smaller angle between the representatives of $\mathbf{u}$ and $\mathbf{v}$ having their tails at the origin.

THEOREM 9.8 If $\mathbf{u}$ and $\mathbf{v}$ are vectors and θ is the angle between them, then

$$\mathbf{u} \cdot \mathbf{v} = |\mathbf{u}||\mathbf{v}| \cos \theta.$$

THEOREM 9.9 The vectors $\mathbf{u}$ and $\mathbf{v}$ (not both $\mathbf{0}$) are orthogonal (perpendicular) if and only if $\mathbf{u} \cdot \mathbf{v} = 0$ (the zero vector is taken to be orthogonal to every vector).

DEFINITION Suppose the nonzero vectors $\mathbf{u}$ and $\mathbf{v}$ are represented by $\overrightarrow{AB}$ and $\overrightarrow{AC}$, respectively. Then the **projection** of $\mathbf{u}$ on $\mathbf{v}$ is the vector $\mathbf{w}$ represented by $\overrightarrow{AD}$, where D is on the line AC and $BD \perp AC$ (see Figure 2.16, page 67).

THEOREM 9.10 *If* **w** *is the projection of* **u** *on* **v**, *then*

$$|\mathbf{w}| = \frac{|\mathbf{u} \cdot \mathbf{v}|}{|\mathbf{v}|} \quad and \quad \mathbf{w} = \left(\frac{\mathbf{u} \cdot \mathbf{v}}{|\mathbf{v}|}\right)\frac{\mathbf{v}}{|\mathbf{v}|} = \frac{\mathbf{u} \cdot \mathbf{v}}{|\mathbf{v}|^2}\mathbf{v}.$$

Theorems 9.7–9.10 are not stated in terms of the dimensions of the vectors; their proofs are identical to those of Chapter 2. Proofs of Theorems 9.4–9.6 are left to the student (see Problems 29–31).

EXAMPLE 1 Given **u** = 2**i** + **j** − 3**k** and **v** = **i** − 2**j** − **k**, find **u** + **v**, **u** − **v**, and **u** · **v**.

SOLUTION
$$\mathbf{u} + \mathbf{v} = (2 + 1)\mathbf{i} + (1 - 2)\mathbf{j} + (-3 - 1)\mathbf{k} = 3\mathbf{i} - \mathbf{j} - 4\mathbf{k}$$
$$\mathbf{u} - \mathbf{v} = (2 - 1)\mathbf{i} + (1 + 2)\mathbf{j} + (-3 + 1)\mathbf{k} = \mathbf{i} + 3\mathbf{j} - 2\mathbf{k}$$
$$\mathbf{u} \cdot \mathbf{v} = 2 \cdot 1 + 1(-2) + (-3)(-1) = 3$$

EXAMPLE 2 Give in component form the vector **v** that is represented by $\overrightarrow{AB}$, where $A = (4, 3, -1)$ and $B = (-1, 2, -3)$ (see Figure 9.6).

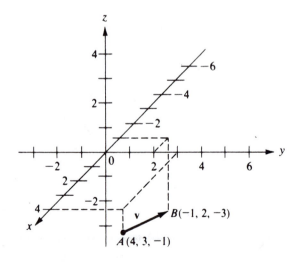

Figure 9.6

SOLUTION
$$\mathbf{v} = (-1 - 4)\mathbf{i} + (2 - 3)\mathbf{j} + (-3 + 1)\mathbf{k} = -5\mathbf{i} - \mathbf{j} - 2\mathbf{k}$$

EXAMPLE 3 Find the endpoints of the representative $\overrightarrow{AB}$ of **v** if **v** = $2\mathbf{i} - 4\mathbf{j} + \mathbf{k}$ and $(2, -3, 5)$ is the midpoint of AB (see Figure 9.7).

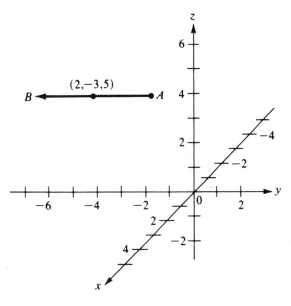

Figure 9.7

SOLUTION If $A = (x_1, y_1, z_1)$ and $B = (x_2, y_2, z_2)$, then, by Theorem 9.5,

$$x_2 - x_1 = 2, \qquad y_2 - y_1 = -4, \qquad \text{and} \qquad z_2 - z_1 = 1.$$

By Theorem 9.3,

$$\frac{x_1 + x_2}{2} = 2, \qquad \frac{y_1 + y_2}{2} = -3, \qquad \text{and} \qquad \frac{z_1 + z_2}{2} = 5.$$

Solving simultaneously, we have

$$A = \left(1, -1, \frac{9}{2}\right) \qquad \text{and} \qquad B = \left(3, -5, \frac{11}{2}\right).$$

EXAMPLE 4 Determine whether the vectors $\mathbf{u} = 3\mathbf{i} + \mathbf{j} - 2\mathbf{k}$ and $\mathbf{v} = 2\mathbf{i} - 4\mathbf{j} + \mathbf{k}$ are orthogonal.

SOLUTION

$$\mathbf{u} \cdot \mathbf{v} = (3)(2) + (1)(-4) + (-2)(1) = 0$$

Thus **u** and **v** are orthogonal.

EXAMPLE 5   Find the projection **w** of $\mathbf{u} = 4\mathbf{i} - \mathbf{j} + \mathbf{k}$ upon $\mathbf{v} = 3\mathbf{i} + \mathbf{j} - 4\mathbf{k}$ (see Figure 9.8).

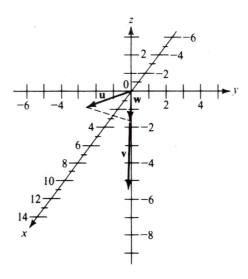

Figure 9.8

SOLUTION

$$\mathbf{u} \cdot \mathbf{v} = (4)(3) + (-1)(1) + (1)(-4) = 7$$

and

$$|\mathbf{v}| = \sqrt{3^2 + 1^2 + (-4)^2} = \sqrt{26}.$$

Thus

$$\mathbf{w} = \left(\frac{\mathbf{u} \cdot \mathbf{v}}{|\mathbf{v}|}\right)\frac{\mathbf{v}}{|\mathbf{v}|}$$

$$= \frac{7}{\sqrt{26}}\frac{3\mathbf{i} + \mathbf{j} - 4\mathbf{k}}{\sqrt{26}}$$

$$= \frac{21}{26}\mathbf{i} + \frac{7}{26}\mathbf{j} - \frac{14}{13}\mathbf{k}.$$

PROBLEMS

A   *In Problems 1–4, give in component form the vector **v** that is represented by $\overrightarrow{AB}$.*

1. $A = (2, 3, -5)$, $B = (-4, 1, 2)$ **2.** $A = (3, -2, 4)$, $B = (5, 4, -1)$

3. $A = (5, 0, -2)$, $B = (2, -4, 1)$ **4.** $A = (2, -3, 8)$, $B = (2, 5, 2)$

*In Problems 5–8, give the unit vector in the direction of **v**.*

5. $\mathbf{v} = 4\mathbf{i} + \mathbf{j} - 2\mathbf{k}$ **6.** $\mathbf{v} = \mathbf{i} - 2\mathbf{j} + 2\mathbf{k}$

7. $\mathbf{v} = \mathbf{i} + 5\mathbf{j} - 3\mathbf{k}$ **8.** $\mathbf{v} = 3\mathbf{i} - 4\mathbf{k}$

B   *In Problems 9–14, find the endpoints of the representative $\overrightarrow{AB}$ of **v** from the given information.*

9. $\mathbf{v} = 2\mathbf{i} - \mathbf{j} + 3\mathbf{k}$, $A = (2, 1, 5)$ **10.** $\mathbf{v} = 3\mathbf{i} + \mathbf{j} - 4\mathbf{k}$, $A = (1, 4, 3)$

11. $\mathbf{v} = -\mathbf{i} + 2\mathbf{j} + 5\mathbf{k}$, $B = (2, 3, 8)$ **12.** $\mathbf{v} = 2\mathbf{i} + 5\mathbf{j} - 3\mathbf{k}$, $B = (4, -2, 6)$

13. $\mathbf{v} = 4\mathbf{i} - 2\mathbf{j} + \mathbf{k}$, $(2, 5, -1)$ is the midpoint of AB

14. $\mathbf{v} = 6\mathbf{i} + \mathbf{j} - 4\mathbf{k}$, $(3, 2, -5)$ is the midpoint of AB

In Problems 15–18, find the angle θ between the given vectors.

15. $\mathbf{u} = \mathbf{i} + \mathbf{j} + 2\mathbf{k}$, $\mathbf{v} = 2\mathbf{i} - \mathbf{j} + \mathbf{k}$ **16.** $\mathbf{u} = 2\mathbf{i} - 2\mathbf{j} - \mathbf{k}$, $\mathbf{v} = -\mathbf{i} + 4\mathbf{j} + 2\mathbf{k}$

17. $\mathbf{u} = 5\mathbf{i} - \mathbf{j} + 3\mathbf{k}$, $\mathbf{v} = 4\mathbf{i} + 5\mathbf{j} - 2\mathbf{k}$ **18.** $\mathbf{u} = 2\mathbf{i} + 4\mathbf{j} + 4\mathbf{k}$, $\mathbf{v} = 4\mathbf{i} - 3\mathbf{k}$

In Problems 19–22, find $\mathbf{u} + \mathbf{v}$, $\mathbf{u} - \mathbf{v}$, and $\mathbf{u} \cdot \mathbf{v}$. Indicate whether $\mathbf{u}$ and $\mathbf{v}$ are orthogonal.

19. $\mathbf{u} = \mathbf{i} - 2\mathbf{j} + 5\mathbf{k}$, $\mathbf{v} = 2\mathbf{i} + 4\mathbf{j} + \mathbf{k}$ **20.** $\mathbf{u} = 3\mathbf{i} + \mathbf{j} - 4\mathbf{k}$, $\mathbf{v} = 2\mathbf{i} + 6\mathbf{j} + 3\mathbf{k}$

21. $\mathbf{u} = 2\mathbf{i} - \mathbf{j} + 6\mathbf{k}$, $\mathbf{v} = \mathbf{i} - 4\mathbf{j} - \mathbf{k}$ **22.** $\mathbf{u} = 4\mathbf{i} + 3\mathbf{j} - \mathbf{k}$, $\mathbf{v} = \mathbf{i} + 2\mathbf{j} + 3\mathbf{k}$

In Problems 23–28, find the projection of $\mathbf{u}$ upon $\mathbf{v}$.

23. $\mathbf{u} = 4\mathbf{i} + \mathbf{j} - \mathbf{k}$, $\mathbf{v} = \mathbf{i} + \mathbf{j} + \mathbf{k}$ **24.** $\mathbf{u} = \mathbf{i} - 2\mathbf{j} + 4\mathbf{k}$, $\mathbf{v} = 2\mathbf{j} + 3\mathbf{k}$

25. $\mathbf{u} = 4\mathbf{i} - 2\mathbf{j} - \mathbf{k}$, $\mathbf{v} = \mathbf{i} - 2\mathbf{j} + \mathbf{k}$ **26.** $\mathbf{u} = 2\mathbf{i} + \mathbf{j}$, $\mathbf{v} = \mathbf{j} - 2\mathbf{k}$

27. $\mathbf{u} = 3\mathbf{i} + \mathbf{j} - 2\mathbf{k}$, $\mathbf{v} = \mathbf{i} + 2\mathbf{j} + \mathbf{k}$ **28.** $\mathbf{u} = 2\mathbf{i} - 3\mathbf{j} + 2\mathbf{k}$, $\mathbf{v} = 3\mathbf{i} + 4\mathbf{k}$

C   **29.** Prove Theorem 9.4.   **30.** Prove Theorem 9.5.   **31.** Prove Theorem 9.6.

32. Prove that $(\mathbf{u} + \mathbf{v}) \cdot (\mathbf{u} + \mathbf{v}) = |\mathbf{u}|^2 + 2\mathbf{u} \cdot \mathbf{v} + |\mathbf{v}|^2$.

33. Prove the triangle inequality $|\mathbf{u} + \mathbf{v}| \leq |\mathbf{u}| + |\mathbf{v}|$.

34. Prove $|\mathbf{u} - \mathbf{v}| \geq ||\mathbf{u}| - |\mathbf{v}||$.

DIRECTION ANGLES, COSINES, AND NUMBERS

In plane geometry, we used the inclination and slope to give the direction of a line. The corresponding terms in solid analytic geometry are direction angles and direction cosines. They are most easily defined for vectors.

DEFINITION *If **v** is a vector, then the ordered set $\{\alpha, \beta, \gamma\}$ is the set of **direction angles*** for **v** if α is the angle between **v** and **i**, β is the angle between **v** and **j**, and γ is the angle between **v** and **k**.*

The direction angles for a vector **v** are illustrated in Figure 9.9. Note that the direction angles are not necessarily in the coordinate planes. The angle α is in the plane determined by **v** and **i**; β and γ are in planes determined by **v** and **j** and by **v** and **k**, respectively.

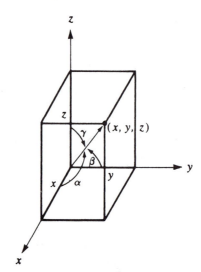

Figure 9.9

The angle between two vectors was defined in the previous section. Note again that this is not a directed angle; in fact, we have given no convention for positive and negative angles in space. Thus direction angles are never negative and never greater than 180°.

Just as the inclination of a line is less convenient than the slope, so the

---

*We shall, for simplicity, use the term "set of direction angles" or simply "direction angles" rather than the more cumbersome "ordered set of direction angles." This same convention will be used for direction cosines and direction numbers.

direction angles for a vector **v** are relatively inconvenient. More convenient to use are the cosines of the direction angles.

DEFINITION *If* **v** *is a vector, then the ordered set* $\{l, m, n\}$ *is the set of* **direction cosines** *for* **v** *if* $l = \cos\alpha$, $m = \cos\beta$, *and* $n = \cos\gamma$, *where* $\{\alpha, \beta, \gamma\}$ *is the set of direction angles for* **v**.

If $\overrightarrow{OP}$ is a representative of **v** with O the origin and P the point (x, y, z) (see Figure 9.9), then

$$l = \cos\alpha = \frac{x}{\rho}, \quad m = \cos\beta = \frac{y}{\rho}, \quad n = \cos\gamma = \frac{z}{\rho},$$

where $\rho = \sqrt{x^2 + y^2 + z^2}$. It is this relation to the coordinates of a point on the line which allows us to prove the next theorem.

THEOREM 9.11 *If* $\{l, m, n\}$ *is a set of direction cosines for a vector, then*

$$l^2 + m^2 + n^2 = 1.$$

PROOF From the relations above, we have

$$l^2 + m^2 + n^2 = \frac{x^2}{\rho^2} + \frac{y^2}{\rho^2} + \frac{z^2}{\rho^2}$$

$$= \frac{x^2 + y^2 + z^2}{\rho^2} = 1. \ \blacksquare$$

For most purposes, an even more convenient set of numbers is a set of direction numbers for a line.

DEFINITION $\{a, b, c\}$ *is a set of* **direction numbers** *for the vector* **v** *if there is a nonzero constant* k *such that* $a = kl$, $b = km$, *and* $c = kn$, *where* $\{l, m, n\}$ *is the set of direction cosines for* **v**.

Theorem 9.11 may be used to find direction cosines of a vector if its direction numbers are known. Of course, additional information is needed to get the proper signs.

EXAMPLE 1 Given that a vector **v** has direction numbers $\{4, 1, -2\}$ and is directed upward, find its direction cosines.

$$\{-4, -1, 2\}$$

SOLUTION
$$a^2 + b^2 + c^2 = k^2l^2 + k^2m^2 + k^2n^2$$
$$= k^2(l^2 + m^2 + n^2) = k^2$$

Thus,
$$k^2 = 16 + 1 + 4 = 21 \quad \text{and} \quad k = \pm\sqrt{21}.$$

Since **v** is directed upward, $\gamma < 90°$ and $n > 0$. Thus $k = -\sqrt{21}$, and

$$l = -\frac{4}{\sqrt{21}}, \quad m = -\frac{1}{\sqrt{21}}, \quad \text{and} \quad n = \frac{2}{\sqrt{21}}.$$

The following theorem is a direct consequence of the foregoing definitions.

THEOREM 9.12 *The vector* $\mathbf{v} = a\mathbf{i} + b\mathbf{j} + c\mathbf{k}$ *has direction numbers* $\{a, b, c\}$.

EXAMPLE 2 Give a set of direction numbers for the vector **v** represented by $\overrightarrow{AB}$, where $A = (3, -1, 2)$ and $B = (5, 2, -1)$.

SOLUTION By Theorem 9.5,

$$\mathbf{v} = (5 - 3)\mathbf{i} + (2 + 1)\mathbf{j} + (-1 - 2)\mathbf{k} = 2\mathbf{i} + 3\mathbf{j} - 3\mathbf{k}.$$

By Theorem 9.12, one set of direction numbers for **v** is $(2, 3, -3\}$.

Direction angles, cosines, and numbers of vectors carry over directly to lines.

DEFINITION *The statement "A vector* **v** *is directed along a line l" means that a representative of* **v** *is on l.*

DEFINITION *A set of direction angles, cosines, or numbers for a line l is any set of direction angles, cosines, or numbers, respectively, for any vector* **v** *directed along l.*

Note that every line has two sets of direction angles and two sets of direction cosines, corresponding to the two possible directions on the line. It is easily seen that if $\{l, m, n\}$ is one set of direction cosines for a line, then the other is $\{-l, -m, -n\}$.

EXAMPLE 3 Find the two sets of direction cosines and direction angles for a line if $\{1, 2, 2\}$ is a set of direction numbers for it.

SOLUTION Since the three numbers given are a number k times the direction cosines, we have the following.

$$kl = 1, \quad km = 2, \quad \text{and} \quad kn = 2$$
$$k^2l^2 + k^2m^2 + k^2n^2 = 1 + 4 + 4 = 9$$
$$k^2(l^2 + m^2 + n^2) = 9$$
$$k^2 = 9 \quad \text{(by Theorem 9.11)}$$
$$k = \pm 3$$

Thus, the two possible sets of direction cosines are $\{1/3, 2/3, 2/3\}$ and $\{-1/3, -2/3, -2/3\}$, and they give approximate direction angles $\{71°, 48°, 48°\}$ and $\{109°, 132°, 132°\}$, respectively.

EXAMPLE 4 Suppose a line has direction numbers $\{2, -4, 1\}$ and contains the point $(1, 3, 4)$. Find another point on the line.

SOLUTION By Theorems 9.5 and 9.12, we have

$$a = x_2 - x_1 \qquad b = y_2 - y_1, \qquad \text{and} \qquad c = z_2 - z_1,$$

or

$$x_2 = x_1 + a, \qquad y_2 = y_1 + b, \qquad \text{and} \qquad z_2 = z_1 + c.$$

Thus,

$$x_2 = 1 + 2 = 3, \qquad y_2 = 3 - 4 = -1, \qquad \text{and} \qquad z_2 = 4 + 1 = 5,$$

which give the point $(3, -1, 5)$ (see Figure 9.10). Of course, any nonzero multiple of the given direction numbers gives another set of direction numbers; these may be used to find other points on the line.

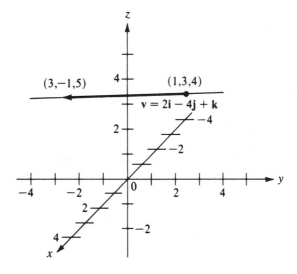

Figure 9.10

Perhaps you feel a bit uneasy about this last example. How do we know that, of all possible sets of direction numbers, the one we were given is *that* one—that is, the set of direction numbers determined by the method of Example 2? The answer is simple—all of them are. Here is the reason. No matter what pair of points we choose on a line, we must get one of the two possible sets of direction cosines. Taking one of those two sets, we can get from it any set of direction numbers by multiplying by the proper number k. In particular, if we take a set of direction cosines for the line of Example 4 and multiply it by the proper number k, we get the direction numbers $\{2, -4, 1\}$. Now there are two points on the line at a distance $|k|$ from the point $(1, 3, 4)$. By choosing the proper one of those two points, we see that that point, together with $(1, 3, 4)$, would give the direction numbers $\{2, -4, 1\}$. Of course, this argument could be repeated for any point and set of direction numbers.

It might be noted that we could, say, double the direction numbers of Example 4 to give another set of direction numbers, $\{4, -8, 2\}$. This, together with the point $(1, 3, 4)$, gives the point $(5, -5, 6)$, which is also on the line. This could be repeated indefinitely to get as many points on the line as we choose.

It is clear that if two lines are parallel and directed the same way, they must have the same set of direction angles and, thus, the same set of direction cosines. If they are parallel and have opposite directions, their direction angles are supplementary and one set of direction cosines is the negative of the other. Thus, any set of direction numbers for one line is proportional to a set of direction numbers for the other. Furthermore, this chain of reasoning can be reversed to show that if two lines have proportional sets of direction numbers, they are parallel.

THEOREM 9.13   *Two distinct lines are parallel if and only if sets of direction numbers for the two lines are proportional.*

Suppose that lines l_1 and l_2 have direction numbers $\{a_1, b_1, c_1\}$ and $\{a_2, b_2, c_2\}$, respectively. Then vectors $\mathbf{v}_1$ and $\mathbf{v}_2$ directed along lines l_1 and l_2, respectively, may be represented by

$$\mathbf{v}_1 = a_1\mathbf{i} + b_1\mathbf{j} + c_1\mathbf{k} \quad \text{and} \quad \mathbf{v}_2 = a_2\mathbf{i} + b_2\mathbf{j} + c_2k.$$

By Theorem 9.8, $\mathbf{v}_1$ and $\mathbf{v}_2$ are orthogonal if and only if

$$\mathbf{v}_1 \cdot \mathbf{v}_2 = a_1a_2 + b_1b_2 + c_1c_2 = 0.$$

This gives the following theorem for perpendicularity of lines.

THEOREM 9.14   *Two lines with direction numbers $\{a_1, b_1, c_1\}$ and $\{a_2, b_2, c_2\}$ are perpendicular if and only if*

$$a_1a_2 + b_1b_2 + c_1c_2 = 0.$$

EXAMPLE 5   Line l_1 contains the points $(1, 2, 5)$ and $(3, -3, 1)$; l_2 contains $(2, 1, -2)$ and $(0, 6, 2)$. Are l_1 and l_2 parallel, perpendicular, coincident, or none of these?

SOLUTION   For l_1 a set of direction numbers is

$$a = 1 - 3 = -2, \qquad b = 2 + 3 = 5, \qquad \text{and} \qquad c = 5 - 1 = 4.$$

For l_2, we have

$$a = 2 - 0 = 2, \qquad b = 1 - 6 = -5, \qquad \text{and} \qquad c = -2 - 2 = -4.$$

Since the two sets of direction numbers are proportional, the lines are either coincident or parallel. Now let us take the point $(1, 2, 5)$ from l_1 and $(2, 1, -2)$ from l_2. If $l_1 = l_2$, these points must give another set of direction numbers proportional to those above. However,

$$a = 1 - 2 = -1, \qquad b = 2 - 1 = 1, \qquad \text{and} \qquad c = 5 + 2 = 7,$$

showing that $l_1 \neq l_2$. Therefore l_1 and l_2 must be parallel.

PROBLEMS

A   *In Problems 1–6, find the set of direction angles for the vector described.*

1. Direction numbers $\{1, 4, 8\}$; directed to the right of the xz plane
2. Direction numbers $\{4, -4, 2\}$; directed to the right of the xz plane
3. Direction numbers $\{1, 2, -4\}$; directed behind the yz plane
4. Direction numbers $\{2, -1, -3\}$; directed above the xy plane
5. Direction numbers $\{1, 1, 1\}$; directed behind the yz plane
6. Direction numbers $\{1, -1, 0\}$; directed to the right of the xz plane

In Problems 7–12, find a set of direction numbers for the lines containing the two given points.

7. $(1, 4, 3)$ and $(5, 2, -1)$
8. $(2, 0, -4)$ and $(-1, 2, 3)$
9. $(2, 2, 1)$ and $(0, 0, 3)$
10. $(3, 5, -2)$ and $(-1, 4, 4)$
11. $(0, 0, 0)$ and $(5, 1, -2)$
12. $(-1, 4, 5)$ and $(3, -4, 0)$

B   *In Problems 13–18, find two more points on the line.*

13. Direction numbers $\{1, 5, 2\}$; containing $(2, 3, -1)$
14. Direction numbers $\{1, 4, 0\}$; containing $(-2, 1, 1)$
15. Direction numbers $\{2, 1, 2\}$; containing $(1, 3, 3)$
16. Direction numbers $\{1, 1, 1\}$; containing $(2, 4, -1)$

17. Direction numbers $\{4, 0, -1\}$; containing $(1, 3, -1\}$

18. Direction numbers $\{4, 4, 3\}$; containing $\{-4, -4, -3\}$

19. Give the direction angles and direction cosines for the coordinate axes with their usual directions.

20. Give a set of formulas for finding all points on the line described in Example 4. [*Hint:* Consider the two paragraphs following Example 4.]

In Problems 21–31, two lines are described by a pair of points on each. Indicate whether the lines are parallel, perpendicular, coincident, or none of these.

21. $(3, 4, 1), (4, 8, -1); (2, 3, -5), (0, -5, -1)$

22. $(2, 1, 5), (3, 3, -1); (4, 2, 10), (1, -4, 5)$

23. $(4, 1, -4), (3, 2, 1); (4, 1, -4), (11, 3, -3)$

24. $(4, 2, -1), (7, 6, 2); (5, 10, 3), (-4, -2, -6)$

25. $(2, 1, 4), (4, -3, 12); (1, 3, 0), (6, -7, 20)$

26. $(4, 5, 1), (3, 2, -4); (4, 1, 2), (5, -1, 3)$

27. $(2, 3, 1), (4, -2, 2); (1, 0, 3), (3, -3, 1)$

28. $(3, 1, 4), (4, 3, 3); (5, 5, 2), (0, -5, 7)$

29. $(2, 1, 3), (5, -1, 1); (3, 4, -1), (5, 3, 3)$

30. $(4, 4, -3), (1, 3, -1); (2, 1, 5), (8, 3, 1)$

31. $(4, -2, 1), (10, -4, 9); (7, -3, 5), (-2, 0, -7)$

9.4

THE LINE

Recall that in Chapter 8 (pages 309–310) we derived a parametric representation for a line containing two given points. The result was basically the point-of-division formulas with r as the parameter. We used a vector argument there. The same argument holds in three dimensions; but instead of using two points, let us consider the line l with direction numbers $\{a, b, c\}$ and containing the point $P_0 = (x_0, y_0, z_0)$. Let $P = (x, y, z)$ be any point on l. If $\mathbf{u}$ and $\mathbf{w}$ are the vectors represented by $\overrightarrow{OP_0}$ and $\overrightarrow{OP}$, respectively (see Figure 9.11), it is seen that

$$\mathbf{u} = x_0\mathbf{i} + y_0\mathbf{j} + z_0\mathbf{k} \qquad \text{and} \qquad \mathbf{w} = x\mathbf{i} + y\mathbf{j} + z\mathbf{k}.$$

Since $\{a, b, c\}$ is a set of direction numbers for l,

$$\mathbf{v} = a\mathbf{i} + b\mathbf{j} + c\mathbf{k}$$

is a vector lying along l. But $\overrightarrow{P_0P}$ represents another vector lying along l; it must be a scalar multiple, $t\mathbf{v}$, of $\mathbf{v}$. Since

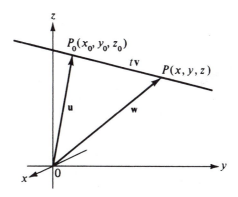

Figure 9.11

$$\mathbf{w} = \mathbf{u} + t\mathbf{v},$$

it follows that

$$x = x_0 + ta,$$

$$y = y_0 = tb,$$

and

$$z = z_0 + tc.$$

THEOREM 9.15 *A parametric representation of the line containing (x_0, y_0, z_0) and having direction numbers $\{a, b, c\}$ is*

$$x = x_0 + at, \qquad y = y_0 + bt, \qquad z = z_0 + ct.$$

The following nonvector argument may also be used to establish this result. Given a set of direction numbers for a line and one point on that line, we can find another simply by adding. Furthermore, we can find other sets of direction numbers by taking a multiple of the original set. Thus if the point given is (x_0, y_0, z_0) and the set of direction numbers given is $\{a, b, c\}$, then any point (x, y, z) such that

$$x = x_0 + at,$$

$$y = y_0 + bt,$$

$$z = z_0 + ct,$$

where t is a real number, is on the given line. Furthermore, if (x, y, z) is a point on the line different from (x_0, y_0, z_0), then a set of direction numbers for the line is $\{x - x_0, y - y_0, z - z_0\}$. These must be a multiple of the given set of direction numbers $\{a, b, c\}$; that is, for some t,

$$x - x_0 = at,$$
$$y - y_0 = bt,$$
$$z - z_0 = ct.$$

These equations hold not only for every point on the line different from (x_0, y_0, z_0) but also for (x_0, y_0, z_0). Thus a point is on the given line if and only if it satisfies the set of equations given above.

EXAMPLE 1 Find a parametric representation for the line containing $(1, 3, -2)$ and having direction numbers $\{3, 2, -1\}$.

SOLUTION
$$x = 1 + 3t, \qquad y = 3 + 2t, \qquad z = -2 - t$$

EXAMPLE 2 Find a parametric representation of the line containing $(4, 2, -1)$ and $(0, 2, 3)$.

SOLUTION A set of direction numbers is $\{4 - 0, 2 - 2, -1 - 3\} = \{4, 0, -4\}$. Thus the line is

$$x = 4 + 4t, \qquad y = 2, \qquad z = -1 - 4t.$$

Once we have the direction numbers, we may use them with either of the two given points. Thus, another representation of the line in Example 2 is

$$x = 4s, \qquad y = 2, \qquad z = 3 - 4s.$$

Although this does not look much like the first representation, it easily seen that they are the same. For instance, $t = 0$ gives the point $(4, 2, -1)$, as does $s = 1$; $t = -1$ gives the point $(0, 2, 3)$, as does $s = 0$, and so forth.

In fact,

$$
\begin{aligned}
x &= 4 + 4t & y &= 2, & z &= -1 - 4t \\
&= 4(t + 1) & & & &= 3 - 4 - 4t \\
&= 4s, & & & &= 3 - 4(t + 1) \\
& & & & &= 3 - 4s,
\end{aligned}
$$

where $s = t + 1$. Thus, whatever point we get using a value of t can be found by choosing $s = t + 1$.

A simpler set of direction numbers can also be found. Since the ones we have are all multiples of 4, we can multiply through by 1/4 to get another set of direction numbers, $\{1, 0, -1\}$. Using these with the first point gives

$$x = 4 + u, \qquad y = 2, \qquad z = -1 - u.$$

Again, we see that $4t = u$, so the two representations are equivalent.

Perhaps you wonder what is needed to be able to say that two parametric representations are equivalent. If a value of t and another of s both give the same point, then, for those values of t and s, the three coordinates must be equal. Eliminating x, y, and z between the two parametric representations gives three equations in t and s (in some of these, the parameters may both be absent, as they are in the representation of y here). If all give the same result when they are solved for one parameter in terms of the other, and if the domain and range are the same, then the representations are equivalent.

Suppose we eliminate the parameter in the representation given by Theorem 9.15. If none of the direction numbers is zero, we can solve each equation for t and set them equal to each other. This gives

$$\frac{x - x_0}{a} = \frac{y - y_0}{b} = \frac{z - z_0}{c}.$$

Actually, this is just a shorter way of writing the three equations

$$\frac{x - x_0}{a} = \frac{y - y_0}{b},$$

$$\frac{y - y_0}{b} = \frac{z - z_0}{c},$$

and

$$\frac{x - x_0}{a} = \frac{z - z_0}{c}.$$

But these three equations are not independent—the last can be found from the first two. Let us discard it and consider only the first two, which, as we shall see in the next section, represent planes. Any point that satisfies both equations is on both planes and therefore on the intersection of the two planes, which is a line. Thus this representation of a line gives it as the intersection of two planes. It might be noted that the equation we discarded is also a plane containing the same line. These three planes can be seen in Figure 9.12. The line PQ is projected upon each of the three coordinate planes. In each case, this projection—together with the original line—determines a plane. The plane

$$\frac{x - x_0}{a} = \frac{y - y_0}{b}$$

is the plane determined by PQ and the projection, $P_1 Q_1$, of PQ on the xy plane.

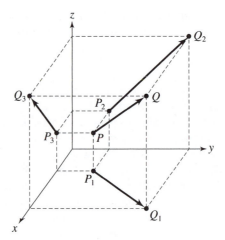

Figure 9.12

The z axis is either on or parallel to this plane. Similarly,

$$\frac{y - y_0}{b} = \frac{z - z_0}{c}$$

is the plane determined by PQ and the projection, P_2Q_2, of PQ on the yz plane; and

$$\frac{x - x_0}{a} = \frac{z - z_0}{b}$$

is the plane determined by PQ and the projection, P_3Q_3, of PQ on the xz plane.

What, now, if one of the direction numbers is zero? Let us suppose that $a = 0$. Then the line in parametric form is

$$x = x_0, \qquad y = y_0 + bt, \qquad z = z_0 + ct.$$

We do not have to eliminate the parameter from the first equation—it is already gone. By eliminating t between the last two equations as before, we have

$$\frac{y - y_0}{b} = \frac{z - z_0}{c}.$$

This together with $x = x_0$ (or $x - x_0 = 0$) gives the line as the intersection of two planes. In this case, plane PQQ_1P_1 and plane PQQ_3P_3 of Figure 9.12 are identical; they are both the plane $x = x_0$, which is parallel to (or on) the yz plane. Thus the line is parallel to the yz plane.

If two of the direction numbers are zero, we have two equations in which the parameter is missing. The parameter in the third equation cannot be eliminated, because there is no other equation with which to combine it. But it is not necessary to eliminate it! The two equations without the parameter already give us the necessary two planes. In this case the line is parallel to (or on) one of the coordinate axes. To illustrate this, suppose that the line PQ of Figure 9.12 is parallel to the y axis. Then $P_3 = Q_3$, and this point lies on the line PQ; hence no plane is determined by this projection. On the other hand, PQQ_1P_1 is a plane perpendicular to the xy plane and parallel to the y axis, while PQQ_2P_2 is a plane perpendicular to the yz plane and parallel to the y axis.

THEOREM 9.16   *If a line contains the point (x_0, y_0, z_0) and has direction numbers $\{a, b, c\}$, then it can be represented by*

(*a*)
$$\frac{x - x_0}{a} = \frac{y - y_0}{b} = \frac{z - z_0}{c}$$

if none of the direction numbers is zero;

(*b*)
$$x - x_0 = 0 \quad and \quad \frac{y - y_0}{b} = \frac{z - z_0}{c}$$

if $a = 0$ and neither b nor c is zero (similar results follow if $b = 0$ or $c = 0$);

(*c*)
$$x - x_0 = 0 \quad and \quad y - y_0 = 0$$

*if $a = 0$ and $b = 0$ (again, similar results follow some other pair of direction numbers equaling zero). These are called **symmetric equations** of the line.*

EXAMPLE 3   Find symmetric equations of the line containing $(4, 1, -2)$ and having direction numbers $\{1, 3, -2\}$.

SOLUTION
$$\frac{x - 4}{1} = \frac{y - 1}{3} = \frac{z + 2}{-2}$$

EXAMPLE 4   Find symmetric equations for the line containing $(4, 1, 3)$ and $(2, 1, -2)$.

SOLUTION   A set of direction numbers is $\{4 - 2, 1 - 1, 3 + 2\} = \{2, 0, 5\}$. Since $b = 0$, we have (using the first point)

$$\frac{x - 4}{2} = \frac{z - 3}{5} \quad and \quad y - 1 = 0.$$

EXAMPLE 5   Find the point of intersection (if any) of the lines

$$x = 3 + 2t, \qquad y = 2 - t, \qquad z = 5 + t$$

and

$$x = -3 - s, \qquad y = 7 + s, \qquad z = 16 + 3s.$$

SOLUTION   Let us assume that there is a point of intersection. Then there is a value of t and a value of s which yield the same values of x, y, and z. For these particular values of t and s, we have

$$x = 3 + 2t = -3 - s$$
$$y = 2 - t = 7 + s$$
$$z = 5 + t = 16 + 3s$$

or

$$2t + s = -6, \qquad t + s = -5, \qquad t - 3s = 11.$$

If we solve the first pair simultaneously, we get

$$t = -1 \qquad \text{and} \qquad s = -4.$$

We see that they also satisfy the third equation. Thus there is a point of intersection which corresponds to $t = -1$ (or $s = -4$). It is $(1, 3, 4)$.

Note that this method requires that the lines be given in parametric form. If they are given as symmetric equations, they must first be changed to a parametric representation. A comparison of Theorems 9.15 and 9.16 makes this easy.

It might be noted that there are three possibilities. One is the situation in which there is a value of t and a value of s satisfying all three of the equations in t and s, as above. This results in a single point of intersection. In a second possibility, there is no value for t or s satisfying all three of the equations; that is, the values of t and s that satisfy the first two equations fail to satisfy the third. Thus there is no point of intersection. The third possibility is that any two of the three equations in t and s are dependent; that is, any pair of values for t and s that satisfies one of them satisfies all three. In this case we have two different representations for the same line (see the discussion following Example 2).

PROBLEMS

A   *In Problems 1–16, represent the given line in parametric form and in symmetric form.*

1. Containing $(5, 1, 3)$; direction numbers $\{3, -2, 4\}$

2. Containing $(2, -4, 2)$; direction numbers $\{2, 3, 1\}$

3. Containing $(5, -2, 1)$; direction numbers $\{4, 1, -2\}$

4. Containing $(2, 0, 3)$; direction numbers $\{4, -1, 3\}$

5. Containing $(1, 1, 1)$; direction numbers $\{2, 0, 1\}$

6. Containing $(1, 0, 5)$; direction numbers $\{3, 1, 0\}$

7. Containing $(4, 4, 1)$; direction numbers $\{0, 0, 1\}$

8. Containing $(3, 1, 2)$; direction numbers $\{1, 0, 0\}$

9. Containing $(4, 0, 5)$ and $(2, 3, 1)$

10. Containing $(3, 3, 1)$ and $(4, 0, 2)$

11. Containing $(8, 4, 1)$ and $(-2, 0, 3)$

12. Containing $(-4, 2, 0)$ and $(3, 1, 2)$

13. Containing $(5, 1, 3)$ and $(5, 2, 4)$

14. Containing $(2, 2, 4)$ and $(1, 2, 7)$

15. Containing $(1, -2, 3)$ and $(1, 4, 3)$

16. Containing $(2, 4, -5)$ and $(5, 4, -5)$

B *In Problems 17–24, find the point of intersection (if any) of the given lines.*

17. $x = 4 + t, y = -8 - 2t, z = 12t;$ $x = 3 + 2s, y = -1 + s, z = -3 - 3s$

18. $x = 2 - t, y = 3 + 2t, z = 4 + t;$ $x = 1 + s, y = -2 + s, z = 5 - 4s$

19. $x = 3 + t, y = 4 - 2t, z = 1 + 5t;$ $x = 5 - s, y = 3 + 2s, z = 8 + 4s$

20. $x = 3 - t, y = 5 + 3t, z = -1 - 4t;$ $x = 8 + 2s, y = -6 - 4s, z = 5 + s$

21. $\dfrac{x - 2}{1} = \dfrac{y - 3}{-2} = \dfrac{z + 1}{1}; \quad \dfrac{x - 3}{2} = \dfrac{y - 1}{-4} = \dfrac{z}{2}$

22. $\dfrac{x - 5}{1} = \dfrac{y + 2}{-2} = \dfrac{z - 3}{5}; \quad \dfrac{x - 4}{-2} = \dfrac{y - 2}{1} = \dfrac{z - 4}{3}$

23. $\dfrac{x - 3}{1} = \dfrac{y + 3}{-4}, z + 1 = 0; \quad \dfrac{x}{-2} = \dfrac{y - 2}{1} = \dfrac{z - 3}{4}$

24. $\dfrac{x - 2}{1} = \dfrac{y - 3}{-2} = \dfrac{z}{4}; \quad x - 4 = 0, \dfrac{y - 2}{1} = \dfrac{z - 3}{-1}$

In Problems 25–32, indicate whether the two given lines are parallel, perpendicular, coincident, or none of these.

25. $x = 3 + 5t, y = -1 - 2t, z = 4 + t;$ $x = 3, y = 4 + 2s, z = -2 + 4s$

26. $x = 4 - t, y = 3 + 2t, z = 1 + t;$ $x = 1 + 2t, y = 4 - 4t, z = 3 - 2t$

27. $x = 2 + t, y = 5 - 3t, z = 1 + 4t;$ $x = 4 - t, y = 2 + 2t, z = 3t$

28. $x = 2 + t, y = 5 - 3t, z = -1 + 2t;$ $x = 4 - 3t, y = -1 + 9t, z = 3 - 6t$

29. $\dfrac{x + 3}{1} = \dfrac{y - 4}{3} = \dfrac{z + 2}{-2}; \quad \dfrac{x - 5}{-3} = \dfrac{y + 3}{-9} = \dfrac{z - 1}{6}$

30. $\dfrac{x-1}{2} = \dfrac{z+3}{4}$, $y - 5 = 0$; $\dfrac{x+2}{6} = \dfrac{y-5}{3} = \dfrac{z}{2}$

31. $\dfrac{x+1}{3} = \dfrac{y-2}{4} = \dfrac{z}{-2}$; $\dfrac{x-1}{2} = \dfrac{y+4}{-2} = \dfrac{z-7}{-1}$

32. $\dfrac{x-4}{2} = \dfrac{y+7}{4} = \dfrac{z-5}{-1}$; $\dfrac{x-4}{-3} = \dfrac{y+7}{-6} = \dfrac{z-5}{2}$

33. Give equations for each of the coordinate axes.

GRAPHING CALCULATOR

34. Suppose we have the following line in space:

$$x = 2 - 3t, \qquad y = 5 + t, \quad z = -3 + 2t.$$

Graph, using pairs of these parametric equations. That is, find the graph represented by x and y above, by x and z, and by y and z. What do these three graphs represent?

9.5

THE CROSS PRODUCT

Let us now look at the other product of two vectors—the cross product.

DEFINITION *If $\mathbf{u} = a_1\mathbf{i} + b_1\mathbf{j} + c_1\mathbf{k}$ and $\mathbf{v} = a_2\mathbf{i} + b_2\mathbf{j} + c_2\mathbf{k}$, then the cross product (vector product, outer product) of $\mathbf{u}$ and $\mathbf{v}$ is*

$$\mathbf{u} \times \mathbf{v} = (b_1c_2 - c_1b_2)\mathbf{i} + (c_1a_2 - a_1c_2)\mathbf{j} + (a_1b_2 - b_1a_2)\mathbf{k}.$$

Some obvious questions arise. Why do we want to define a cross product this way? What is it good for? What are its properties? In some ways, all answers are the same. We define the cross product in this way to establish some interesting properties that are useful for certain applications. In a way, this is approaching the problem backward. It would be more logical to define the cross product of two vectors as that one having the desired properties and then show that such a vector must take the form given. The reason for our way of doing it is that it is by far the simpler approach. Before looking at some properties, let us consider a simpler form for the cross product.

THEOREM 9.17 *If $\mathbf{u} = a_1\mathbf{i} + b_1\mathbf{j} + c_1\mathbf{k}$ and $\mathbf{v} = a_2\mathbf{i} + b_2\mathbf{j} + c_2\mathbf{k}$, then*

$$\mathbf{u} \times \mathbf{v} = \begin{vmatrix} \mathbf{i} & \mathbf{j} & \mathbf{k} \\ a_1 & b_1 & c_1 \\ a_2 & b_2 & c_2 \end{vmatrix}.$$

This theorem follows directly from the definition if we expand the above determinant by minors along the first row.

EXAMPLE 1   If $\mathbf{u} = 3\mathbf{i} + \mathbf{j} - 2\mathbf{k}$ and $\mathbf{v} = \mathbf{i} + 2\mathbf{j} + \mathbf{k}$, find $\mathbf{u} \times \mathbf{v}$ and $\mathbf{v} \times \mathbf{u}$.

SOLUTION

$$\mathbf{u} \times \mathbf{v} = \begin{vmatrix} \mathbf{i} & \mathbf{j} & \mathbf{k} \\ 3 & 1 & -2 \\ 1 & 2 & 1 \end{vmatrix} = 5\mathbf{i} - 5\mathbf{j} + 5\mathbf{k}$$

$$\mathbf{v} \times \mathbf{u} = \begin{vmatrix} \mathbf{i} & \mathbf{j} & \mathbf{k} \\ 1 & 2 & 1 \\ 3 & 1 & -2 \end{vmatrix} = -5\mathbf{i} + 5\mathbf{j} - 5\mathbf{k}$$

Note that $\mathbf{u} \times \mathbf{v} \neq \mathbf{v} \times \mathbf{u}$!

Again we are not multiplying numbers; there is no reason to assume that the cross product of two vectors has the same properties as the product of two numbers. We have already seen one difference in Example 1. The cross product has the following properties.

THEOREM 9.18   *If* $\mathbf{u}$, $\mathbf{v}$, *and* $\mathbf{w}$ *are vectors and* a *is a scalar, then the following properties hold.*

(*a*) $\mathbf{u} \times \mathbf{v} = -(\mathbf{v} \times \mathbf{u})$

(*b*) $\mathbf{u} \times (\mathbf{v} + \mathbf{w}) = \mathbf{u} \times \mathbf{v} + \mathbf{u} \times \mathbf{w}$

(*c*) $\mathbf{u} \times \mathbf{0} = \mathbf{0} \times \mathbf{u} = \mathbf{0}$

(*d*) *If* $\mathbf{u} = a\mathbf{v}$, *then* $\mathbf{u} \times \mathbf{v} = \mathbf{0}$ (*that is, the cross product of parallel vectors is* $\mathbf{0}$)

(*e*) $(\mathbf{u} \times \mathbf{v}) \cdot \mathbf{w} = \mathbf{u} \cdot (\mathbf{v} \times \mathbf{w})$

PROOF   (*a*) Suppose $\mathbf{u} = a_1\mathbf{i} + b_1\mathbf{j} + c_1\mathbf{k}$, $\mathbf{v} = a_2\mathbf{i} + b_2\mathbf{j} + c_2\mathbf{k}$, and $\mathbf{w} = a_3\mathbf{i} + b_3\mathbf{j} + c_3\mathbf{k}$. Since, by Theorem 9.17, $\mathbf{u} \times \mathbf{v}$ and $\mathbf{v} \times \mathbf{u}$ are given by determinants that are identical except for the reversal of the second and third rows, it follows that

$$\mathbf{u} \times \mathbf{v} = -(\mathbf{v} \times \mathbf{u}).$$

(*b*) Since $\mathbf{v} + \mathbf{w} = (a_2 + a_3)\mathbf{i} + (b_2 + b_3)\mathbf{j} + (c_2 + c_3)\mathbf{k}$,

$$\mathbf{u} \times (\mathbf{v} + \mathbf{w}) = \begin{vmatrix} \mathbf{i} & \mathbf{j} & \mathbf{k} \\ a_1 & b_1 & c_1 \\ a_2 + a_3 & b_2 + b_3 & c_2 + c_3 \end{vmatrix}$$

$$= \begin{vmatrix} \mathbf{i} & \mathbf{j} & \mathbf{k} \\ a_1 & b_1 & c_1 \\ a_2 & b_2 & c_2 \end{vmatrix} + \begin{vmatrix} \mathbf{i} & \mathbf{j} & \mathbf{k} \\ a_1 & b_1 & c_1 \\ a_3 & b_3 & c_3 \end{vmatrix}$$

$$= (\mathbf{u} \times \mathbf{v}) + (\mathbf{u} \times \mathbf{w}). \quad \blacksquare$$

The proofs of the remaining three parts are left to the student (see Problems 30 and 31).

It might be noted that the definition of the cross product was stated in terms of three-dimensional vectors. In fact, we must have a three-dimensional vector space, for $\mathbf{u} \times \mathbf{v}$ is not in the plane determined by $\mathbf{u}$ and $\mathbf{v}$, as shown in the next theorem.

THEOREM 9.19   *If* $\mathbf{u}$ *and* $\mathbf{v}$ *are nonzero vectors, then* $\mathbf{u} \times \mathbf{v}$ *is perpendicular to both* $\mathbf{u}$ *and* $\mathbf{v}$.

PROOF

$$\mathbf{u} \cdot (\mathbf{u} \times \mathbf{v}) = (\mathbf{u} \times \mathbf{u}) \cdot \mathbf{v} \qquad \text{(why?)}$$
$$= \mathbf{0} \cdot \mathbf{v} \qquad \text{(why?)}$$
$$= 0 \qquad \text{(why?)}$$

Thus $\mathbf{u}$ and $\mathbf{u} \times \mathbf{v}$ are perpendicular. A similar argument shows that $\mathbf{u} \times \mathbf{v}$ and $\mathbf{v}$ are perpendicular. ■

This property of the cross product gives us its principal use. Certain problems in three-dimensional analytic geometry that were relatively difficult without the use of the cross product are easier now.

EXAMPLE 2   Find a set of direction numbers for a line perpendicular to the plane containing

$$x = 1, \qquad y = 3 + 2t, \qquad z = 4 + t$$

and

$$x = 1 + 4s, \qquad y = 3 + 2s, \qquad z = 4 + 2s.$$

SOLUTION   Any line perpendicular to a given plane is perpendicular to any line in that plane. This suggests the use of the cross product. Vectors directed along the given lines are

$$\mathbf{u} = 2\mathbf{j} + \mathbf{k} \qquad \text{and} \qquad \mathbf{v} = 4\mathbf{i} + 2\mathbf{j} + 2\mathbf{k}.$$

Since $\mathbf{u} \times \mathbf{v} = 2\mathbf{i} + 4\mathbf{j} - 8\mathbf{k}$ (see Figure 9.13), we have $\{2, 4, -8\}$ as one set of direction numbers for the desired line; $\{1, 2, -4\}$ is a simpler set.

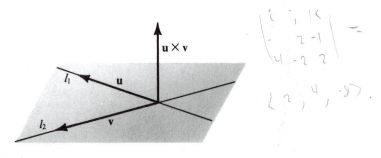

Figure 9.13

EXAMPLE 3   Find equations for the line containing $(1, 4, 3)$ and perpendicular to

$$\frac{x-1}{2} = \frac{y+3}{1} = \frac{z-2}{4} \quad \text{and} \quad \frac{x+2}{3} = \frac{y-4}{2} = \frac{z+1}{-2}.$$

SOLUTION   Again, vectors along the two given lines are

$$\mathbf{u} = 2\mathbf{i} + \mathbf{j} + 4\mathbf{k} \quad \text{and} \quad \mathbf{v} = 3\mathbf{i} + 2\mathbf{j} - 2\mathbf{k};$$

and $\mathbf{u} \times \mathbf{v} = -10\mathbf{i} + 16\mathbf{j} + \mathbf{k}$ is perpendicular to both of them. The desired line is, therefore,

$$\frac{x-1}{-10} = \frac{y-4}{16} = \frac{z-3}{1}.$$

EXAMPLE 4   Find the distance between the lines

$$x = 1 - 4t, \qquad y = 2 + t, \qquad z = 3 + 2t$$

and

$$x = 1 + s, \qquad y = 4 - 2s, \qquad z = -1 + s.$$

SOLUTION   The desired distance is to be measured along a line perpendicular to both of the given lines. Again, vectors along the given lines are $\mathbf{u} = -4\mathbf{i} + \mathbf{j} + 2\mathbf{k}$ and $\mathbf{v} = \mathbf{i} - 2\mathbf{j} + \mathbf{k}$ (see Figure 9.14). Thus the distance is to be measured along

$$\mathbf{u} \times \mathbf{v} = 5\mathbf{i} + 6\mathbf{j} + 7\mathbf{k}.$$

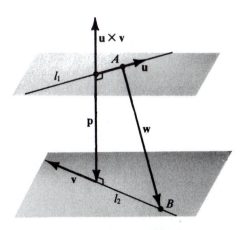

Figure 9.14

The point $A = (1, 2, 3)$ is on the first line, and $B = (1, 4, -1)$ is on the second. The vector represented by $\overrightarrow{AB}$ is

$$\mathbf{w} = 2\mathbf{j} - 4\mathbf{k}.$$

We want a vector whose representatives are all perpendicular to both of the given lines and with one representative having its head on one line and its tail on the other. All of the representatives of $\mathbf{u} \times \mathbf{v}$ are perpendicular to both lines and one of the representatives of $\mathbf{w}$ has its endpoints on the given lines. Thus, the projection $\mathbf{p}$ of $\mathbf{w}$ on $\mathbf{u} \times \mathbf{v}$ has the desired properties and its length is the distance between the given lines.

$$|\mathbf{p}| = \frac{|\mathbf{w} \cdot (\mathbf{u} \times \mathbf{v})|}{|\mathbf{u} \times \mathbf{v}|} = \frac{|0 \cdot 5 + 2 \cdot 6 - 4 \cdot 7|}{\sqrt{25 + 36 + 49}} = \frac{16}{\sqrt{110}}$$

Up to this point we have been dealing exclusively with the direction of $\mathbf{u} \times \mathbf{v}$. Its length also has some interesting properties.

THEOREM 9.20   *If $\mathbf{u}$ and $\mathbf{v}$ are vectors and θ is the angle between them, then*

$$|\mathbf{u} \times \mathbf{v}| = |\mathbf{u}|\,|\mathbf{v}| \sin \theta.$$

PROOF   Since $\cos \theta = \mathbf{u} \cdot \mathbf{v}/(|\mathbf{u}|\,|\mathbf{v}|)$ by Theorem 9.8,

$$|\mathbf{u}|\,|\mathbf{v}| \sin \theta = |\mathbf{u}|\,|\mathbf{v}| \sqrt{1 - \cos^2 \theta} \qquad \text{(see Note 1)}$$

$$= |\mathbf{u}|\,|\mathbf{v}| \sqrt{1 - \frac{(\mathbf{u} \cdot \mathbf{v})^2}{|\mathbf{u}|^2 |\mathbf{v}|^2}}$$

$$= \sqrt{|\mathbf{u}|^2 |\mathbf{v}|^2 - (\mathbf{u} \cdot \mathbf{v})^2}.$$

If we let $\mathbf{u} = a_1\mathbf{i} + b_1\mathbf{j} + c_1\mathbf{k}$ and $\mathbf{v} = a_2\mathbf{i} + b_2\mathbf{j} + c_2\mathbf{k}$, then

$$|\mathbf{u}|\,|\mathbf{v}| \sin \theta = \sqrt{(a_1^2 + b_1^2 + c_1^2)(a_2^2 + b_2^2 + c_2^2) - (a_1a_2 + b_1b_2 + c_1c_2)^2}$$

$$= \sqrt{(b_1c_2 - c_1b_2)^2 + (c_1a_2 - a_1c_2)^2 + (a_1b_2 - b_1a_2)^2}$$

$$= |\mathbf{u} \times \mathbf{v}|. \qquad \text{(see Note 2)}$$

Note 1: By the definition of the angle between two vectors, $0° \leq \theta \leq 180°$; and $\sin \theta \geq 0$.

Note 2: The algebra here is routine but tedious. It is left to the student. ∎

Note the similarity between this theorem and the first part of Theorem 9.8. One consequence of this theorem is given in Problem 25.

It appears that Theorems 9.19 and 9.20 give a geometric description of $\mathbf{u} \times \mathbf{v}$, the first giving its direction and the second, its length. Actually, this is not quite true. There are two vectors of a given length which are perpendicular to both $\mathbf{u}$ and $\mathbf{v}$; they have opposite orientations (that is, one is the negative of the other). It can be shown that $\mathbf{u} \times \mathbf{v}$ is the one that gives the system $\{\mathbf{u}, \mathbf{v}, \mathbf{u} \times \mathbf{v}\}$ a right-hand orientation; that is, if the index and second fingers of the right hand point in the directions of $\mathbf{u}$ and $\mathbf{v}$, respectively, then the thumb points in the direction of $\mathbf{u} \times \mathbf{v}$ (see Figure 9.15). This is summarized in the next theorem.

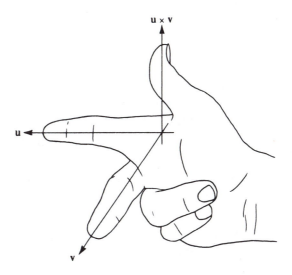

Figure 9.15

THEOREM 9.21   *If $\mathbf{u}$ and $\mathbf{v}$ are vectors, θ is the angle between them, and $\mathbf{n}$ is the unit vector perpendicular to both $\mathbf{u}$ and $\mathbf{v}$ such that $\{\mathbf{u}, \mathbf{v}, \mathbf{n}\}$ forms a right-hand system, then*

$$\mathbf{u} \times \mathbf{v} = |\mathbf{u}|\,|\mathbf{v}| \sin \theta\, \mathbf{n}.$$

Some authors take this as the definition of $\mathbf{u} \times \mathbf{v}$. Another direct result from Theorem 9.20 follows.

THEOREM 9.22   *If $\mathbf{u}$ and $\mathbf{v}$ are two nonzero vectors, then $\mathbf{u} \times \mathbf{v} = \mathbf{0}$ if and only if $\mathbf{u} = k\mathbf{v}$ for some scalar k.*

This extends Theorem 9.18d, which gives part of this theorem. The proof is left to the student (see Problem 37).

If might be noted that Examples 2 and 3 can also be solved without the use of vectors. We simply use direction numbers instead of vectors.

EXAMPLE 2   Find a set of direction numbers for a line perpendicular to the plane containing

$$x = 1, \qquad y = 3 + 2t, \qquad z = 4 + t$$

and

$$x = 1 + 4s, \qquad y = 3 + 2s, \qquad z = 4 + 2s.$$

ALTERNATE SOLUTION   A line perpendicular to the given plane must be perpendicular to any line in the plane. Thus the desired line with direction numbers $\{a, b, c\}$ is perpendicular to the two given lines with direction numbers $\{0, 2, 1\}$ and $\{4, 2, 2\}$. This gives the equations

$$2b + c = 0$$
$$4a + 2b + 2c = 0.$$

Since we have only two equations in three unknowns, we cannot solve for all of them; however, we can solve for two of them in terms of the third. By subtracting the first equation from the second, we get

$$4a + c = 0$$

or $c = -4a$. Doubling the first equation and subtracting from the second, we have

$$4a - 2b = 0$$

or $b = 2a$. Thus $\{a, b, c\} = \{a, 2a, -4a\}$ are direction numbers for the desired line. A second set of direction numbers can be found by multiplying by $1/a$, giving $\{1, 2, -4\}$.

EXAMPLE 3   Find equations for the line containing $(1, 4, 3)$ and perpendicular to

$$\frac{x - 1}{2} = \frac{y + 3}{1} = \frac{z - 2}{4} \qquad \text{and} \qquad \frac{x + 2}{3} = \frac{y - 4}{2} = \frac{z + 1}{-2}.$$

ALTERNATE SOLUTION   Again, direction numbers for the two given lines are $\{2, 1, 4\}$ and $\{3, 2, -2\}$; for the desired line, $\{a, b, c\}$. Since the desired line is perpendicular to both of the given lines, we have

$$2a + b + 4c = 0$$
$$3a + 2b - 2c = 0.$$

Doubling the second equation and adding to the first gives

$$8a + 5b = 0$$

or $b = -8a/5$. Doubling the first equation and subtracting the second, we have

$$a + 10c = 0$$

or $c = -a/10$. Thus the direction numbers for the desired line are $\{a, -8a/5, -a/10\}$ or, multiplying by $-10/a$, $\{-10, 16, 1\}$. Using these with the given point, we see that symmetric equations for the line are

$$\frac{x - 1}{-10} = \frac{y - 4}{16} = \frac{z - 3}{1}.$$

line ⊥
to 2
things!

Example 4 can also be solved without the use of vectors; however, the solution involves planes, which we have not yet considered.

PROBLEMS

In Problems 1–6, find $\mathbf{u} \times \mathbf{v}$. *Use the dot product to verify that your result is perpendicular to both* $\mathbf{u}$ *and* $\mathbf{v}$.

A

1. $\mathbf{u} = 3\mathbf{i} - \mathbf{j} + 4\mathbf{k}$, $\mathbf{v} = 2\mathbf{i} + \mathbf{j} + \mathbf{k}$
2. $\mathbf{u} = \mathbf{i} + \mathbf{j} + \mathbf{k}$, $\mathbf{v} = 2\mathbf{i} - \mathbf{j} - 4\mathbf{k}$
3. $\mathbf{u} = 2\mathbf{i} + 3\mathbf{j} - \mathbf{k}$, $\mathbf{v} = -\mathbf{i} + 2\mathbf{j}$
4. $\mathbf{u} = 4\mathbf{i} + 2\mathbf{j}$, $\mathbf{v} = 3\mathbf{i} - \mathbf{j}$
5. $\mathbf{u} = 3\mathbf{i} + \mathbf{k}$, $\mathbf{v} = -\mathbf{i} + \mathbf{j}$
6. $\mathbf{u} = 2\mathbf{i} + \mathbf{j} - \mathbf{k}$, $\mathbf{v} = -\mathbf{i} - \mathbf{j} + 3\mathbf{k}$

B *In Problems 7–12, find direction numbers for the line described.*

7. Perpendicular to the plane containing $(4, 1, 2)$, $(2, -1, 1)$, and $(3, 0, 4)$
8. Perpendicular to the plane containing $(2, 2, 3)$, $(-1, 4, 1)$, and $(0, 1, 2)$
9. Perpendicular to the plane containing $x = 2 + t$, $y = 3 - 2t$, $z = -t$ and $x = 2 - 2s$, $y = 3 + s$, $z = -s$
10. Perpendicular to the plane containing $x = 3 + 4t$, $y = 1 - t$, $z = 3$ and $x = 3 - 2s$, $y = 1 + 2s$, $z = 3 - s$
11. Perpendicular to the plane containing $x = 4 + t$, $y = -1 + 2t$, $z = 2t$ and $x = 2 + s$, $y = 4 + 2s$, $z = 1 + 2s$
12. Perpendicular to the plane containing $x = 2 + 2t$, $y = 3 - t$, $z = -1 + t$ and $x = 4 + 2s$, $y = 2 - s$, $z = 4 + s$

In Problems 13–18, find equations for the line described.

13. Containing $(3, 2, 1)$ and perpendicular to $x = 1 - 2t$, $y = 3 + t$, $z = 4 - t$ and $x = 2 + s$, $y = -1 + 2s$, $z = 3 - s$
14. Containing $(4, -1, 0)$ and perpendicular to $x = 3 + t$, $y = 2 - t$, $z = 2t$ and $x = 4$, $y = 2 + s$, $z = -1 + s$

15. Containing $(2, 3, 1)$ and perpendicular to the plane determined by $(2, 3, 1)$ and the line $x = 0$, $y = 2t$, $z = t$

16. Containing $(0, 4, -2)$ and perpendicular to the plane determined by $(0, 4, -2)$ and the line $x = -2 + 2t$, $y = 8t$, $z = -1 + t$

17. Containing $(2, 0, 5)$ and perpendicular to and containing a point of $x = 4 + t$, $y = 3 - 2t$, $z = 1 + t$

18. Containing $(1, 1, 2)$ and perpendicular to and containing a point of $x = 1 - t$, $y = 2 + 2t$, $z = 4t$

In Problems 19–24, find the distance between the given lines.

19. $x = 1 + t$, $y = -2 + 3t$, $z = 4 + t$ and $x = 2 - s$, $y = 3 + 2s$, $z = 1 + s$

20. $x = 2 + t$, $y = 1 - t$, $z = 4t$ and $x = 2 + s$, $y = 4 - 2s$, $z = 1 + 3s$

21. $x = 1 + t$, $y = 1 - 5t$, $z = 2 + t$ and $x = 4 + s$, $y = 5 + 2s$, $z = -3 + 4s$

22. $x = 2 + t$, $y = -4 + t$, $z = 1 - 3t$ and $x = 3 - s$, $y = 4 + 2s$, $z = 2 + s$

23. $x = 2 + 3t$, $y = 5 + t$, $z = -1 - 2t$ and $x = 2 + 3s$, $y = 3 + s$, $z = 5 - 2s$

24. $x = 4t$, $y = 1 + t$, $z = -2 - t$ and $x = 9 + 4s$, $y = 1 + s$, $z = -2 - s$

25. Suppose the vectors **u** and **v** are represented by $\overrightarrow{AB}$ and $\overrightarrow{AC}$, respectively. Show that the area of $\triangle ABC$ is $|\mathbf{u} \times \mathbf{v}|/2$. (Equivalently, the parallelogram determined by AB and AC has area $|\mathbf{u} \times \mathbf{v}|$.) [*Hint:* Use Theorem 9.20.]

In Problems 26–29, use the result of Problem 25 to find the area of the triangles with the given vertices.

26. $(1, 0, 4)$, $(2, -1, 2)$, $(4, 4, 1)$ 27. $(3, -2, 1)$, $(-1, 2, 0)$, $(4, 4, 2)$

28. $(2, 4, 3)$, $(1, 0, 1)$, $(-2, 2, 4)$ 29. $(4, 2)$, $(3, -1)$, $(-1, 0)$

C   30. Prove parts (c) and (d) of Theorem 9.18.

31. Show that if $\mathbf{u} = a_1\mathbf{i} + b_1\mathbf{j} + c_1\mathbf{k}$, $\mathbf{v} = a_2\mathbf{i} + b_2\mathbf{j} + c_2\mathbf{k}$, and $\mathbf{w} = a_3\mathbf{i} + b_3\mathbf{j} + c_3\mathbf{k}$, then

$$\mathbf{u} \cdot (\mathbf{v} \times \mathbf{w}) = \begin{vmatrix} a_1 & b_1 & c_1 \\ a_2 & b_2 & c_2 \\ a_3 & b_3 & c_3 \end{vmatrix}.$$

Use this result to prove Theorem 9.18(e).

32. Given that $\overrightarrow{AB}$, $\overrightarrow{AC}$, and $\overrightarrow{AD}$ represent the vectors **u**, **v**, and **w**, respectively, show that the parallelepiped determined by AB, AC, and AD (see Figure 9.16) has volume $|\mathbf{u} \cdot (\mathbf{v} \times \mathbf{w})|$. [*Hint:* By Problem 25, the area of the base is $|\mathbf{v} \times \mathbf{w}|$.]

In Problems 33–36, use the result of Problem 32 to find the volume of the parallelepiped determined by AB, AC, and AD.

33. $A = (0, 0, 0)$, $B = (2, 1, 3)$, $C = (5, 3, 1)$, $D = (2, -1, 4)$

34. $A = (1, 3, 2)$, $B = (4, 1, 5)$, $C = (1, 5, 2)$, $D = (0, 5, -1)$

35. $A = (1, -2, 5)$, $B = (-2, 3, 1)$, $C = (4, -2, 3)$, $D = (-1, -3, 4)$

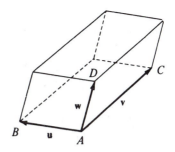

Figure 9.16

36. $A = (2, 5, 1)$, $B = (3, -2, 3)$, $C = (4, -3, 5)$, $D = (-2, 4, 1)$

37. Prove Theorem 9.22.

9.6

THE PLANE

Let us now consider the plane. Perhaps the simplest way of determining a plane is by three noncollinear points. But, for the purpose of determining its equation, it is better to describe it by a single point and a line perpendicular to it.

Let p be a plane in space, containing the point $P_0 = (x_0, y_0, z_0)$ (see Figure 9.17); and let l, with direction numbers $\{a, b, c\}$, be a line perpendicular to p. In order to determine an equation of p, we consider any point $P = (x, y, z)$ lying in the plane. The directed line segment $\overrightarrow{P_0P}$ represents a vector

$$\mathbf{v} = (x - x_0)\mathbf{i} + (y - y_0)\mathbf{j} + (z - z_0)\mathbf{k}$$

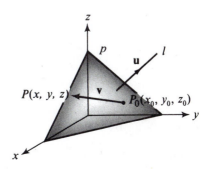

Figure 9.17

in the plane p. Since l is perpendicular to p, it is perpendicular to any line in this plane; in particular, it is perpendicular to P_0P. Since

$$\mathbf{u} = a\mathbf{i} + b\mathbf{j} + c\mathbf{k}$$

is a vector lying along l, we have the following.

$$\mathbf{u} \cdot \mathbf{v} = 0$$
$$[a\mathbf{i} + b\mathbf{j} + c\mathbf{k}] \cdot [(x - x_0)\mathbf{i} + (y - y_0)\mathbf{j} + (z - z_0)\mathbf{k}] = 0$$
$$a(x - x_0) + b(y - y_0) + c(z - z_0) = 0.$$

Furthermore, the argument may be traced backward to show that any point (x, y, z) that satisfies the last equation must lie in the plane p. Thus we have proved the following theorem.

THEOREM 9.23   *A point is on a plane containing (x_1, y_1, z_1) and perpendicular to a line with direction numbers $\{A, B, C\}$ if and only if it satisfies the equation*

$$A(x - x_1) + B(y - y_1) + C(z - z_1) = 0.$$

This theorem can also be proved using a nonvector argument. The vectors are simply replaced by direction numbers. Suppose p is the plane in space, containing the point $P_0 = (x_0, y_0, z_0)$; and l, with direction numbers $\{a, b, c\}$, is a line perpendicular to p. Then if $P = (x, y, z)$ is any other point of p, a set of direction numbers for the line PP_0 is $\{x - x_0, y - y_0, z - z_0\}$. Since this line in the plane p must be perpendicular to l, we have

$$a(x - x_0) + b(y - y_0) + c(z - z_0) = 0.$$

EXAMPLE 1   Find the equation of the plane containing $(1, 3, -2)$ and perpendicular to the line through $(2, 5, 1)$ and $(0, 1, -3)$.

SOLUTION   A set of direction numbers for the given line is $\{2, 4, 4\}$ or $\{1, 2, 2\}$. Thus the desired plane is

$$1(x - 1) + 2(y - 3) + 2(z + 2) = 0$$
$$x + 2y + 2z - 3 = 0.$$

EXAMPLE 2   Find an equation of the plane containing the two lines

$$x = 1, \ y = 3 + 2t, \ z = 4 + t \quad \text{and} \quad x = 1 + 4s, \ y = 3 + 2s, \ z = 4 + 2s.$$

SOLUTION   These lines clearly intersect at $(1, 3, 4)$. All we need, then, is a set of direction numbers for a line perpendicular to the desired plane. This was done in Example 2 of the previous section by using the cross product of two vectors or (as shown in the alternate solution) by using the direction numbers. One such set is $\{1, 2, -4\}$. By Theorem 9.23, the corresponding plane is

$$1(x - 1) + 2(y - 3) - 4(z - 4) = 0$$
$$x + 2y - 4z + 9 = 0.$$

EXAMPLE 3   Find an equation of the plane containing the two lines

$$x = 1 + t, \quad y = 3 - 2t, \quad z = -2 + 2t$$

and

$$x = 4 + s, \quad y = 2 - 2s, \quad z = -1 + 2s.$$

SOLUTION   The two lines are parallel since they both have direction numbers $\{1, -2, 2\}$. Since $(1, 3, -2)$ is on the first line and $(4, 2, -1)$ is on the second, the line through these two points intersects both of the given lines and lies in the desired plane (see Figure 9.18). Its direction numbers are $\{4 - 1, 2 - 3, -1 + 2\}$ or $\{3, -1, 1\}$. We now have two intersecting lines with direction numbers $\{1, -2, 2\}$ and $\{3, -1, 1\}$ and lying in the desired plane. By the methods of the previous section, a line perpendicular to both of these lines (and therefore perpendicular to the plane containing them) has direction numbers $\{0, 1, 1\}$. Thus, by Theorem 9.23, the desired plane is

$$0(x - 1) + 1(y - 3) + 1(z + 2) = 0$$
$$y + z - 1 = 0.$$

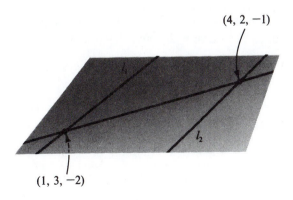

(4, 2, −1)

(1, 3, −2)

Figure 9.18

The following theorem is a direct consequence of Theorem 9.23.

THEOREM 9.24   *Any plane can be represented by an equation of the form*

$$Ax + By + Cz + D = 0,$$

where $\{A, B, C\}$ is a set of direction numbers for a line normal to (that is, perpendicular to) the plane. Conversely, an equation of the above form (where A, B, and C are not all zero) represents a plane with $\{A, B, C\}$ a set of direction numbers for a normal line.

EXAMPLE 4   Find an equation of the plane containing the points $P_1 = (1, 0, 1)$, $P_2 = (-1, -4, 1)$, and $P_3 = (-2, -2, 2)$.

SOLUTION   This problem may be solved using either Theorem 9.24 or Theorem 9.23. Let us do it both ways.

By Theorem 9.24, the equation we seek is, for the proper choices of A, B, C, and D,

$$Ax + By + Cz + D = 0.$$

We get an equation in A, B, C, and D from each of the three given points.

$$
\begin{aligned}
P_1 = (1, 0, 1): & \quad A \qquad\quad + C + D = 0, \\
P_2 = (-1, -4, 1): & \quad -A - 4B + C + D = 0, \\
P_3 = (-2, -2, 2): & \quad -2A - 2B + 2C + D = 0.
\end{aligned}
$$

Although we cannot solve for A, B, C, and D directly, since we have only three equations in four unknowns, we can solve for three of them in terms of the other one. If we take A to be fixed and solve for the other three, we have $B = -A/2$, $C = 2A$, and $D = -3A$. We may give A any nonzero value we want; let us choose $A = 2$. Then $B = -1$, $C = 4$, and $D = 6$; the resulting equation is

$$2x - y + 4z - 6 = 0.$$

ALTERNATE SOLUTION   We now solve the same problem using Theorem 9.23. We let $\overrightarrow{P_1P_2}$ and $\overrightarrow{P_1P_3}$ represent the vectors $\mathbf{u}$ and $\mathbf{v}$, respectively. Then

$$\mathbf{u} = (-1 - 1)\mathbf{i} + (-4 - 0)\mathbf{j} + (1 - 1)\mathbf{k} = -2\mathbf{i} - 4\mathbf{j}$$

$$\mathbf{v} = (-2 - 1)\mathbf{i} + (-2 - 0)\mathbf{j} + (2 - 1)\mathbf{k} = -3\mathbf{i} - 2\mathbf{j} + \mathbf{k}.$$

Since $\mathbf{u}$ and $\mathbf{v}$ lie in the desired plane, their cross product is perpendicular to it.

$$\mathbf{u} \times \mathbf{v} = \begin{vmatrix} \mathbf{i} & \mathbf{j} & \mathbf{k} \\ -2 & -4 & 0 \\ -3 & -2 & 1 \end{vmatrix} = -4\mathbf{i} + 2\mathbf{j} - 8\mathbf{k}$$

Thus $\{-4, 2, -8\}$ is a set of direction numbers for a line perpendicular to the desired plane. A simpler set is $\{2, -1, 4\}$. Using this, together with the point $(1, 0, 1)$ in Theorem 9.23, we have

$$2(x - 1) - y + 4(z - 1) = 0$$
$$2x - y + 4z - 6 = 0.$$

EXAMPLE 5   Sketch $x + 2y + 3z = 6$.

SOLUTION   By Theorem 9.24, we know that this equation represents a plane. Knowing this, we merely need to find three points to determine the plane. The simplest points to find are the intercepts (the points where the plane crosses the coordinate axes), which are $(6, 0, 0)$, $(0, 3, 0)$, and $(0, 0, 2)$. Thus we have the plane shown in Figure 9.19.

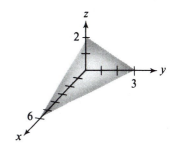

Figure 9.19

EXAMPLE 6   Sketch $x + 2y = 4$.

SOLUTION   This equation represents a line if we are considering only the xy plane (for which $z = 0$). But we get the same line when $z = 1$ or $z = 2$, and so forth. Thus the result is a plane that is parallel to the z axis (see Figure 9.20).

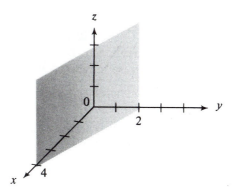

Figure 9.20

Of course the two planes

$$A_1x + B_1y + C_1z + D_1 = 0$$

and

$$A_2x + B_2y + C_2z + D_2 = 0$$

are parallel if and only if their normal lines are parallel. Similarly, the planes are perpendicular if and only if their normal lines are perpendicular. Thus from Theorems 9.13 and 9.14, we have the following theorem for planes.

THEOREM 9.25   *The planes*

$$A_1x + B_1y + C_1z + D_1 = 0$$

and

$$A_2x + B_2y + C_2z + D_2 = 0$$

are **parallel** (*or* **coincident**) *if and only if there is a number k such that*

$$A_2 = kA_1, \qquad B_2 = kB_1, \qquad C_2 = kC_1;$$

they are **perpendicular** *if and only if*

$$A_1A_2 + B_1B_2 + C_1C_2 = 0.$$

EXAMPLE 7   Show that $3x + y - 4z = 2$ and $6x + 2y - 8z = 3$ are parallel planes and that $3x - y + 2z = 5$ is perpendicular to both of them.

SOLUTION   The coefficients of x, y, and z in the three equations are

$$A_1 = 3, \qquad A_2 = 6, \qquad A_3 = 3,$$
$$B_1 = 1, \qquad B_2 = 2, \qquad B_3 = -1,$$
$$C_1 = -4, \qquad C_2 = -8, \qquad C_3 = 2.$$

Since $A_2 = 2A_1$, $B_2 = 2B_1$, and $C_2 = 2C_1$, the first and second planes are either parallel or coincident. But since $D_2 = -3 \neq 2D_1 = 2(-2)$, they are not equivalent equations—the planes are parallel. Since

$$A_1A_3 + B_1B_3 + C_1C_3 = 3 \cdot 3 + 1(-1) - 4 \cdot 2 = 0,$$

the first and third planes are perpendicular. Of course, the third plane must then be perpendicular to the second as well; but this may also be checked using Theorem 9.25.

$$A_2A_3 + B_2B_3 + C_2C_3 = 6 \cdot 3 + 2(-1) - 8 \cdot 2 = 0$$

In the previous section, we used the cross product and the projection of one vector upon another to find the distance between a pair of lines in space. Similar methods can be used to find the distance between a point and a plane or between a point and a line.

EXAMPLE 8

Find the distance between $(3, -4, 1)$ and $x - 2y + 2z + 4 = 0$.

SOLUTION

A vector perpendicular to the given plane is

$$\mathbf{v} = \mathbf{i} - 2\mathbf{j} + 2\mathbf{k}$$

(see Figure 9.21). We now choose an arbitrary point on $x - 2y + 2z + 4 = 0$, say $(0, 0, -2)$, and let $\mathbf{u}$ be the vector represented by the directed line segment from $(3, -4, 1)$ to $(0, 0, -2)$.

$$\mathbf{u} = (0 - 3)\mathbf{i} + (0 + 4)\mathbf{j} + (-2 - 1)\mathbf{k}$$
$$= -3\mathbf{i} + 4\mathbf{j} - 3\mathbf{k}$$

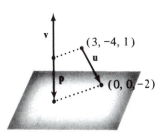

Figure 9.21

Now the distance we want is the length of the projection $\mathbf{p}$ of $\mathbf{u}$ upon $\mathbf{v}$.

$$d = \frac{|\mathbf{u} \cdot \mathbf{v}|}{|\mathbf{v}|}$$
$$= \frac{|(-3)(1) + (4)(-2) + (-3)(2)|}{\sqrt{1 + 4 + 4}} = \frac{17}{3}$$

Exactly the same method can be used to find the distance between the point (x_1, y_1, z_1) and the plane $Ax + By + Cz + D = 0$. We obtain the following.

THEOREM 9.26   *The distance between the point (x_1, y_1, z_1) and the plane $Ax + By + Cz + D = 0$ is*

$$d = \frac{|Ax_1 + By_1 + Cz_1 + D|}{\sqrt{A^2 + B^2 + C^2}}.$$

The proof is left to the student (see Problem 59). Notice that this formula is similar to the one on page 106 for the distance between a point and a line in two dimensions. With this formula, the distance of Example 1 is

$$d = \frac{|Ax_1 + By_1 + Cz_1 + D|}{\sqrt{A^2 + B^2 + C^2}}.$$
$$= \frac{|1 \cdot 3 - 2(-4) + 2 \cdot 1 + 4|}{\sqrt{1^2 + (-2)^2 + 2^2}}$$
$$= \frac{17}{3}.$$

EXAMPLE 9   Find the distance between $P = (5, 1, 3)$ and the line $x = 3$, $y = 7 + t$, $z = 1 + t$.

SOLUTION   A vector directed along the given line is

$$\mathbf{u} = \mathbf{j} + \mathbf{k}$$

(see Figure 9.22), and the point $Q = (3, 7, 1)$ is on the line. Letting $\mathbf{v}$ be the vector represented by $\overrightarrow{QP}$,

$$\mathbf{v} = (5 - 3)\mathbf{i} + (1 - 7)\mathbf{j} + (3 - 1)\mathbf{k}$$
$$= 2\mathbf{i} - 6\mathbf{j} + 2\mathbf{k}.$$

From Figure 9.22, the distance is

$$d = |\mathbf{v}| \sin \theta.$$

Figure 9.22

But by Theorem 9.20,

$$|\mathbf{u} \times \mathbf{v}| = |\mathbf{u}| \, |\mathbf{v}| \sin \theta.$$

Therefore

$$d = |\mathbf{v}| \sin \theta = \frac{|\mathbf{u} \times \mathbf{v}|}{|\mathbf{u}|} = \frac{|8\mathbf{i} + 2\mathbf{j} - 2\mathbf{k}|}{|\mathbf{j} + \mathbf{k}|} = \frac{\sqrt{64 + 4 + 4}}{\sqrt{2}}$$
$$= \frac{\sqrt{72}}{\sqrt{2}} = \sqrt{36} = 6.$$

PROBLEMS

A   *In Problems 1–6, sketch the plane.*

1. $2x + 3y + z = 6$
2. $3x - y + z = 9$
3. $2x + y - 4z + 4 = 0$
4. $x - y - 4z = 8$
5. $x + 2y = 3$
6. $y - 5 = 0$

In Problems 7–12, find an equation(s) of the plane(s) satisfying the given conditions.

7. Containing $(3, 2, -5)$ and perpendicular to a line with direction numbers $\{3, -4, 1\}$
8. Containing $(4, 2, 3)$ and perpendicular to a line with direction numbers $\{-2, 5, 1\}$
9. Containing $(4, 1, -3)$ and perpendicular to the line $x = 2 + 3t, y = 4 - t, z = 3 - 2t$
10. Containing $(3, 2, 5)$ and perpendicular to the line $x = 1 + t, y = 3t, z = 4 + t$
11. Containing $(3, 5, 1)$ and parallel to $3x - 4y + 2z = 3$
12. Containing $(4, -1, 2)$ and parallel to $x + y - 2z = 4$

In Problems 13–18, find the distance between the plane and point given.

13. $2x - 4y + 4z + 3 = 0; (1, 3, -2)$
14. $4x + y - 8z + 1 = 0; (2, 0, 3)$
15. $x + y - 2z - 4 = 0; (3, 3, 1)$
16. $2x - y + z + 5 = 0; (1, 0, 2)$
17. $x + z - 5 = 0; (3, 3, 1)$
18. $y + 7 = 0; (1, 3, 1)$

B   19. If the distance between $(1, 4, z)$ and $8x - y + 4z - 3 = 0$ is 1, find z.

20. If the distance between $(2, y, 3)$ and $4x - 4y + 2z - 5 = 0$ is 3/2, find y.

In Problems 21–32, find an equation(s) of the plane(s) satisfying the given conditions.

21. Containing $(1, 1, 0)$, $(1, 3, 2)$, and $(2, -1, 1)$
22. Containing $(2, -2, -2)$, $(1, -3, 5)$, and $(-1, 4, 1)$
23. Containing $(1, 4, 2)$, $(2, 3, -1)$, and $(5, 0, 2)$
24. Containing $(3, 1, -4)$, $(2, 3, 1)$, and $(7, 4, -2)$
25. Containing $x = 4 + t, y = 2 - t, z = 1 + 2t$ and $x = 4 - 3s, y = 2 + 2s, z = 1 - s$
26. Containing $x = 2 + 2t, y = -1 + t, z = 4 - t$ and $x = 2 - s, y = -1 - 2s, z = 4 + 3s$
27. Containing $(4, 1, 2)$ and $x = 4 - t, y = 1 + 2t, z = 3 - t$
28. Containing $(-2, 3, -4)$ and $x = 1 + t, y = 3 - 2t, z = -2 + t$
29. Containing $x = 3 + 2t, y = 4 - t, z = 1 + t$ and $x = -1 + 2s, y = 3 - s, z = 4 + s$
30. Containing $x = 4 + t, y = 2t, z = 5$ and $x = 1 + s, y = 3 + 2s, z = -2$
31. Containing $(1, 5, -2)$ and perpendicular to $3x + 2y - z + 1 = 0$ and $x - y + 2z = 0$
32. Containing $(3, 0, -4)$ and perpendicular to $2x - 5y + z = 1$ and $x - 2y - z = 3$

In Problems 33–38, find the distance between the point and line given.

33. $(1, 3, -2); x = 4, y = -3 + 4t, z = 11 + 5t$

34. $(4, 3, 3)$; $x = 2 + 2t, y = 5 - 5t, z = -1 - t$

35. $(2, 4, -1)$; $x = 5 + t, y = -2 + 3t, z = 3 + t$

36. $(-1, 0, 5)$; $\dfrac{x - 2}{2} = \dfrac{y - 1}{1} = \dfrac{z + 2}{3}$

37. $(1, 4, 2)$; $\dfrac{x - 1}{3} = \dfrac{y + 2}{1} = \dfrac{z - 4}{-2}$

38. $(3, -1, 4)$; $\dfrac{x + 2}{1} = \dfrac{y}{-2} = \dfrac{z + 4}{-2}$

In Problems 39–42, find the distance between the parallel planes.

39. $2x - y + 2z = 9, 2x - y + 2z = -12$

40. $x - 4y - 2z = 5, x - 4y - 2z = 10$

41. $x + 2y = 6, x + 2y = 1$

42. $x - y - z = 4, 2x - 2y - 2z = -3$

In Problems 43–46, find equations of the given line.

43. Containing $(2, 5, -1)$ and perpendicular to $2x - y + 3z + 2 = 0$

44. Containing $(4, -2, 3)$ and perpendicular to $3x + 2y - z + 6 = 0$

45. Containing $(2, -4, 5)$ and parallel to $x - y + 3z = 4$ and $3x - 3y + 2z = 5$

46. Containing $(4, 0, 5)$ and parallel to $2x - 5y + z + 1 = 0$ and $x + 2y - z + 2 = 0$

In Problems 47–52, indicate whether the given planes are parallel, perpendicular, coincident, or none of these.

47. $3x + y - 5z = 2, x + 2y + z = 4$

48. $4x - 2y + 2z = 6, 2x - y + z = 3$

49. $4x + y - z = 5, x - y + 2z = 2$

50. $4x - 2y + z = 1, x + y - 2z = 0$

51. $2x - y + 3z = 4, 6x - 3y + 9z = 5$

52. $x + 3y - z = 4, 2x - y + z = 3$

In Problems 53–56, find the point of intersection of the plane and the line.

53. $3x - 2y + z = 4$; $x = 2 + t, y = 1 - 2t, z = 2 - 5t$

54. $x + 2y - 4z = 12$; $x = 1 + t, y = -2 + 3t, z = -2t$

55. $3x - y + 4z = 7$; $x = 2 - t, y = 5 + 2t, z = 1 + t$

56. $2x + 3y - z = 2$; $x = 1 + t, y = -2 + 3t, z = 6 - t$

57. Does the line $x = 1, y = 5 + 4t, z = 2 + t$ lie in the plane $2x - y + 4z = 5$?

58. Give equations of each of the coordinate planes. Compare with Problem 33, page 342.

C   **59.** Prove Theorem 9.26.

9.7

CYLINDERS AND SPHERES

We now turn our attention to more complex surfaces, beginning with the cylinder. A cylinder is formed by a line (**generatrix**) moving along a curve (**directrix**) while remaining parallel to a fixed line. If the generatrix is parallel to one of the coordinate axes, the equation of the cylinder is quite simple.

THEOREM 9.27   *A nonlinear equation of the form*

$$f(x, y) = 0$$

is a cylinder with generatrix parallel to the z axis and directrix $f(x, y) = 0$ in the xy plane. Similar statements hold when one of the other variables is absent.

It is a simple matter to see why this is so. If $x = x_0$ and $y = y_0$ satisfies the equation $f(x, y) = 0$, then any point of the form (x_0, y_0, z), for any choice of z, is on the surface. But the set of all such points is a line parallel to the z axis. Thus any point on the curve $f(x, y) = 0$ in the xy plane determines a line parallel to the z axis in space. The result is then a cylinder.

EXAMPLE 1   Sketch $x^2 + y^2 = 4$.

SOLUTION   The surface is a cylinder with generatrix parallel to the z axis and directrix a circle in the xy plane. A portion of the cylinder is given in Figure 9.23.

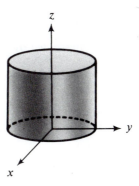

Figure 9.23

EXAMPLE 2   Sketch $y = z^2$.

SOLUTION   The surface is a cylinder with generatrix parallel to (or on) the x axis and directrix a parabola in the yz plane. A portion of the cylinder is given in Figure 9.24.

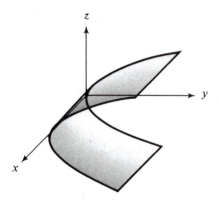

Figure 9.24

Another relatively simple surface is the sphere. The following theorems concerning the sphere are analogous to those for a circle and are proved in much the same way.

THEOREM 9.28   *A point (x, y, z) is on the sphere of radius r and center at (h, k, l) if and only if it satisfies the equation*

$$(x - h)^2 + (y - k)^2 + (z - l)^2 = r^2.$$

THEOREM 9.29   *Any sphere can be represented by an equation of the form*

$$Ax^2 + Ay^2 + Az^2 + Gx + Hy + Iz + J = 0,$$

where $A \neq 0$.

THEOREM 9.30   *An equation of the form*

$$Ax^2 + Ay^2 + Az^2 + Gx + Hy + Iz + J = 0,$$

where $A \neq 0$, represents a sphere, a point, or no locus.

The proofs are left to the student.

EXAMPLE 3   Give an equation for the sphere with center $(1, 3, -2)$ and radius 3.

SOLUTION   By Theorem 9.28, the equation is

$$(x - 1)^2 + (y - 3)^2 + (z + 2)^2 = 9$$

or

$$x^2 + y^2 + z^2 - 2x - 6y + 4z + 5 = 0.$$

EXAMPLE 4   Describe the locus of $x^2 + y^2 + z^2 + 2x - 4y - 8z + 5 = 0$.

SOLUTION   Let us put the equation into the form of Theorem 9.28 by completing squares.

$$x^2 + 2x \quad + y^2 - 4y \quad + z^2 - 8z \quad = -5$$
$$x^2 + 2x + 1 + y^2 - 4y + 4 + z^2 - 8z + 16 = -5 + 1 + 4 + 16$$
$$(x + 1)^2 + (y - 2)^2 + (z - 4)^2 = 16$$

This represents a sphere with center $(-1, 2, 4)$ and radius 4.

EXAMPLE 5   Describe the locus of

$$2x^2 + 2y^2 + 2z^2 - 2x + 6y - 4z + 7 = 0.$$

SOLUTION

$$x^2 + y^2 + z^2 - x + 3y - 2z + \frac{7}{2} = 0$$

$$x^2 - x \quad + y^2 + 3y \quad + z^2 - 2z \quad = -\frac{7}{2}$$

$$x^2 - x + \frac{1}{4} + y^2 + 3y + \frac{9}{4} + z^2 - 2z + 1 = -\frac{7}{2} + \frac{1}{4} + \frac{9}{4} + 1$$

$$\left(x - \frac{1}{2}\right)^2 + \left(y + \frac{3}{2}\right)^2 + (z - 1)^2 = 0$$

The equation represents the point $(1/2, -3/2, 1)$.

Electronic graphing can be a considerable aid in visualizing these surfaces in three dimensions. However, there are certain problems associated with it. The first problem is that, while most of the computer programs for graphing include three-dimensional graphs, the graphing calculators do not. Furthermore, after giving the equation—and it must again be solved for z as a function of x and y—you must describe the point from which it is to be viewed. If you have no idea of what the surface looks like, this can pose something of a problem; the view from some points may not be much help in visualizing the surface. Thus, a certain amount of trial and error may be necessary.

If you have a calculator that cannot graph in three dimensions, you can still use it to find parallel cross sections, which can help you visualize the surface. Nevertheless, this has its own problems. Suppose, for example, that we want cross sections on and parallel to the xy plane. We must assign a particular value to z, then solve the result for y as a function of x before entering the resulting equation. This must be repeated several times to get several cross sections. It is a tedious chore to do all of this, and it is not recommended for graphing spheres. However, a two-dimensional plot can be used to determine the directrix of a cylinder.

PROBLEMS

A In Problems 1–10, sketch the given surface.

1. $y^2 + z^2 = 1$

2. $x^2 + z^2 = 4$

3. $y = x^2$

4. $x^2 - z^2 = 1$

5. $xy = 4$

6. $x^2 + z^2 + 2x = 0$

7. $z = 4 - y^2$

8. $x = \sin z$

9. $y = x^3$

10. $z = \dfrac{y}{1 + y}$

In Problems 11–20, identify the equation as representing a sphere, a point, or no locus. If it is a sphere, give its center and radius. If it is a point, give its coordinates.

11. $x^2 + y^2 + z^2 - 2x + 4z - 4 = 0$

12. $x^2 + y^2 + z^2 + 6x - 10y + 2z + 19 = 0$

13. $x^2 + y^2 + z^2 - 8x + 4y - 10z + 46 = 0$

14. $x^2 + y^2 + z^2 + 6x - 8y - 2z + 22 = 0$

15. $2x^2 + 2y^2 + 2z^2 + 2x - 6y + 4z - 1 = 0$

16. $2x^2 + 2y^2 + 2z^2 - 2x + 2y - 10z + 13 = 0$

17. $9x^2 + 9y^2 + 9z^2 - 6x + 6y + 12z - 2 = 0$

18. $3x^2 + 3y^2 + 3z^2 + 4x - 2y - 8z + 7 = 0$

19. $4x^2 + 4y^2 + 4z^2 - 8x - 4y + 16z + 21 = 0$

20. $6x^2 + 6y^2 + 6z^2 - 6x - 4y - 3z = 0$

B *In Problems* 21–29, *find an equation(s) in the general form of the sphere(s) described.*

21. Center (4, 1, −2) and radius 3

22. Center (3, 1, 1) and containing the origin

23. Center (2, 4, 7) and tangent to $4x − 8y + z = 1$

24. Center (4, 1, −3) and tangent to $2x − y − 2z = 4$

25. Tangent to $x − 3y + 4z + 23 = 0$ at (1, 4, −3) with radius $\sqrt{26}$

26. Tangent to $x + 2y + 2z − 17 = 0$ at (1, 4, 4) with radius 3

27. Containing (3, 1, −1), (2, 5, 2), (−3, 0, 1), and (−1, 0, 0)

28. Containing (4, 1, 0), (−2, −1, 0), (0, 2, 1), and (1, 1, 1)

29. Center on the line $x = 5 + 3t$, $y = −1 + t$, $z = −2t$ and tangent to the three coordinate planes.

C 30. Prove Theorem 9.28

31. Prove Theorem 9.29

32. Prove Theorem 9.30

9.8

QUADRIC AND OTHER SURFACES

In the plane, a second-degree equation represents a parabola, an ellipse, a hyperbola, or a degenerate case of one of these. There are far more variations in space, where we have already seen that certain cylinders and the sphere are represented by second-degree equations. The **traces** in the coordinate planes of a given surface are simply the intersections of the surface with the coordinate planes. The traces in the coordinate planes of quadric surfaces, represented by second-degree equations, are conics or degenerate conics. We say that a quadric surface is in the standard position if its traces in the coordinate planes are in the standard position—that is, they have their center or vertex at the origin and axes along the coordinate axes. In the following discussion, all surfaces are assumed to be in the standard position.

The **ellipsoid** (see Figure 9.25 on page 366) is represented by an equation of the form

$$\frac{x^2}{a^2} + \frac{y^2}{b^2} + \frac{z^2}{c^2} = 1.$$

Its traces in the coordinate planes are ellipses (or circles). We have already seen a special case of this, an which $a = b = c$. In that case, we have a sphere. There are two other special cases. One is the **prolate spheroid**. Here, two of the denominators are equal and both are less than the third. It has the shape of a football and may be generated by rotating an ellipse about its major axis. The other case is the **oblate spheroid**, in which two of the denominators are equal and

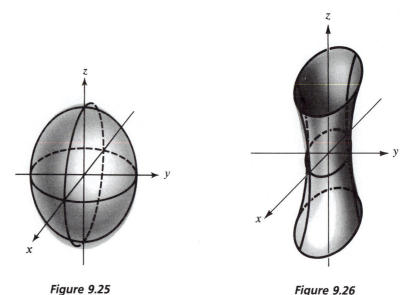

Figure 9.25   **Figure 9.26**

both greater than the third. It has the shape of a doorknob and may be generated by rotating an ellipse about its minor axis.

The **hyperboloid of one sheet** (see Figure 9.26) is represented by an equation of the form

$$\frac{x^2}{a^2} + \frac{y^2}{b^2} - \frac{z^2}{c^2} = 1.$$

Its traces in the xz plane and yz plane are hyperbolas; in the xy plane, it is an ellipse. If $a = b$, it may be generated by rotating a hyperbola about its conjugate axis.

The **hyperboloid of two sheets** (see Figure 9.27) is represented by an equation of the form

$$\frac{x^2}{a^2} - \frac{y^2}{b^2} - \frac{z^2}{c^2} = 1.$$

Its traces in the xy plane and xz plane are hyperbolas. It has no trace in the yz plane; however, if $|x| > a$, its intersection with a plane parallel to the yz plane is an ellipse. If $b = c$, it may be generated by rotating a hyperbola about its transverse axis.

The **elliptic paraboloid** (see Figure 9.28) is represented by an equation of the form

$$\frac{x^2}{a^2} + \frac{y^2}{b^2} = \frac{z}{c}.$$

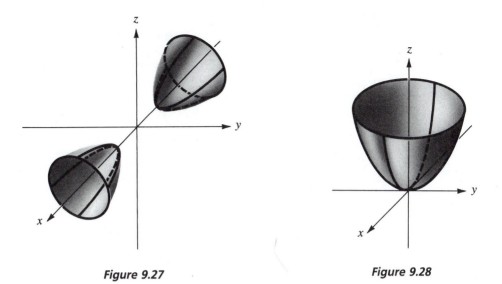

Figure 9.27

Figure 9.28

Its traces in the xz plane and yz plane are parabolas. Its trace in the xy plane is a single point. If $c > 0$, then its intersection with a plane parallel to and above the xy plane is an ellipse; below the xy plane there is no intersection. This situation is reversed if $c < 0$. In Figure 9.28, $c > 0$. If $a = b$, the elliptic paraboloid is generated by rotating a parabola about its axis.

The **hyperbolic paraboloid** (see Figure 9.29) is represented by an equation of the form

$$\frac{x^2}{a^2} - \frac{y^2}{b^2} = \frac{z}{c}.$$

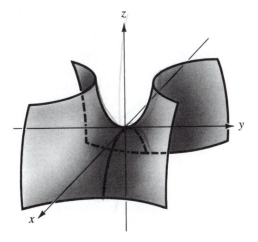

Figure 9.29

Its traces in the xz plane and yz plane are parabolas, one opening upward and the other down. Its trace in the xy plane is a pair of lines intersecting at the origin (a degenerate hyperbola). Its intersection with a plane parallel to the xy plane is a hyperbola. If $c > 0$, those hyperbolas above the xy plane have the transverse axis parallel to the x axis, while those below have it parallel to the y axis. If $c < 0$, this situation is reversed. In Figure 9.29, $c < 0$.

The **elliptic cone** (see Figure 9.30) is represented by an equation of the form

$$\frac{x^2}{a^2} + \frac{y^2}{b^2} - \frac{z^2}{c^2} = 0.$$

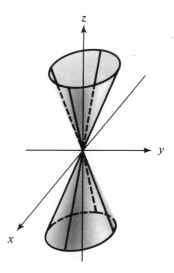

Figure 9.30

Its trace in the xz plane is a pair of lines intersecting at the origin. Its trace in the yz plane is also a pair of lines intersecting at the origin. Its trace in the xy plane is a single point at the origin. Its intersection with a plane parallel to the xy plane is an ellipse. If $a = b$, it is a circular cone.

EXAMPLE 1   Describe and sketch

$$9x^2 + 9y^2 - 4z^2 = 36.$$

SOLUTION   Dividing by 36, we have a hyperboloid of one sheet in the standard form.

$$\frac{x^2}{4} + \frac{y^2}{4} - \frac{z^2}{9} = 1$$

Since the denominators of the x^2 and y^2 terms are equal, the trace in the xy plane, as well as in any plane parallel to it, is a circle. The surface is shown in Figure 9.31.

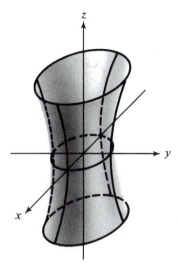

Figure 9.31

Although graphing calculators can graph functions of only a single variable—that is, two-dimensional graphs—they still can be a great help in graphing three-dimensional surfaces. They help by graphing the surface's traces in the coordinate planes and planes parallel to them. Let us see how this is done using the equation of Example 1.

EXAMPLE 2　Using the surface $9x^2 + 9y^2 - 4z^2 = 36$ of Example 1, find the traces in the three coordinate planes as well as in the planes $z = \pm 1$, $z = \pm 2$, $z = \pm 3$. Use these to sketch the surface.

SOLUTION　Let us begin by finding the intersections of the given surface with $z = 0$, $z = \pm 1$, $z = \pm 2$, and $z = \pm 3$. Substituting $z = 0$ into the given equation gives $9x^2 + 9y^2 = 36$ or $y = \pm\sqrt{4 - x^2}$. Repeating for the other values of z gives

$$z = \pm 1: \qquad 9x^2 + 9y^2 - 4 = 36 \quad \text{or} \quad y = \pm\frac{\sqrt{40 - 9x^2}}{3},$$

$$z = \pm 2: \qquad 9x^2 + 9y^2 - 16 = 36 \quad \text{or} \quad y = \pm\frac{\sqrt{52 - 9x^2}}{3},$$

$$z = \pm 3: \qquad 9x^2 + 9y^2 - 36 = 36 \quad \text{or} \quad y = \pm\sqrt{8 - x^2}.$$

These equations represent circles with radii 2, $2\sqrt{10}/3 = 2.11$, $2\sqrt{13}/3 = 2.40$, and $2\sqrt{2} = 2.83$. Their graphs are shown in Figure 9.32. The innermost circle is the intersection of the surface with the *xy* plane. The other three circles represent the intersection with planes 1, 2, and 3 units above the *xy* plane. As we go up or down from the *xy* plane the circles get larger, by amounts that increase as we get farther away from the *xy* plane. This is in keeping with what we have already seen in Example 1.

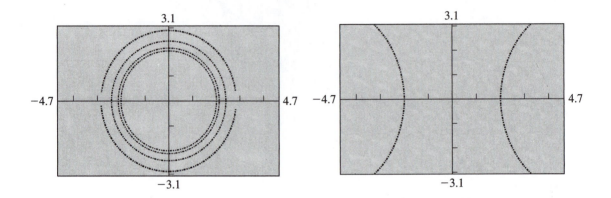

Figure 9.32   *Figure 9.33*

Now let us consider the intersections with the other two coordinate planes. The *xy* plane is $y = 0$ and the *yz* plane is $x = 0$. Substituting these values into the original equation, we get $9x^2 - 4z^2 = 36$ and $9y^2 - 4z^2 = 36$. Solving for *z* we get

$$z = \pm\frac{3}{2}\sqrt{x^2 - 4}$$

and

$$z = \pm\frac{3}{2}\sqrt{y^2 - 4}.$$

They are the same curves with different orientations. Their graphs are identical and given in Figure 9.33. The vertical axis here is the *z* axis (although it was represented as the *y* axis by the calculator). Now imagine the circles of Figure 9.32 in planes that are perpendicular to the plane of this graph and lying on or parallel to the *y* axis. The result is the surface of Figure 9.34.

EXAMPLE 3   Describe and sketch

$$9x^2 + 9y^2 - 4z^2 = 0.$$

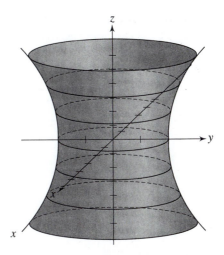

Figure 9.34

SOLUTION Again dividing by 36, we have

$$\frac{x^2}{4} + \frac{y^2}{4} - \frac{z^2}{9} = 0,$$

which is a circular cone with its axis the z axis. The cone is given in Figure 9.35.

Let us find traces in the same planes that were considered in Example 2. The trace of this surface with the xy plane ($z = 0$) is easily seen to be a single point at

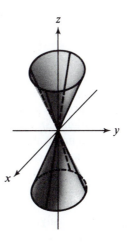

Figure 9.35

the origin. Using planes parallel to the xy plane, we have

$$z = \pm 1: \quad x^2 + y^2 = \frac{4}{9},$$

$$z = \pm 2: \quad x^2 + y^2 = \frac{16}{9},$$

$$z = \pm 3: \quad x^2 + y^2 = 4.$$

Even without a graphing calculator, these equations are easily seen to be circles with radii 2/3, 4/3, and 2. The trace in the yz plane ($x = 0$) is

$$\frac{y^2}{4} - \frac{z^2}{9} = 0$$

or

$$y = \pm \frac{2}{3} z$$

which is easily seen to be two lines that intersect at the origin. The trace in the xz plane is similar. This verifies what we have already seen in Figure 9.35.

Finding traces in the coordinate planes and in planes parallel to them can be used with more complex surfaces. Let us consider some examples.

EXAMPLE 4   Sketch the surface given by $z = (x^2 + y^2)(x^2 + y^2 - 4)$.

SOLUTION   We begin by taking the traces in the coordinate planes. In the xy plane ($z = 0$) we have

$$(x^2 + y^2)(x^2 + y^2 - 4) = 0$$

giving

$$x^2 + y^2 = 0 \quad \text{or} \quad x^2 + y^2 = 4.$$

These equations represent a single point at the origin and the circle of radius 2 with center at the origin. Since the point is represented by an equation in the form of a circle, cross sections above and below the xy plane are likely to be either single circles or pairs of circles. Let us consider the traces in the other two coordinate planes to see if they give us some hint.

The trace in the xz plane ($y = 0$) is $z = x^2(x^2 - 4)$. This is easily sketched with or without a graphing calculator. However, a graphing calculator allows us to locate the minimum easily. Thus we have the graph of Figure 9.36. By using CALC, we appear to have minima at $(\pm\sqrt{2}, -4)$. The trace in the yz plane is identical. This supports the idea that cross sections parallel to the xy plane are

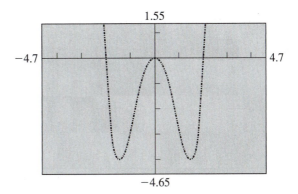

Figure 9.36

either circles (above the xy plane) or pairs of circles (below it). Let us see if we can find some of these circles.

We consider the plane $z = 1$. Substituting this into the given equation gives us

$$1 = (x^2 + y^2)(x^2 + y^2 - 4).$$

It is not at all clear from this equation that we have circles. But let us make the substitution $C = x^2 + y^2$, which gives

$$1 = C(C - 4)$$

$$C^2 - 4C - 1 = 0$$

$$C = \frac{4 \pm \sqrt{16 + 4}}{2} = 2 \pm \sqrt{5}.$$

Replacing C by $x^2 + y^2$ gives us $x^2 + y^2 = 2 \pm \sqrt{5}$. Since $2 - \sqrt{5}$ is negative, we are left with the single circle $x^2 + y^2 = 2 + \sqrt{5}$.

If we repeat the preceding argument using $z = -1$, we get the two circles $x^2 + y^2 = 2 \pm \sqrt{3}$. Thus our original conjecture is verified; these cross sections are either circles or pairs of circles, and the graph is as shown in Figure 9.37.

EXAMPLE 5   Sketch the surface determined by

$$z = \frac{4y}{x^2 + y^2 + 1}.$$

SOLUTION   The traces in the coordinate planes are: In the xy plane ($z = 0$), the line $y = 0$; in the xz plane ($y = 0$), the line $z = 0$; and in the yz plane ($x = 0$), we have the equation

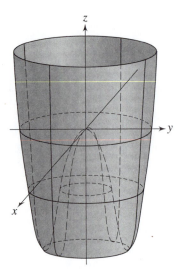

Figure 9.37

$$z = \frac{4y}{y^2 + 1}$$

with the graph shown in Figure 9.38. It has a maximum at $(1, 2)$ and a minimum at $(-1, -2)$.

Now let us put these traces together. This is done in Figure 9.39. Note that $y = 0$ in the xy plane and $z = 0$ in the xz plane represent the same line. Nevertheless, we get the additional information that the surface touches no other point of the xy plane and no other point of the xz plane. We need to find traces in planes parallel to one or more coordinate plane. What appears to be the best orientation to take? Planes parallel to the xy plane are likely to reveal little. Planes parallel to

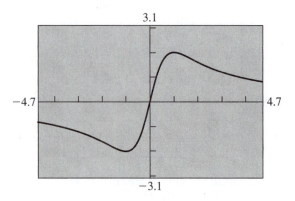

Figure 9.38

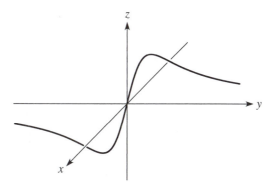

Figure 9.39

the other two coordinate planes are the only candidates. Let us consider planes parallel to the xz plane, starting with $x = \pm1$ and $x = \pm2$. This results in the equations

$$z = \frac{4y}{y^2 + 2} \quad \text{and} \quad z = \frac{4y}{y^2 + 5}.$$

Graphing these, together with the previously graphed

$$z = \frac{4y}{y^2 + 1},$$

we have the result shown in Figure 9.40.

Similarly, $y = \pm1$ and $y = \pm2$ give equations

$$z = \frac{\pm4}{x^2 + 2} \quad \text{and} \quad z = \frac{\pm8}{x^2 + 5}$$

with the graphs shown in Figure 9.41.

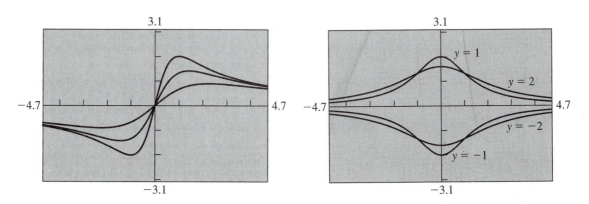

Figure 9.40　　　　　　　　　　　　　　　　　　**Figure 9.41**

Putting these graphs together gives us a reasonable approximation of the surface, as shown in Figure 9.42.

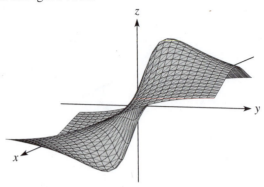

Figure 9.42

Picturing these three-dimensional surfaces is not a simple matter, as you can see. There are computer programs that do it for us quite rapidly. As of 1995, graphing calculators graph functions in only two dimensions. We can expect this to change in the near future.

PROBLEMS

B *Describe and sketch the quadric surfaces in Problems 1–25.*

1. $36x^2 + 9y^2 + 4z^2 = 36$

2. $4x^2 + 9y^2 + 9z^2 = 36$

3. $x - y^2 - z^2 = 0$

4. $4x^2 + 4y^2 - z^2 + 16 = 0$

5. $x^2 - y^2 + z^2 = 0$

6. $x^2 - y - z^2 = 0$

7. $x^2 + 2y + z^2 = 0$

8. $x^2 - 4y^2 - 4z^2 = 0$

9. $25x^2 - 4y^2 + 25z^2 + 100 = 0$

10. $16x^2 - 9y^2 - 9z^2 + 144 = 0$

11. $x^2 + 4y - z^2 = 0$

12. $25x^2 + 16y^2 + 25z^2 = 400$

13. $36x^2 - 9y^2 + 16z^2 + 144 = 0$

14. $16x^2 - 9y + 16z^2 = 0$

15. $16x^2 - 9y^2 - 9z^2 = 0$

16. $36x^2 - 4y - 9z^2 = 0$

17. $9x^2 - 36y^2 + 16z^2 + 144 = 0$

18. $x^2 + y^2 - 4z = 0$

19. $x^2 + y^2 - 4z^2 = 4$

20. $x^2 + y^2 + 4z^2 = 4$

21. $x^2 - y^2 - 9z = 0$

22. $9x^2 - y^2 - z^2 = 9$

23. $9x^2 - y^2 - z^2 = 0$

24. $x^2 + y^2 + 2z = 0$

25. $25x^2 - 4y^2 + 25z^2 = 100$

C *If the coordinate axes are translated with the origin moved to (h, k, l), the equations of translation are*

$$x' = x - h, \qquad y' = y - k, \qquad z' = z - l.$$

In many cases, translations may be carried out by completing the square. In Problems 26–30, translate and sketch.

26. $z = 4 - x^2 - 2y^2$

27. $z = 1 + x^2 + y^2$

28. $z = x^2 + y^2 + 2x + 4y + 7$

29. $x^2 + 4y^2 + 9z^2 + 2x + 16y - 18z - 10 = 0$

30. $x^2 + 4y^2 - z^2 - 2x - 24y - 8z + 17 = 0$

GRAPHING CALCULATOR

In Problems 31–34, use the traces in the coordinate planes and in planes parallel to them to sketch the surface represented by the given equation.

31. $z(x^2 + y^2) = 1$

32. $z = \dfrac{4(x^2 + y^2)}{(x^2 + y^2)^2 + 1}$

33. $8z = y^3(x^2 + 1)$

34. $z = \dfrac{x^2}{x^2 + y^2 + 1}$

APPLICATIONS

35. The hypotenuse of a right triangle is related to the other two sides by the Theorem of Pythagoras, $z^2 = x^2 + y^2$. The numbers that satisfy this equation lie on what surface?

36. Suppose a gas satisfies the ideal gas law, $PV = RT$, where P is the pressure in atmospheres, V is the volume in liters, T is the absolute temperature, and R is the ideal gas constant, 0.082. What does the graph of this equation look like? Note that none of the quantities P, V, and T can be negative.

37. The reflector of an automobile headlight is a circular paraboloid that is 8 inches in diameter and 2 inches deep. Give its equation with the vertex at the origin and the z axis on the axis of the paraboloid. Where is its focus?

38. The mirror of a reflecting telescope is a circular paraboloid that is 4 meters in diameter and 0.3 meter deep. Give its equation if the axes are placed with the origin at the vertex and the z axis on the axis of the paraboloid. Where is its focus?

39. The ceiling of a room is an ellipsoid that is 50 feet long, 20 feet wide, and 6 feet high. Give an equation for it with axes placed with the origin at the center and x, y, and z axes placed along the length, width, and height, respectively, of the ceiling.

40. A Cassegrain focus telescope has the dimensions shown in Figure 5.25 on page 184. Give equations for both mirrors. Place the z axis on the axis of the telescope, with any convenient placement of the origin. You need not have the same placement of axes for both surfaces.

41. Give equations of the mirrors of Problem 40 using the same placement of axes for both surfaces, with the z axis on the axis of the telescope and the origin at (**a**) the vertex of the paraboloid and (**b**) the center of the hyperboloid.

9.9

CYLINDRICAL AND SPHERICAL COORDINATES

Here we look at two other coordinate systems in space that are useful. The first of these is called a **cylindrical coordinate system**, which is especially convenient when the z axis is a line of symmetry. In this system, a point P with projection Q on the xy plane (see Figure 9.43) is represented by (r, θ, z), where (r, θ) is a polar representation of Q and z is the (directed) distance of P from the xy plane.

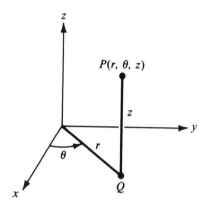

Figure 9.43

The relations between rectangular and cylindrical coordinates are the same as the relations between rectangular and polar coordinates as given by the following theorem.

THEOREM 9.31   *If (x, y, z) and (r, θ, z) represent the same point in rectangular and cylindrical coordinates, then*

$$x = r\cos\theta, \qquad y = r\sin\theta, \qquad z = z$$

or

$$r = \pm\sqrt{x^2 + y^2}, \qquad \theta = \arctan\frac{y}{x}, \qquad z = z.$$

EXAMPLE 1   Express in rectangular coordinates the point having cylindrical coordinates $(4, 30°, -2)$.

SOLUTION

$$x = r\cos\theta = 4\cos 30° = 4\cdot\frac{\sqrt{3}}{2} = 2\sqrt{3}$$

$$y = r \sin \theta = 4 \sin 30° = 4 \cdot \frac{1}{2} = 2$$

$$z = -2$$

Thus, the point is $(2\sqrt{3}, 2, -2)$.

EXAMPLE 2   Express in cylindrical coordinates the point having rectangular coordinates $(2, 2, 4)$.

SOLUTION

$$r = \pm\sqrt{x^2 + y^2} = \pm\sqrt{8} = \pm 2\sqrt{2}$$

$$\theta = \arctan\frac{y}{x} = \arctan 1 = \frac{\pi}{4} + \pi n$$

$$z = 4$$

There are two choices for r and infinitely many for θ. The choices we make are not independent of each other—the choice of one of them puts restrictions on the other. If we choose θ to be $\pi/4$ (or any first-quadrant angle), we must choose $r = 2\sqrt{2}$; if we choose θ to be $5\pi/4$ (or any third-quadrant angle), we must choose $r = -2\sqrt{2}$. Thus, two possible representations are

$$(2\sqrt{2}, \pi/4, 4) = (-2\sqrt{2}, 5\pi/4, 4).$$

EXAMPLE 3   Express in cylindrical coordinates the rectangular coordinate equation $x + y + z = 1$.

SOLUTION   Substituting $x = r \cos \theta$ and $y = r \sin \theta$, we have

$$r \cos \theta + r \sin \theta + z = 1.$$

EXAMPLE 4   Express in rectangular coordinates the cylindrical coordinate equation $r = z \sin \theta$.

SOLUTION   Multiplying both sides by r, we have

$$r^2 = z \cdot r \sin \theta \qquad \text{or} \qquad x^2 + y^2 = yz.$$

Another useful system for representing points in space uses **spherical coordinates**. In this system, a point is represented by (ρ, θ, φ), where ρ is the distance of the point from the origin, θ has the same meaning as in cylindrical coordinates, and φ is the angle between the positive end of the z axis and the segment joining the origin to the given point (see Figure 9.44). Since φ is undirected, it is never negative. Furthermore

$$0° \leq \varphi \leq 180°.$$

Likewise, we restrict $\rho : \rho \geq 0$.

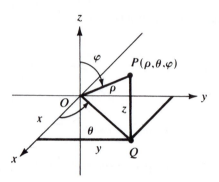

Figure 9.44

From Figure 9.44 we see that $\overline{OQ} = \rho \sin \varphi$. Thus

$$x = \overline{OQ} \cos \theta = \rho \sin \varphi \cos \theta,$$
$$y = \overline{OQ} \sin \theta = \rho \sin \varphi \sin \theta,$$

and

$$z = \rho \cos \varphi.$$

We can easily see from these that

$$\rho^2 = x^2 + y^2 + z^2.$$

Thus we have the following theorem.

THEOREM 9.32   *If (x, y, z) and (ρ, θ, φ) represent the same point in rectangular and spherical coordinates, respectively, then*

$$x = \rho \sin \varphi \cos \theta,$$
$$y = \rho \sin \varphi \sin \theta,$$
$$z = \rho \cos \varphi,$$

and

$$\rho^2 = x^2 + y^2 + z^2.$$

EXAMPLE 5

Express in rectangular coordinates the point having spherical coordinates $(2, \pi/6, 2\pi/3)$.

SOLUTION

$$x = \rho \sin \varphi \cos \theta$$
$$= 2 \sin \frac{2\pi}{3} \cdot \cos \frac{\pi}{6} = 2 \cdot \frac{\sqrt{3}}{2} \cdot \frac{\sqrt{3}}{2} = \frac{3}{2}$$
$$y = \rho \sin \varphi \sin \theta$$
$$= 2 \sin \frac{2\pi}{3} \sin \frac{\pi}{6} = 2 \cdot \frac{\sqrt{3}}{2} \cdot \frac{1}{2} = \frac{\sqrt{3}}{2}$$
$$z = \rho \cos \varphi$$
$$= 2 \cos \frac{2\pi}{3} = 2\left(-\frac{1}{2}\right) = -1$$

Thus the point is $(3/2, \sqrt{3}/2, -1)$.

EXAMPLE 6

Express in spherical coordinates the point having rectangular coordinates $(\sqrt{3}, \sqrt{3}, -\sqrt{2})$.

SOLUTION

$$\rho^2 = x^2 + y^2 + z^2$$
$$= 3 + 3 + 2 = 8$$
$$\rho = 2\sqrt{2}$$
$$z = \rho \cos \varphi$$
$$-\sqrt{2} = 2\sqrt{2} \cos \varphi$$
$$\cos \varphi = -\frac{1}{2}$$
$$\varphi = 120°$$
$$x = \rho \sin \varphi \cos \theta$$
$$\sqrt{3} = 2\sqrt{2} \cdot \frac{\sqrt{3}}{2} \cos \theta$$
$$\cos \theta = \frac{1}{\sqrt{2}}$$
$$\theta = 45°$$

The point is $(2\sqrt{2}, 45°, 120°)$.

EXAMPLE 7

Express in spherical coordinates the rectangular coordinate equation $x^2 + y^2 - z^2 = 0$.

SOLUTION

$$\rho^2 \sin^2\varphi \cos^2\theta + \rho^2 \sin^2\varphi \sin^2\theta - \rho^2 \cos^2\varphi = 0$$
$$\rho^2[\sin^2\varphi(\cos^2\theta + \sin^2\theta) - \cos^2\varphi] = 0$$
$$\rho^2(\sin^2\varphi - \cos^2\varphi) = 0$$
$$\sin^2\varphi - \cos^2\varphi = 0 \qquad \text{or} \qquad \rho = 0$$
$$\cos 2\varphi = 0$$
$$2\varphi = \frac{\pi}{2}, \frac{3\pi}{2}$$
$$\varphi = \frac{\pi}{4}, \frac{3\pi}{4}$$

Thus, we have $\varphi = \pi/4$, $\varphi = 3\pi/4$, and $\rho = 0$. $\varphi = \pi/4$ gives the top half of the cone, $\varphi = 3\pi/4$ gives the bottom half, and $\rho = 0$ gives a single point—the origin. Since $\rho = 0$ is included in both of the others, we may drop it. The final result is

$$\varphi = \frac{\pi}{4} \qquad \text{and} \qquad \varphi = \frac{3\pi}{4}.$$

EXAMPLE 8

Express in rectangular coordinates the spherical coordinate equation $\rho^2 \sin\varphi \cos\varphi \cos\theta = 1$.

SOLUTION

$$\rho^2 \sin\varphi \cos\varphi \cos\theta = 1$$
$$(\rho \sin\varphi \cos\theta)(\rho \cos\varphi) = 1$$
$$xz = 1$$

PROBLEMS

A

1. The following points are given in cylindrical coordinates. Express them in rectangular coordinates.
   - (a) $(2, 45°, 1)$
   - (b) $(3, 2\pi/3, -2)$
   - (c) $(1, 0°, 2)$
   - (d) $(0, \pi/4, -3)$

2. The following points are given in rectangular coordinates. Express them in cylindrical coordinates.
   - (a) $(1, 1, 3)$
   - (b) $(0, 2, -2)$
   - (c) $(-1, \sqrt{3}, 4)$
   - (d) $(-2\sqrt{3}, -2, 3)$

3. The following points are given in spherical coordinates. Express them in rectangular coordinates.

(a) $(3, 45°, 30°)$ (b) $(1, \pi/6, 0)$
(c) $(1, 90°, 45°)$ (d) $(2, 5\pi/6, 3\pi/4)$

4. The following points are given in rectangular coordinates. Express them in spherical coordinates.

(a) $(2, 2, 0)$ (b) $(2, 1, -2)$
(c) $(2, -\sqrt{3}, 4)$ (d) $(1, 1, \sqrt{3})$

B   **5.** The following points are given in cylindrical coordinates. Express them in spherical coordinates.

(a) $(3, 30°, 4)$ (b) $(2, \pi/4, -2)$
(c) $(0, 45°, 3)$ (d) $(2, \pi/2, -4)$

6. The following points are given in spherical coordinates. Express them in cylindrical coordinates.

(a) $(4, 45°, 30°)$ (b) $(2, 2\pi/3, \pi/2)$
(c) $(2, 210°, 135°)$ (d) $(3, \pi/6, 2\pi/3)$

In Problems 7–14, express the given rectangular coordinate equations in cylindrical and spherical coordinates.

7. $x^2 + y^2 = 4$ **8.** $x^2 + y^2 + z^2 = 4$
9. $x^2 + y^2 = z$ **10.** $x^2 - y^2 = z$
11. $x^2 - y^2 - z^2 = 1$ **12.** $x^2 - y^2 + z^2 = 1$
13. $x^2 + y^2 - z^2 = 1$ **14.** $4x^2 + 9y^2 + 9z^2 = 1$

In Problems 15–22, express the given equations in rectangular coordinates.

15. $z = r^2 \sin 2\theta$ **16.** $z = r^2 \cos 2\theta$
17. $z = r^2$ **18.** $z = 1 + \sin \theta$
19. $\rho \sin \varphi \tan \varphi \sin 2\theta = 2$ **20.** $\rho \sin \varphi = 1$
21. $\rho = \sin \varphi \cos \theta$ **22.** $\rho^2 \sin \varphi \cos \varphi = 1$

APPLICATIONS

23. The mirror of a reflecting telescope is a circular paraboloid that is 4 meters in diameter and 0.2 meter deep. Give equations for it in both cylindrical and spherical coordinates, with the origin at the vertex and z axis along the axis of the paraboloid.

24. Give cylindrical and spherical coordinate equations for the ceiling of Problem 39 on page 377.

25. Give cylindrical and spherical coordinate equations for the mirrors of Problem 40 on page 377.

REVIEW PROBLEMS

A   1. Find the point 1/3 of the way from $(-2, 1, 5)$ to $(1, 4, -4)$.

2. Find the projection of $\mathbf{u} = \mathbf{i} - 3\mathbf{j} + \mathbf{k}$ upon $\mathbf{v} = 4\mathbf{i} + 2\mathbf{j} + 3\mathbf{k}$.

3. Find the set of direction angles for the vector $\mathbf{v}$ with direction numbers $\{1, -1, \sqrt{2}\}$ if $\mathbf{v}$ is directed upward.

4. Suppose the line l_1 contains $(2, -2, 4)$ and $(5, 3, 0)$, while l_2 contains $(4, -3, 1)$ and $(3, -4, -1)$. Are l_1 and l_2 parallel, perpendicular, coincident, or none of these?

5. Suppose the line l_1 contains $(5, 1, -2)$ and $(2, -3, 1)$, while l_2 contains $(3, 8, 1)$ and $(-3, 0, 7)$. Are l_1 and l_2 parallel, perpendicular, coincident, or none of these?

In Problems 6 and 7, represent the given line in parametric form and in symmetric form.

6. Containing $(4, 0, 3)$ and $(5, 1, 3)$

7. Containing $(2, -3, 5)$; direction numbers $\{3, -2, 1\}$

In Problems 8 and 9, find the point of intersection (if any) of the given lines.

8. $\dfrac{x}{2} = \dfrac{y - 6}{-4} = \dfrac{z - 3}{1}$; $\dfrac{x - 2}{-1} = \dfrac{y - 4}{3} = \dfrac{z - 7}{1}$

9. $x = 2 + t,\ y = 3 - 2t,\ z = 3 + 5t$; $x = -3 + 2s,\ y = -1 + 3s,\ z = -s$

10. Find the distance between the lines

$$x = 4 - t,\ y = 3 + 2t,\ z = 4 + t \quad \text{and} \quad x = 3 + 2s,\ y = 5 - s,\ z = -1 + 4s.$$

In Problems 11 and 12, find an equation(s) of the plane(s) satisfying the given conditions.

11. Containing $(4, 0, -2)$ and perpendicular to $x = 4 - t,\ y = 3 + 2t,\ z = 1$

12. Containing $(2, 5, 1)$, $(3, -2, 4)$, and $(1, 0, 2)$

13. Find the distance between $(4, 3, 1)$ and $2x - 2y + z = 4$.

14. Find the distance between $x - 4y + z = 3$ and $x - 4y + z = 8$.

15. Find the distance between $(2, 5, -2)$ and the line $x = 3 + t,\ y = 4 - t,\ z = 2 + 2t$.

16. Find the distance between the lines

$$x = 1 + 2t,\ y = 4 + 3t,\ z = 2 - t \quad \text{and} \quad x = 4 + 2s,\ y = 2 + 3s,\ z = 5 - s.$$

17. Find the acute angle between the planes $2x - 5y + z = 3$ and $x + y - 2z = 1$.

In Problems 18 and 19, describe the locus of the given equation.

18. $x^2 + y^2 + z^2 - 2x - 4y + 8z + 5 = 0$

19. $2x^2 + 2y^2 + 2z^2 - 2x - 4y + 6z + 7 = 0$

In Problems 20–23, sketch and describe.

20. $9x^2 + 9y^2 - 4z^2 = 36$

21. $z^2 - x^2 = 4$

22. $9x^2 - 36y^2 + 4z^2 = 0$

23. $x^2 + y - z^2 = 0$

24. The following points are given in rectangular coordinates. Express them in cylindrical and spherical coordinates.

(a) $(2, 2, 1)$ (b) $(-1, 1, -2)$

25. Express $z = r \cos \theta$ in rectangular coordinates.

26. Express $z = 4x^2 + 4y^2$ in cylindrical and spherical coordinates.

27. Express $\rho = \sin \varphi \, (2 \cos \theta - \sin \theta)$ in rectangular coordinates.

B   **28.** The point $(5, 3, -2)$ is a distance 3 from the midpoint of the segment joining $(5, 7, 2)$ and $(1, 1, z)$. Find z.

29. Find the endpoints of the representative $\overrightarrow{AB}$ of $\mathbf{v}$ if $\mathbf{v} = 3\mathbf{i} - \mathbf{j} + 4\mathbf{k}$ and the midpoint of AB is $(2, 4, -3)$.

30. Suppose $\mathbf{u} = 2\mathbf{i} - \mathbf{j} + 3\mathbf{k}$ and $\mathbf{v} = \mathbf{i} + 4\mathbf{j} - \mathbf{k}$. Express $\mathbf{u}$ in the form $\mathbf{u} = \mathbf{p} + \mathbf{q}$, where $\mathbf{p} = k\mathbf{v}$ and $\mathbf{q} \cdot \mathbf{v} = 0$. Interpret geometrically.

In Problems 31 and 32, find equations for the line described.

31. Containing $(3, 5, -2)$ and perpendicular to the plane containing

$$x = 5 - t, y = 2 + 3t, z = 4 + t \quad \text{and} \quad x = 3 - s, y = 5 + 3s, z = -1 + s$$

32. Perpendicular at $(4, 2, 3)$ to the plane containing $(4, 2, 3)$, $(-1, 3, 1)$, and $(2, 5, -3)$.

33. Find an equation(s) of the plane(s) containing $(4, 2, -3)$ and perpendicular to

$$2x - y + 4z = 5 \quad \text{and} \quad x + 3y - z = 2.$$

34. Find, in parametric form, the line of intersection of

$$2x - y + 3z = 2 \quad \text{and} \quad x + 3y + 2z = 4.$$

35. Sketch and describe $x^2 + y^2 + 4z^2 - 2x + 4y + 1 = 0$.

C   **36.** Prove or show to be false: $\mathbf{u} \times (\mathbf{v} \times \mathbf{w}) = (\mathbf{u} \times \mathbf{v}) \times \mathbf{w}$.

Appendix **A** Trigonometry Review

The trigonometric functions are periodic, or repeating; they are very useful in the study of other periodic functions. A review of the trigonometric functions here will probably be useful.

An *angle* consists of a pair of rays (half-lines) OA and OB with a common end point O, called the *vertex* of the angle. An angle is in *standard position* if its vertex is at the origin and one of the rays, the *initial side*, is on the right half of the x axis. The other ray, the *terminal side*, is looked upon as a rotation of the initial side in either a clockwise or counterclockwise direction. The amount of this rotation determines the measure of the angle. A counterclockwise rotation has a positive measure and a clockwise rotation has a negatuive measure. The two most common ways of measuring angles are in *degrees* and *radians*. While measuring angles in degrees is convenient for many purposes, radian measure is more natural when we differentiate and integrate functions. If the vertex of the angle is at the center of a circle (see Figure 1), then the radian measure of the angle θ is the length s of the arc subtended by θ* divided by the radius of the circle:

$$\theta = \frac{s}{r}.$$

Since a complete rotation of 360° subtends an arc whose length is the circumference of the circle, it follows that 360° is equivalent to

$$\theta = \frac{s}{r} = \frac{2\pi r}{r} = 2\pi \text{ radians}.$$

With this result we can find a formula for the area of a sector (a pie-shaped region). If the central angle of

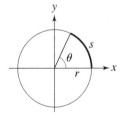

Figure 1

---

* While an angle (geometric) and its measure (arithmetic) are not the same, we follow the usual practice of using a single symbol θ to represent both. There should be no confusion between the two ideas because the context will make it clear which one we are referring to.

a sector is 2π radinas, the sector becomes a circle with area πr^2. Assuming that the area of a sector is proportional to the measure of the central angle, we compare the sector with central angle 2π and area πr^2 with one having central angle θ and area A. This gives the ratio

$$\frac{A}{\pi r^2} = \frac{\theta}{2\pi};$$

therefore the area of the sector is

$$A = \frac{1}{2}r^2\theta.$$

Of course, θ is the *radian* measure of the angle.

The six common trigonometric functions are defined in terms of the coordinates of a point other than the origin on the terminal side (see Figure 2). Thus we have:

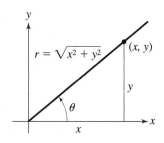

$$\sin\theta = \frac{y}{r}, \qquad \tan\theta = \frac{y}{x}, \qquad \sec\theta = \frac{r}{x},$$

$$\cos\theta = \frac{x}{r}, \qquad \cot\theta = \frac{x}{y}, \qquad \csc\theta = \frac{r}{y}.$$

Figure 2

EXAMPLE 1   If $\sin\theta = 4/5$ and θ is not a first-quadrant angle, find the values of the other five trigonometric functions.

Since $\sin\theta = y/r$ and r is always positive, it follows that $\sin\theta$ is positive when y is positive—that is, for the first- and second-quadrant angles. But we are given that θ is not a first quadrant angle; then it must be in the second quadrant (see Figure 3). Using the theorem of Pythagoras on the triangle of Figure 3 with $y = 4$ and $r = 5$, it is easily seen that $x = -3$. We now use the definitions to find the other five trigonometric functions:

$$\tan\theta = \frac{y}{x} = -\frac{4}{3}, \qquad \sec\theta = \frac{r}{x} = -\frac{5}{3},$$

$$\cos\theta = \frac{x}{r} = -\frac{3}{5}, \qquad \cot\theta = \frac{x}{y} = -\frac{3}{4}, \qquad \csc\theta = \frac{r}{y} = \frac{5}{4}.$$

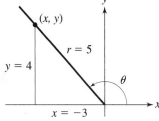

Figure 3

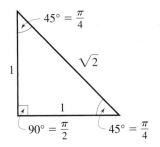
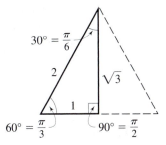

Figure 4

The definitions of the trigonometric functions together with the two familiar triangles of Figure 4 allow us to evaluate the trigonometric functions for any multiple of $30° = \pi/6$ or $45° = \pi/4$.

EXAMPLE 2

Find the exact value of $\cos(3\pi/4)$ and $\tan(-\pi/3)$.

By placing the triangles of Figure 4 on the coordinate axes as shown in Figure 5 and noting which coordinates are negative, we may refer to the definitions to get the desired results:

$$\cos\frac{3\pi}{4} = \frac{x}{r} = \frac{-1}{\sqrt{2}} = -\frac{1}{\sqrt{2}} \quad \text{and} \quad \tan-\frac{\pi}{3} = \frac{y}{x} = \frac{-\sqrt{3}}{1} = -\sqrt{3}.$$

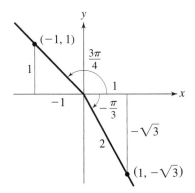

Figure 5

The definitions of the trigonometric functions dictate the following simple identities.

$$\cot \theta = \frac{1}{\tan \theta}, \qquad \sec \theta = \frac{1}{\cos \theta}, \qquad \csc \theta = \frac{1}{\sin \theta},$$

$$\tan \theta = \frac{\sin \theta}{\cos \theta}, \qquad \cot \theta = \frac{\cos \theta}{\sin \theta}.$$

Only slightly more difficult is the following derivation of a very important identity.

By the Pythagorean theorem,

$$x^2 + y^2 = r^2.$$

Dividing by r^2 and using our definitions, we have

$$\frac{x^2}{r^2} + \frac{y^2}{r^2} = 1$$

$$\left(\frac{x}{r} \right)^2 + \left(\frac{y}{r} \right)^2 = 1$$

$$\cos^2 \theta + \sin^2 \theta = 1.$$

Once we have this identity, we may find two other useful ones by dividing through by $\cos^2 \theta$ and by $\sin^2 \theta$.

$$\frac{\cos^2 \theta}{\cos^2 \theta} + \frac{\sin^2 \theta}{\cos^2 \theta} = \frac{1}{\cos^2 \theta},$$

$$1 + \tan^2 \theta = \sec^2 \theta.$$

Similarly, $\cot^2 \theta + 1 = \csc^2 \theta$.

Other common identities are given in Appendix B.

The graphs of the six common trigonometric functions are given in Figure 6. Note that all of them are periodic (repeating). To say that a function f has period p means that $f(x + p) = f(x)$ for all x; in other words, the function repeats itself after every p units. The period of $y = \tan x$ and $y = \cot x$ is π; the other four functions have period 2π. When considering algebraic equations of the form $y = P(x)/Q(x)$, we find vertical asymptotes at those values of x that make $Q(x) = 0$. If there is no denominator $[Q(x) = 1]$, there is no vertical asymptote. Nevertheless, four of the graphs of the trigonometric functions have vertical asymptotes even though there is no denominator. The reason for this is that we have defined the trigonometric functions as *ratios* of certain lengths; there really are denominators which are not immediately apparent.

Since the graphs of $y = \sin x$ and $y = \cos x$ are waves, they are the most important of the six. Let us see how they are altered by changing certain constants. The equation

$$y = A \sin B(x - C)$$

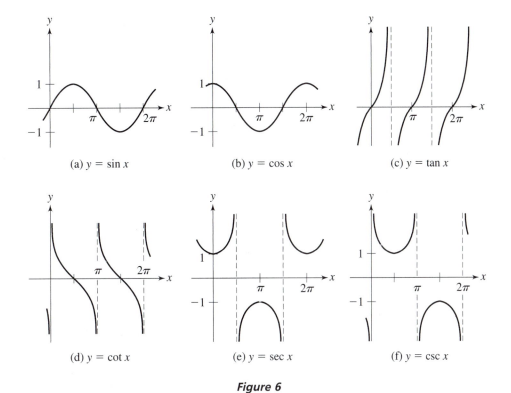

(a) $y = \sin x$ (b) $y = \cos x$ (c) $y = \tan x$

(d) $y = \cot x$ (e) $y = \sec x$ (f) $y = \csc x$

Figure 6

has a graph of the form shown in Figure 7. The amplitude is $|A|$, with the curve inverted if A is negative. The period is $2\pi/|B|$, again with the curve inverted if B is negative. The displacement is C. If C is positive, the displacement is to the right; if it is negative, there is a negative (or left) displacement. Note, however, that we have $x - C$ in the given equation. Thus $y = \sin(x + 3)$ has a negative displacement; $C = -3$. We have the same results with the cosine curve except that a negative value of B does not invert the curve.

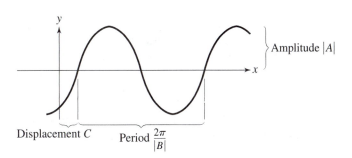

Displacement C Period $\dfrac{2\pi}{|B|}$

Figure 7

EXAMPLE 3

Sketch $y = 3 \sin 2x$.

First of all, note that $-1 \leq \sin x \leq 1$. Thus the factor of 3 in $3 \sin 2x$ alters this range by a factor of 3. The fact that we have $\sin 2x$ instead of $\sin x$ does not alter the range. Now it takes one complete cycle for whatever we are taking the sine of to go from 0 to 2π; that is, we have one complete cycle for

$$0 \leq 2x \leq 2\pi \qquad \text{or} \qquad 0 \leq x \leq \pi.$$

We have a sine curve with an amplitude of 3, a period of $2\pi/2 = \pi$, and no displacement. The result is shown in Figure 8.

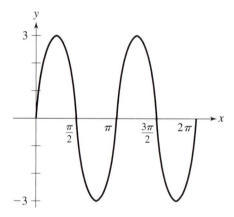

Figure 8

EXAMPLE 4

Sketch $y = 4 \cos\left(2x + \dfrac{\pi}{2}\right)$.

First let us write the equation in the form

$$y = 4 \cos 2\left(x + \frac{\pi}{4}\right).$$

Now we see that we have a cosine curve with an amplitude of 4, a period of $2\pi/2 = \pi$, and a displacement of $-\pi/4$, as shown in Figure 9.

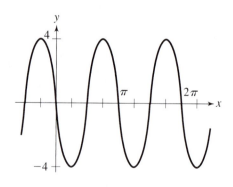

Figure 9

EXAMPLE 5   Sketch $y = \cos x + \sin 2x$.

The method of addition of ordinates is useful for equations of the form $y = f(x) + g(x)$. We note that $y = y_1 + y_2$, where $y_1 = f(x)$ and $y_2 = g(x)$. Now we sketch both $y_1 = f(x)$ and $y_2 = g(x)$ on the same coordinate axes; and for each value of x we add the y values y_1 and y_2 from the two graphs to get the y of the original equation. Thus we sketch $y = \cos x$ and $y = \sin 2x$ and add the ordinates; the result is shown as the solid curve in Figure 10.

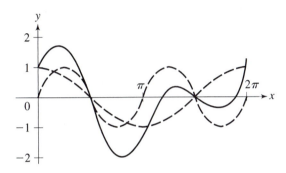

Figure 10

EXAMPLE 6   Sketch $y = x + \sin x$.

Perhaps you wonder how we can add x and $\sin x$ when x is an angle and $\sin x$ is a number. Actually, both x and $\sin x$ are numbers. Although it is convenient to talk about the sine of an angle, we actually take trigonometric functions, not of angles, but of numbers. The numbers are simply the *measures* of angles. It is quite possible to consider trigonometric functions of numbers idependently of any angular interpretations; but if we do want to impose such an interpretation, the value of x is the measure of an angle in radians. Again, addition of ordinates works very well and Figure 11 is self-explanatory.

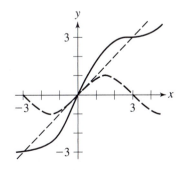

Figure 11

PROBLEMS

A   **1.** Express the following degree measures in radian measure.

$45°$, $-210°$, $270°$, $30°$, $-180°$, $-60°$, $135°$, $150°$.

2. Express the following radian measures in degree measure.

$\pi/3$, π, $3\pi/4$, $-\pi/2$, $5\pi/6$, $-2\pi/3$, $3\pi/2$, $10\pi/6$,

3. If $\tan \theta = 5/12$ and $\sin \theta$ is negative, find the values of all six trigonometric functions.

4. If $\sec \theta = -4/3$ and θ is not in the third quadrant, find the values of the remaining five trigonometric functions.

In Problems 5–8, find exact values of the given trigonometric functions.

5. $\tan 225°$ **6.** $\sin(-120°)$

7. $\cos(7\pi/6)$ **8.** $\sin(-3\pi/4)$

B   *In Problems 9–18, sketch one complete cycle.*

9. $y = 3 \sin x$.   **10.** $y = 2 \cos 4x$.

11. $y = 4 \sin \pi x$.   **12.** $y = 2 \sin(2x + \pi)$.

13. $y = -\cos\left(x - \dfrac{\pi}{3}\right)$.   **14.** $y = 3 \cos\left(2\pi x + \dfrac{\pi}{2}\right)$.

15. $y = 2 \sin x + \sin 2x$.   **16.** $y = \cos x - \sin 2x$.

17. $y = 3 \cos x + \sin x$.   **18.** $y = 4 \sin x + 2 \sin 2x - \sin 4x$.

In Problems 19 and 20, sketch the graph.

19. $y = x - \sin x$.   **20.** $y = x^2 + \sin x$.

ELEMENTARY ALGEBRA

(1) Quadratic formula

The solutions of the quadratic equation $ax^2 + bx + c = 0$ $(a \neq 0)$ are

$$x = \frac{-b \pm \sqrt{b^2 - 4ac}}{2a}.$$

(2) Binomial theorem (*n* a positive integer)

$$(a + b)^n = a^n + na^{n-1}b + \frac{n(n-1)}{2!}a^{n-2}b^2 + \frac{n(n-1)(n-2)}{3!}a^{n-3}b^3 + \cdots$$

$$+ \frac{n(n-1)(n-2)\cdots(n-r+1)}{r!}a^{n-r}b^r + \cdots + nab^{n-1} + b^n.$$

(3) Exponents

$$a^m \cdot a^n = a^{m+n}.$$

$$\frac{a^m}{a^n} = a^{m-n} \qquad (a \neq 0).$$

$$(a^m)^n = a^{mn}.$$

$$(ab)^n = a^n b^n.$$

$$a^{-n} = \frac{1}{a^n} \qquad (a \neq 0).$$

$$a^{p/q} = \sqrt[q]{a^p} = (\sqrt[q]{a})^p \qquad (a > 0).$$

(4) Logarithms

$$\log_a x + \log_a y = \log_a xy.$$

$$\log_a x - \log_a y = \log_a \frac{x}{y}.$$

$$n \log_a x = \log_a x^n.$$

$$\frac{1}{n} \log_a x = \log_a \sqrt[n]{x}$$

$$\log_a x = \frac{\log_b x}{\log_b a}.$$

$$\log_a 1 = 0, \qquad \log_a a = 1.$$

$$a^{\log_a x} = x.$$

(5) Determinants

$$\begin{vmatrix} a & b \\ c & d \end{vmatrix} = ad - bc.$$

$$\begin{vmatrix} a_1 & b_1 & c_1 \\ a_2 & b_2 & c_2 \\ a_3 & b_3 & c_3 \end{vmatrix} = a_1 \begin{vmatrix} b_2 & c_2 \\ b_3 & c_3 \end{vmatrix} - b_1 \begin{vmatrix} a_2 & c_2 \\ a_3 & c_3 \end{vmatrix} + c_1 \begin{vmatrix} a_2 & b_2 \\ a_3 & b_3 \end{vmatrix}.$$

B.2

TRIGONOMETRY

(1) Trigonometric identities

$$\tan x = \frac{\sin x}{\cos x}, \qquad \cot x = \frac{\cos x}{\sin x},$$

$$\sec x = \frac{1}{\cos x}, \qquad \csc x = \frac{1}{\sin x}.$$

$$\sin^2 x + \cos^2 x = 1$$

$$\tan^2 x + 1 = \sec^2 x.$$

$$1 + \cot^2 x = \csc^2 x.$$

$$\sin(x + y) = \sin x \cos y + \cos x \sin y.$$

$$\sin(x - y) = \sin x \cos y - \cos x \sin y.$$

$$\cos(x + y) = \cos x \cos y - \sin x \sin y.$$

$$\cos(x - y) = \cos x \cos y + \sin x \sin y.$$

$$\tan(x + y) = \frac{\tan x + \tan y}{1 - \tan x \tan y}.$$

$$\tan(x - y) = \frac{\tan x - \tan y}{1 + \tan x \tan y}.$$

$$\sin 2x = 2 \sin x \cos x.$$

$$\begin{aligned}
\cos 2x &= \cos^2 x - \sin^2 x \\
&= 1 - 2 \sin^2 x \\
&= 2 \cos^2 x - 1.
\end{aligned}$$

$$\tan 2x = \frac{2 \tan x}{1 - \tan^2 x}.$$

$$\sin \frac{x}{2} = \pm \sqrt{\frac{1 - \cos x}{2}}.$$

$$\cos \frac{x}{2} = \pm \sqrt{\frac{1 + \cos x}{2}}.$$

$$\tan \frac{x}{2} = \pm \sqrt{\frac{1 - \cos x}{1 + \cos x}} = \frac{1 - \cos x}{\sin x} = \frac{\sin x}{1 + \cos x}.$$

$$\sin x + \sin y = 2 \sin \frac{x + y}{2} \cos \frac{x - y}{2}.$$

$$\sin x - \sin y = 2 \cos \frac{x + y}{2} \sin \frac{x - y}{2}.$$

$$\cos x + \cos y = 2 \cos \frac{x + y}{2} \cos \frac{x - y}{2}.$$

$$\cos x - \cos y = 2 \sin \frac{x + y}{2} \sin \frac{x - y}{2}.$$

(2) Triangles

Law of sines: $\dfrac{a}{\sin A} = \dfrac{b}{\sin B} = \dfrac{c}{\sin C}.$

Law of cosines: $c^2 = a^2 + b^2 - 2ab \cos C.$

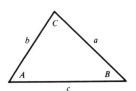

(3) Reduction formulas

$$\sin \alpha = -\cos(\alpha + \pi/2) = -\sin(\alpha + \pi) = +\cos(\alpha + 3\pi/2).$$

$$\sin \alpha = +\cos(\alpha - \pi/2) = -\sin(\alpha - \pi) = -\cos(\alpha - 3\pi/2).$$

$$\cos \alpha = +\sin(\alpha + \pi/2) = -\cos(\alpha + \pi) = -\sin(\alpha + 3\pi/2).$$

$$\cos \alpha = -\sin(\alpha - \pi/2) = -\cos(\alpha - \pi) = +\sin(\alpha - 3\pi/2).$$

(4)   Signs of trigonometric functions

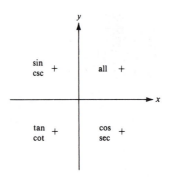

(5)   Trigonometric functions of multiples of 30° and 45°

Degrees	Radians	sin	cos	tan
0	0	0	1	0
30	$\pi/6$	1/2	$\sqrt{3}/2$	$1/\sqrt{3}$
45	$\pi/4$	$1/\sqrt{2}$	$1/\sqrt{2}$	1
60	$\pi/3$	$\sqrt{3}/2$	1/2	$\sqrt{3}$
90	$\pi/2$	1	0	—
120	$2\pi/3$	$\sqrt{3}/2$	$-1/2$	$-\sqrt{3}$
135	$3\pi/4$	$1/\sqrt{2}$	$-1/\sqrt{2}$	-1
150	$5\pi/6$	1/2	$-\sqrt{3}/2$	$-1/\sqrt{3}$
180	π	0	-1	0
210	$7\pi/6$	$-1/2$	$-\sqrt{3}/2$	$1/\sqrt{3}$
225	$5\pi/4$	$-1/\sqrt{2}$	$-1/\sqrt{2}$	1
240	$4\pi/3$	$-\sqrt{3}/2$	$-1/2$	$\sqrt{3}$
270	$3\pi/2$	-1	0	—
300	$5\pi/3$	$-\sqrt{3}/2$	1/2	$-\sqrt{3}$
315	$7\pi/4$	$-1/\sqrt{2}$	$1/\sqrt{2}$	-1
330	$11\pi/6$	$-1/2$	$\sqrt{3}/2$	$-1/\sqrt{3}$
360	2π	0	1	0

Appendix **C** *Theorems from Plane Geometry*

INTERSECTING AND PARALLEL LINES

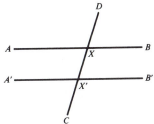

$$\sphericalangle AXC = \sphericalangle BXD$$

$$AB \parallel A'B' \Rightarrow \sphericalangle AXC = \sphericalangle BXD = \sphericalangle A'X'C = \sphericalangle B'X'D$$

$$\sphericalangle AXC = \sphericalangle A'X'C \Rightarrow AB \parallel A'B'$$

TRIANGLES

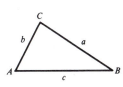 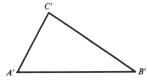

$$a^2 = b^2 + c^2 - 2bc \cos A$$

$$\sphericalangle A + \sphericalangle B + \sphericalangle C = 180°$$

The medians of a triangle are concurrent at a point that is 2/3 of the way from a vertex to the midpoint of the opposite side.

$$\triangle ABC \cong \triangle A'B'C' \Rightarrow \begin{cases} AB = A'B', & BC = B'C', & AC = A'C' \\ \sphericalangle A = \sphericalangle A', & \sphericalangle B = \sphericalangle B', & \sphericalangle C = \sphericalangle C' \end{cases}$$

$$\left.\begin{array}{l} AB = A'B', \quad BC = B'C', \quad AC = A'C' \quad \Rightarrow \\ AB = A'B', \quad AC = A'C', \quad \angle A = \angle A' \quad \Rightarrow \\ AB = A'B', \quad \angle A = \angle A', \quad \angle B = \angle B' \\ \qquad\qquad\qquad (\angle C = \angle C') \end{array}\right\} \Rightarrow \right\} \triangle ABC \cong \triangle A'B'C'$$

$$\triangle ABC \sim \triangle A'B'C' \Rightarrow \begin{cases} \dfrac{AB}{A'B'} = \dfrac{BC}{B'C'} = \dfrac{AC}{A'C'} \\ \angle A = \angle A', \; \angle B = \angle B', \; \angle C = \angle C' \end{cases}$$

$$\left.\begin{array}{l} \dfrac{AB}{A'B'} = \dfrac{BC}{B'C'} = \dfrac{AC}{A'C'} \qquad\qquad\quad \Rightarrow \\[2mm] \dfrac{AB}{A'B'} = \dfrac{AC}{A'C'}, \; \angle A = \angle A' \qquad\qquad \Rightarrow \\[2mm] \angle A = \angle A', \; \angle B = \angle B', \; (\angle C = \angle C') \Rightarrow \end{array}\right\} \triangle ABC \sim \triangle A'B'C'$$

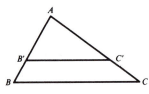

$$B'C' \parallel BC \Rightarrow \triangle AB'C' \sim \triangle ABC$$

RIGHT TRIANGLES

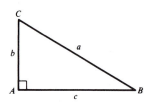

$$\angle A = 90° \Leftrightarrow \begin{cases} 1. \; \angle B + \angle C = 90° \\ 2. \; a^2 = b^2 + c^2 \end{cases}$$

ISOSCELES TRIANGLES

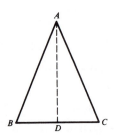

$$AB = AC \Leftrightarrow \angle B = \angle C$$

$$AB = AC, \, AD \perp BC \Rightarrow BD = DC$$

$$AB = AC, \, BD = DC \Rightarrow AD \perp BC$$

EQUILATERAL TRIANGLES

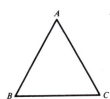

$$AB = BC = AC \Leftrightarrow \angle A = \angle B = \angle C = 60°$$

PARALLELOGRAMS

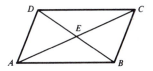

$$ABCD \text{ is a parallelogram} \Leftrightarrow \begin{cases} 1. & AB \parallel CD, AD \parallel BC \\ 2. & AB = CD, AD = BC \\ 3. & \angle A = \angle C, \angle B = \angle D \\ 4. & AE = EC, BE = ED \end{cases}$$

RECTANGLES

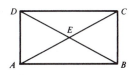

$$ABCD \text{ is a rectangle} \Rightarrow \begin{cases} AB \parallel CD, AD \parallel BC \\ AB = CD, AD = BC \\ \angle A = \angle B = \angle C = \angle D = 90° \\ AE = EC, BE = ED \\ AC = BD \end{cases}$$

$$\left. \begin{array}{l} ABCD \text{ is a parallelogram}, \angle A = 90° \Rightarrow \\ ABCD \text{ is a parallelogram}, AC = BD \Rightarrow \end{array} \right\} ABCD \text{ is a rectangle}$$

RHOMBUSES

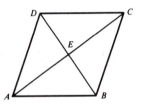

$$ABCD \text{ is a rhombus} \Leftrightarrow \begin{cases} AB \parallel CD, \ BC \parallel AD \\ AB = BC = CD = DA \\ AE = EC, \ BE = ED \\ AC \perp BD \end{cases}$$

$$\left. \begin{array}{l} ABCD \text{ is a parallelogram}, \ AB = BC \Rightarrow \\ ABCD \text{ is a parallelogram}, \ AC \perp BD \Rightarrow \end{array} \right\} ABCD \text{ is a rhombus}$$

SQUARES

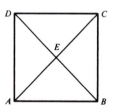

$$ABCD \text{ is a square} \Rightarrow \begin{cases} AB \parallel CD, \ AD \parallel BC \\ AB = BC = CD = DA \\ \angle A = \angle B = \angle C = \angle D = 90° \\ AC = BD, \ AC \perp BD \\ AE = EC, \ BE = ED \end{cases}$$

$$\left. \begin{array}{l} ABCD \text{ is a rectangle}, \ AB = BC \Rightarrow \\ ABCD \text{ is a rectangle}, \ AC \perp BD \Rightarrow \\ ABCD \text{ is a rhombus}, \ \angle A = 90° \Rightarrow \\ ABCD \text{ is a rhombus}, \ AC = BD \Rightarrow \end{array} \right\} ABCD \text{ is a square}$$

CIRCLES

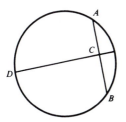

$CD \perp AB$, $AC = CB \Rightarrow$ center is on CD

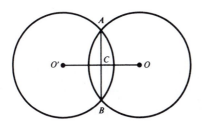

O, O' are centers $\Rightarrow \begin{cases} OO' \perp AB \\ AC = CB \end{cases}$

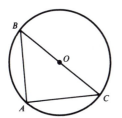

center O is on $BC \Rightarrow AB \perp AC$

Answers to Odd-Numbered Problems

CHAPTER 1

Section 1.2 [pages 9–11]

1. $\sqrt{65}$ **3.** 2   **5.** $\sqrt{37}/2$ **7.** $\sqrt{6}$ **9.** Collinear   **11.** Not collinear

13. Collinear   **15.** Right triangle   **17.** Right triangle   **19.** $-3, 5$ **21.** 2, 3

23. Since $(5, 2)$ is equidistant from A and B $(d = \sqrt{17})$, it is on the perpendicular bisector.

25. Lettering the vertices A, B, C, and D, respectively, it follows that $\overline{AB} = \overline{CD} = 3$ and $\overline{BC} = \overline{AD} = \sqrt{10}$.

27. $(0, -1)$ inside; $(1, 7)$, $(2, 0)$, $(-5, 7)$, $(-5, -1)$, $(-6, 6)$ on; $(-3, 8)$, $(4, 2)$ outside

29. $A = R - A_1 - A_2 - A_3$

$= (x_2 - x_1)(y_3 - y_2) - \tfrac{1}{2}(x_2 - x_1)(y_1 - y_2) - \tfrac{1}{2}(x_2 - x_3)(y_3 - y_2) - \tfrac{1}{2}(x_3 - x_1)(y_3 - y_1)$

$= \tfrac{1}{2}(2x_2y_3 - 2x_2y_2 - 2x_1y_3 + 2x_1y_2 - x_2y_1 + x_2y_2 + x_1y_1 - x_1y_2 - x_2y_3 + x_2y_2$

$\quad + x_3y_3 - x_3y_2 - x_3y_3 + x_3y_1 + x_1y_3 - x_1y_1)$

$= \tfrac{1}{2}(x_1y_2 + x_2y_3 + x_3y_1 - x_1y_3 - x_2y_1 - x_3y_2)$

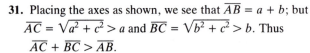

If the points are numbered in clockwise order, rather than the counterclockwise order given, the result will be the negative of the one above. Thus the absolute value is needed.

31. Placing the axes as shown, we see that $\overline{AB} = a + b$; but $\overline{AC} = \sqrt{a^2 + c^2} > a$ and $\overline{BC} = \sqrt{b^2 + c^2} > b$. Thus $\overline{AC} + \overline{BC} > \overline{AB}$.

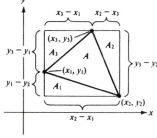

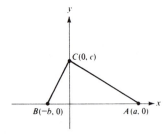

33. 13

Section 1.3 [pages 18–19]

1. (4, 3)　　**3.** (16/5, −9/5)　　**5.** (17, 22)　　**7.** (2, 1)　　**9.** (7/2, 1)　　**11.** (12, −4)

13. (0, −7)　　**15.** (3, 0)　　**17.** (14/5, 7/5)　　**19.** (6, −7)　　**21.** (4/3, 5/3)　　**23.** (2, −1/2)

25. Placing the axes as shown, we see that the midpoints of the three sides are $D = (0, 0)$, $E = (a/2, c/2)$, and $F = (-a/2, c/2)$. Thus $\overline{CE} = \overline{ED} = \overline{DF} = \overline{FC} = \sqrt{a^2 + c^2}/2$.

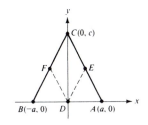

27. 7, −9　　**29.** (13, −2), (−9, 12), (−5, −10)

31. Placing the axes as shown, we see that $\overline{AB} = \overline{CD} = a$ and $\overline{BC} = \overline{DA} = \sqrt{b^2 + c^2}$. Thus $\overline{AB}^2 + \overline{BC}^2 + \overline{CD}^2 + \overline{DA}^2 = 2(a^2 + b^2 + c^2)$. Since $\overline{AC} = \sqrt{(a - b)^2 + c^2}$ and $\overline{BD} = \sqrt{(a + b)^2 + c^2}$, it follows that $\overline{AC}^2 + \overline{BD}^2 = (a - b)^2 + c^2 + (a + b)^2 + c^2 = 2(a^2 + b^2 + c^2)$.

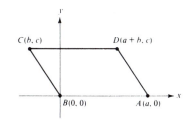

33.

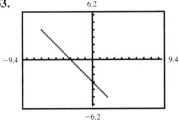

Midpoint = (−2.5, −0.5)

35.

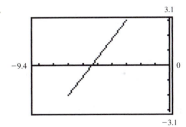

Midpoint = (−5, 0.5)

37. 3.33, H.0　　**39.** B.47, 7.63; 1.2 minutes after second car starts

Section 1.5 [pages 28–30]

1. 5/3, 59°　　**3.** 2/3, 34°　　**5.** No slope, 90°　　**7.** 1, 45°

9.

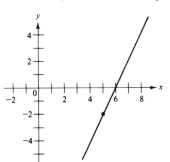

11.

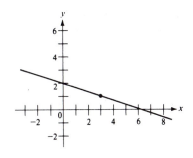

13.

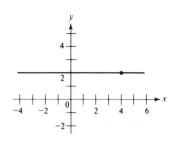

15. Parallel   **17.** None   **19.** Coincident   **21.** Perpendicular   **23.** Parallel

25. 14/3   **27.** 10/7, 11   **29.** 9

31. Labeling the points A, B, C, and D, respectively, we have $m_{AB} = m_{CD} = -1$ and $m_{BC} = m_{AD} = 1$. Thus $AB \parallel CD$ and $BC \parallel AD$, implying that it is a parallelogram. $AB \perp BC$, implying that it is a rectangle. Since m_{AC} does not exist and $m_{BD} = 0$, $AC \perp BD$, implying that it is a square.

33. $m_{AC} = \dfrac{a - 0}{0 - a} = -1;\quad m_{BD} = \dfrac{a - 0}{a - 0} = 1;\quad AC \perp BD.$

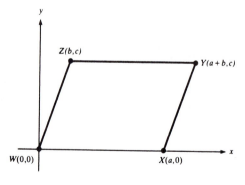

35. Since $WXYZ$ is a rhombus,
$$\overline{WX} = \overline{WZ},$$
$$a = \sqrt{b^2 + c^2},$$
$$a^2 = b^2 + c^2.$$
$$m_{WY} = \frac{c}{a + b}$$
$$m_{XZ} = \frac{c}{b - a} = -\frac{c}{a - b}$$
$$m_{WY} \cdot m_{XZ} = -\frac{c^2}{a^2 - b^2} = -\frac{c^2}{c^2} = -1$$
$$WY \perp XZ$$

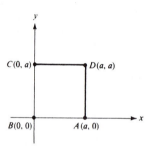

37.

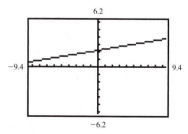

39.

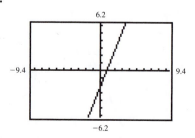

41.

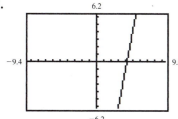

43.

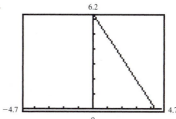

45. 62.5%   **47.** N 21°48′E

Section 1.6 [page 35]

1. 135°   **3.** 135°   **5.** 6°   **7.** 56°   **9.** 12°   **11.** 60°   **13.** 135°   **15.** 27°

17. $-7 - 5\sqrt{2}$ **19.** $1 + \sqrt{2}$ **21.** $-8 + \sqrt{65}$ **23.** 2   **25.** 37°, 72°, 72°   **27.** 1/5

29. 2   **31.** −2   **33.** ±1

Section 1.7 [pages 42–45]

1. Function

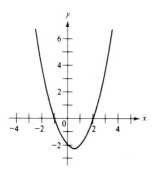

3. Function

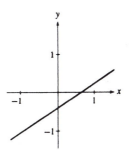

5. Function

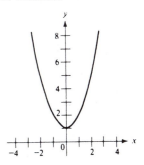

7. Function

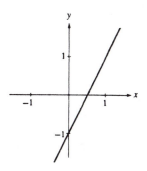

9. Not a function

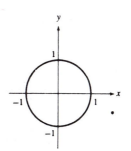

11. Function

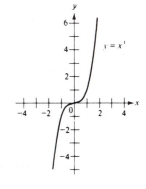

13.

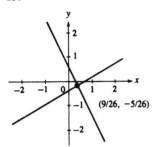

(9/26, −5/26)

15.

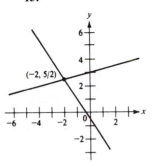

(−2, 5/2)

17. Not a function

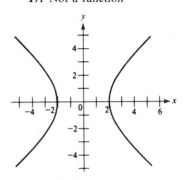

19. Not a function

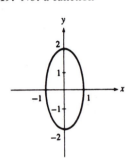

21. Function

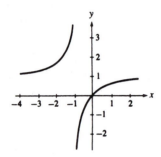

23. Not a function

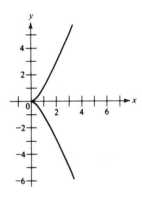

25.

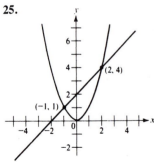

(2, 4)

(−1, 1)

27.

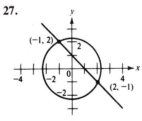

(−1, 2)

(2, −1)

29.

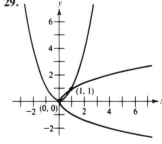

(1, 1)

(0, 0)

31.

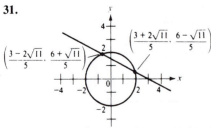

$\left(\dfrac{3 - 2\sqrt{11}}{5}, \dfrac{6 + \sqrt{11}}{5}\right)$

$\left(\dfrac{3 + 2\sqrt{11}}{5}, \dfrac{6 - \sqrt{11}}{5}\right)$

33. (−4, 2) (2, 4)   **35.** (−4, 3), (5, −3)

37. (−2, 2), (3, −3), (−1, 5), (1, 5)

39.

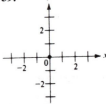

41.

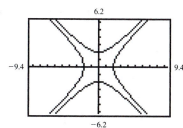

43. $(-1, 1), (2, 4)$

45.

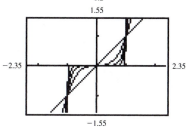

47.

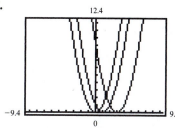

Same shape moved 2 units to the left of the origin. Replacing x by $x - k$ moves the figure k units to the right.

49.

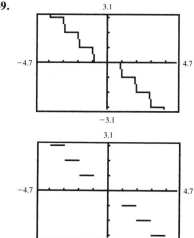

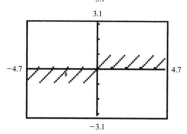

In both cases, the dot mode gives the better representation.

51.

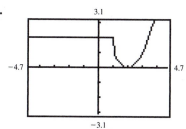

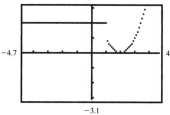

Dot mode gives the better representation near $x = 1$; connected mode is better away from $x = 1$.

53. No. They are identical for $x \le 16$, but the second graph contains points with $x > 16$, which are not on $\sqrt{x} + \sqrt{y} = 4$.

55. 10, 80. Beyond $x = 100$, the difference between the two graphs becomes significant.

57.

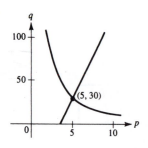

59.

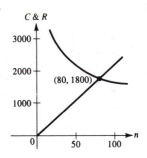

Section 1.8 [pages 48–49]

1. $14x + 8y - 69 = 0$ **3.** $x^2 + y^2 - 10x - 16y + 80 = 0$ **5.** $2x - y - 8 = 0$

7. $5x - 2y - 19 = 0$ **9.** $x^2 + y^2 - 9x + 18 = 0$

11. $x^2 + y^2 - x - 9y + 18 = 0$, except $(2, 5)$ and $(-1, 4)$

13. $y^2 - 8x + 16 = 0$ **15.** $3x^2 - y^2 + 6x - 9 = 0$ **17.** $25x^2 + 9y^2 = 225$

19. $x^2 + y^2 = 16$ **21.** $8x^2 - y^2 = 8$ **23.** $xy = \pm 4$

25. $x^2 - 2xy + y^2 - 2x - 2y - 1 = 0$

Review Problems [pages 49–50]

1. $(22/7, -9/7)$ **2.** Noncollinear (distances: $3\sqrt{5}, \sqrt{157}, \sqrt{34}$; slopes, 2: 11/6)

3. $\sqrt{221}/2, \sqrt{41}, \sqrt{185}/2$

4. Not a right triangle (distances: 5, $2\sqrt{2}, \sqrt{29}$; slopes: $-3/4, 1, -5/2$)

5. 2   **6.** None (slopes; 2, 1/4)   **7.** $-8/5, -1/2, 5$ **8.** $(1/3, -1), (-4/3, 3)$

9. $3x - 2y - 8 = 0$ **10.** $(-2, 0), (5, 1)$

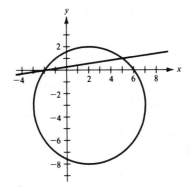

11. $-1, -9$

12. $E = \left(\dfrac{a+b}{2}, 0\right),$

$F = \left(\dfrac{b+d}{2}, \dfrac{c}{2}\right).$

$G = \left(\dfrac{d}{2}, \dfrac{c+e}{2}\right),$

$H = \left(\dfrac{a}{2}, \dfrac{c}{2}\right).$

$m_{EF} = \dfrac{e/2 - 0}{\dfrac{b+d}{2} - \dfrac{a+b}{2}}$

$= \dfrac{e}{d-a}$

$m_{GH} = \dfrac{\dfrac{c+e}{2} - \dfrac{c}{2}}{\dfrac{d}{2} - \dfrac{a}{2}} = \dfrac{e}{d-a}$

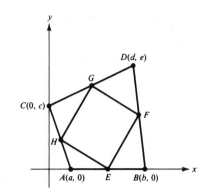

Since $m_{EF} = m_{GH}$, $EF \parallel GH$.

$m_{EH} = \dfrac{c/2 - 0}{\dfrac{a}{2} - \dfrac{a+b}{2}} = -\dfrac{c}{b}, \quad m_{FG} = \dfrac{\dfrac{c+e}{2} - \dfrac{e}{2}}{\dfrac{d}{2} - \dfrac{b+d}{2}} = -\dfrac{c}{b}.$

Since $m_{EH} = m_{FG}$, $EH \parallel FG$.

13. $(2/3, 1)$ **14.** $x^2 - 2y + 1 = 0$ **15.** $(3, -2)$ **16.** $(\sqrt{3}a, \sqrt{3}a), (-\sqrt{3}a, -\sqrt{3}a)$

17. $(1, 4), (4, 6)$ **18.** $(8, 7)$ **19.** $x = y \cot y$

CHAPTER 2

Section 2.1 [pages 60–61]

1. $-6\mathbf{i} - 2\mathbf{j}$ **3.** $7\mathbf{i} + \mathbf{j}$ **5.** $-\mathbf{i} + 5\mathbf{j}$ **7.** $4\mathbf{i} - 3\mathbf{j}$ **9.** $(3/(\sqrt{10})\mathbf{i} - (1/\sqrt{10})\mathbf{j}$

11. $-(1/\sqrt{5})\mathbf{i} + (2/\sqrt{5})\mathbf{j}$ **13.** $(1/\sqrt{5})\mathbf{i} + (2/\sqrt{5})\mathbf{j}$ **15.** $-(4/5)\mathbf{i} + (3/5)\mathbf{j}$

17. $B = (4, 3)$ **19.** $A = (5, 0)$ **21.** $B = (6, 0)$ **23.** $A = (1, 3)$

25. $4\mathbf{i} + \mathbf{j}, 2\mathbf{i} - 3\mathbf{j}$ **27.** $3\mathbf{i} + \mathbf{j}, -\mathbf{i} - 3\mathbf{j}$ **29.** $3\mathbf{i} + \mathbf{j}, -\mathbf{i} - 7\mathbf{j}$

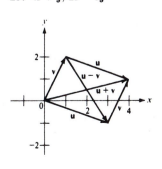

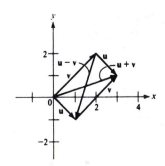

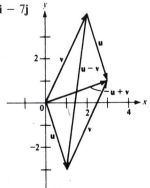

31. $5\mathbf{i} - 3\mathbf{j}$, $\mathbf{i}$, $+ \mathbf{j}$

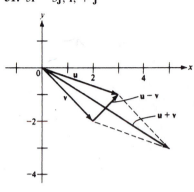

33.

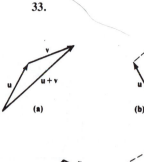

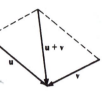

35. $A = (5/2, -3/2)$, $B = (11/2, 7/2)$ **37.** $A = (3/2, -1/2)$, $B = (5/2, 1/2)$ **39.** $8\mathbf{i} + 5\mathbf{j}$

41. $-5\mathbf{i} - 3\mathbf{j}$

43. **(a)** $\overrightarrow{AB} \equiv \overrightarrow{AB}$ follows directly from the definition of equivalence.

   (b) Since any true statement about $\overrightarrow{AB}$ and $\overrightarrow{CD}$ is also a true statement about $\overrightarrow{CD}$ and $\overrightarrow{AB}$, it follows that if $\overrightarrow{AB} \equiv \overrightarrow{CD}$, then $\overrightarrow{CD} \equiv \overrightarrow{AB}$.

   (c) Suppose $\overrightarrow{AB} \equiv \overrightarrow{CD}$ and $\overrightarrow{CD} \equiv \overrightarrow{EF}$. If $\overrightarrow{AB}$ has length 0, then $\overrightarrow{CD}$ has length 0; if $\overrightarrow{CD}$ has length 0, then $\overrightarrow{EF}$ has length 0. Thus $\overrightarrow{AB} \equiv \overrightarrow{EF}$. Suppose AB is not of length 0. Then $\overrightarrow{AB}$ and $\overrightarrow{CD}$ have the same length, both lie on the same or parallel lines, and they are directed in the same way. A similar statement can be made about $\overrightarrow{CD}$ and $\overrightarrow{EF}$. Thus $\overrightarrow{AB}$ and $\overrightarrow{EF}$ have the same length, both lie on the same or parallel lines, and both are directed in the same way. A similar statement can be made about $\overrightarrow{CD}$ and $\overrightarrow{EF}$. Thus $\overrightarrow{AB}$ and $\overrightarrow{EF}$ have the same length, both lie on the same or parallel lines, and both are directed in the same way. Thus $\overrightarrow{AB} \equiv \overrightarrow{EF}$.

45. **(a)** $a_1\mathbf{i} + b_1\mathbf{j}$ is represented by $\overrightarrow{OP_1}$ from $(0, 0)$ to (a_1, b_1). $a_2\mathbf{i} + b_2\mathbf{j}$ is represented by $\overrightarrow{OP_2}$ from $(0, 0)$ to (a_2, b_2), or by $\overrightarrow{P_1P_3}$ from (a_1, b_1) to $(a_1 + a_2, b_1 + b_2)$. Hence the sum is represented by $\overrightarrow{OP_3}$ from $(0, 0)$ to $(a_1 + a_2, b_1 + b_2)$, giving $(a_1\mathbf{i} + b_1\mathbf{j}) + (a_2\mathbf{i} + a_2\mathbf{j}) = (a_1 + a_2)\mathbf{i} + (b_1 + b_2)\mathbf{j}$.

   (b) Let $(a_1\mathbf{i} + b_1\mathbf{j}) - (a_2\mathbf{i} + b_2\mathbf{j}) = a\mathbf{i} + b\mathbf{j}$. Then $a_1\mathbf{i} + b_1\mathbf{j} = (a_2\mathbf{i} + b_2\mathbf{j}) + (a\mathbf{i} + b\mathbf{j}) = (a_2 + a)\mathbf{i} + (b_2 + b)\mathbf{j}$. Thus $a_1 = a_2 + a$ or $a = a_1 - a_2$, and $b_1 = b_2 + b$ or $b = b_1 - b_2$.

   (c) $\mathbf{v} = a\mathbf{i} + b\mathbf{j}$ is represented by $\overrightarrow{OP}$ from $(0, 0)$ to (a, b); $\mathbf{w} = da\mathbf{i} + db\mathbf{j}$ is represented by $\overrightarrow{OQ}$ from $(0, 0)$ to (da, db). Clearly these points lie on the same line; so $\mathbf{w}$ is in the same direction as or opposite direction from $\mathbf{v}$, depending upon the sign of d. Furthermore, $|\mathbf{w}| = \sqrt{d^2a^2 + d^2b^2} = \sqrt{d^2(a^2 + b^2)} = |d|\sqrt{a^2 + b^2} = |d||\mathbf{v}|$.

   (d) $a\mathbf{i} + b\mathbf{j}$ is represented by $\overrightarrow{OP}$ from $(0, 0)$ to (a, b). Its length is $\sqrt{(a - 0)^2 + (b - 0)^2} = \sqrt{a^2 + b^2}$.

47. Clearly if $a = c$ and $b = d$, then $a\mathbf{i} + b\mathbf{j} = c\mathbf{i} + d\mathbf{j}$. Suppose $a\mathbf{i} + b\mathbf{j} = c\mathbf{i} + d\mathbf{j}$, $(a - c)\mathbf{i} + (b - d)\mathbf{j} = \mathbf{0}$. This vector has length 0. Therefore $(a - c)^2 + (b - d)^2 = 0$.
$a - c = 0$ and $b - d = 0$
$a = c$ and $b = d$

Section 2.2 [pages 69–70]

1. Arccos $1/5\sqrt{2} = 82°$ **3.** 90°   **5.** 90°   **7.** Arccos $3/\sqrt{34} = 59°$ **9.** 1, not orthogonal

11. 4, not orthogonal   **13.** -1, not orthogonal   **15.** 3, not orthogonal   **17.** $(1/2)\mathbf{i} + (1/2)\mathbf{j}$

19. $-(6/5)\mathbf{i} + (12/5)\mathbf{j}$ **21.** $(2/5)\mathbf{i} + (1/5)\mathbf{j}$ **23.** $(6/5)\mathbf{i} - (3/5)\mathbf{j}$ **25.** $30\sqrt{2}$

27. 3   **29.** -8 **31.** $-1/2$ **33.** -1 **35.** $4 + 2\sqrt{3}$ **37.** $(8 - 5\sqrt{3})/11$

39. $-(2/17)\mathbf{i} - (8/17)\mathbf{j}, -(19/17)\mathbf{i} - (76/17)\mathbf{j}$ **41.** $-(2/5)\mathbf{i} - (4/5)\mathbf{j}, (3/5)\mathbf{i} + (6/5)\mathbf{j}$

43. Let $\mathbf{u} = a_1\mathbf{i} + b_1\mathbf{j}, \quad \mathbf{v} = a_2\mathbf{i} + b_2\mathbf{j}, \quad \mathbf{w} = a_3\mathbf{i} + b_3\mathbf{j}.$

$$(\mathbf{u} + \mathbf{v}) \cdot \mathbf{w} = [(a_1 + a_2)\mathbf{i} + (b_1 + b_2)\mathbf{j}] \cdot (a_3\mathbf{i} + b_3\mathbf{j})$$
$$= (a_1 + a_2)a_3 + (b_1 + b_2)b_3$$

$$\mathbf{u} \cdot \mathbf{w} + \mathbf{v} \cdot \mathbf{w} = a_1a_3 + b_1b_3 + a_2a_3 + b_2b_3$$

45. $\mathbf{0} \cdot \mathbf{u} = (0\mathbf{i} + 0\mathbf{j}) \cdot (a\mathbf{i} + b\mathbf{j}) = 0 \cdot a + 0 \cdot b = 0$ **47.** $|\cos \theta| = \left|\dfrac{\mathbf{u} \cdot \mathbf{v}}{|\mathbf{u}||\mathbf{v}|}\right| = \dfrac{|\mathbf{u} \cdot \mathbf{v}|}{|\mathbf{u}||\mathbf{v}|} \le 1$

Section 2.3 [pages 81–83]

1. $6\mathbf{i} + 2\mathbf{j}$

3. $5\sqrt{5}$ lb to the right and inclined upward at an angle of 63° with the horizontal

5. $\sqrt{25 - 12\sqrt{2}}$ lb to the right and inclined downward at an angle of 3° with the horizontal

7. $3\sqrt{3}\mathbf{i} + 3\mathbf{j}$ **9.** $3\sqrt{2}/2\mathbf{i} - 3\sqrt{2}/2\mathbf{j}$ **11.** $1.638\mathbf{i} + 1.147\mathbf{j}$ **13.** $\mathbf{j}$

15. $\sqrt{29}$ lb to the right and inclined downward at an angle of 68° with the horizontal

17. $\sqrt{34 + 15\sqrt{2}}$ lb to the left and inclined downward at an angle of 28° with the horizontal

19. $4\sqrt{2}\mathbf{i} + (4\sqrt{2} - 9)\mathbf{j}$ **21.** $(3\sqrt{2} - 8)\mathbf{i} + 3\sqrt{2}\mathbf{j}$ **23.** $(4\sqrt{3} - 5)\mathbf{i} + (5\sqrt{3} - 24)\mathbf{j}$

25. 100 lb, $200/\sqrt{3}$ lb, $100/\sqrt{3}$ lb   **27.** 100 lb, $100\sqrt{3}$ lb, 100 lb

29. 100 lb, $100(\sqrt{3} - 1)$ lb, $50\sqrt{2}(3 - \sqrt{3})$ lb, $50(3 - \sqrt{3})$ lb, $50(\sqrt{3} - 1)$ lb

31. $\mathbf{q} = (1/2)\mathbf{u} + \mathbf{v} + (1/2)\mathbf{w}$

$\mathbf{q} = -(1/2)\mathbf{u} + \mathbf{p} - (1/2)\mathbf{w}$

Adding, we have

$$2\mathbf{q} = \mathbf{p} + \mathbf{v}, \quad \mathbf{q} = \frac{\mathbf{p} + \mathbf{v}}{2}$$

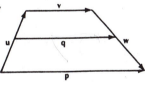

Since $\mathbf{p}$ and $\mathbf{v}$ are parallel, one is a scalar multiple of the other; say, $\mathbf{p} = k\mathbf{v}$. Thus

$$\mathbf{q} = \frac{\mathbf{p} + \mathbf{v}}{2} = \frac{k\mathbf{v} + \mathbf{v}}{2} = \frac{k + 1}{2}\mathbf{v},$$

making $\mathbf{q}$ a scalar multiple of $\mathbf{p}$ and therefore parallel to $\mathbf{p}$. Furthermore,

$$|\mathbf{q}| = \frac{k + 1}{2}|\mathbf{v}| = \frac{k|\mathbf{v}| + |\mathbf{v}|}{2} = \frac{|\mathbf{p}| + |\mathbf{v}|}{2}.$$

33. $\mathbf{a} = \mathbf{u} - \mathbf{v}, \quad \mathbf{b} = \mathbf{u} + \mathbf{v}$

$$\mathbf{a} \cdot \mathbf{b} = (\mathbf{u} - \mathbf{v}) \cdot (\mathbf{u} + \mathbf{v})$$
$$= (\mathbf{u} \cdot \mathbf{u}) + (\mathbf{u} \cdot \mathbf{v}) - (\mathbf{v} \cdot \mathbf{u}) - (\mathbf{v} \cdot \mathbf{v})$$
$$= |\mathbf{u}|^2 - |\mathbf{v}|^2 = 0$$

$$|\mathbf{u}| = |\mathbf{v}|$$

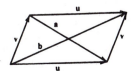

35. $u = a + d = a - b$, $v = a + b$

$|u|^2 + |v|^2 = |a - b|^2 + |a + b|^2$

$= |a|^2 + |-b|^2 - 2|a||-b|\cos\beta$
$+ |a|^2 + |b|^2 - 2|a||b|\cos\alpha$

$= 2|a|^2 + 2|b|^2$

$= |a|^2 + |b|^2 + |c|^2 + |d|^2$

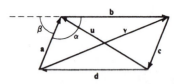

37. $|u| = |v|$, $v = u + w$

$\cos\alpha = \dfrac{u \cdot (-w)}{|u||-w|} = \dfrac{-u \cdot w}{|u||w|}$

$\cos\beta = \dfrac{w \cdot v}{|w||v|} = \dfrac{w \cdot (u + w)}{|w||v|}$

$= \dfrac{w \cdot u + w \cdot w}{|w||u|}$

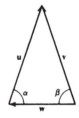

$|w|^2 = |u|^2 + |v|^2 - 2|u||v|\cos(180° - \alpha - \beta)$
$= |u|^2 + |v|^2 - 2u \cdot v = 2|u|^2 - 2u \cdot (u + w)$
$= 2u \cdot u - 2u \cdot u - 2u \cdot w = -2u \cdot w$

$\cos\beta = \dfrac{u \cdot w + w \cdot w}{|w||u|} = \dfrac{u \cdot w - 2u \cdot w}{|w||u|} = -\dfrac{u \cdot w}{|u||w|} = \cos\alpha$

39. $a + b + c = 0$

$u = -a - 1/2c$

$v = a + 1/2b$

$sv = a + ru$

$s(a + 1/2b) = a + r(-a - 1/2c)$, $sa + s/2b = (1 - r)a - r/2c$

$s/2b + r/2c = (1 - r - s)a = (1 - r - s)(-b - c)$

$(s/2 + 1 - r - s)b + (r/2 + 1 - r - s)c = 0$

$1 - r - s/2 = 1 - r/2 - s = 0$, $2r + s = r + 2s = 2$

$2r + s = 2$, $r + 2s = 2$, $r = s = 2/3$

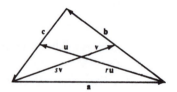

Review Problems [pages 83–84]

1. $\dfrac{2}{\sqrt{13}}i - \dfrac{3}{\sqrt{13}}j$ **2.** $(3, -2)$

3. $A = (-9/2, 5/2)$, $B = (1/2, -1/2)$

4. $u + v = (4, 4)$, $u - v = (2, -8)$, $3u = (9, -6)$, $2u + v = (7, 2)$

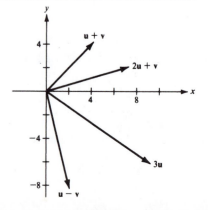

5. 8 **6.** 135° **7.** $-\dfrac{4}{5}\mathbf{i} + \dfrac{18}{5}\mathbf{j}$ **8.** $\dfrac{3}{2}\mathbf{i} - \dfrac{3}{2}\mathbf{j}$

9. 3/2 **10.** 3/4 **11.** $(4\sqrt{3} - 5\sqrt{2})\mathbf{i} + (10 - 5\sqrt{2})\mathbf{j}$

12. $|\mathbf{a}| = |\mathbf{c}|, \mathbf{v} = \mathbf{a} + \mathbf{b}, -\mathbf{u} = \mathbf{b} + \mathbf{c}$

$|\mathbf{v}|^2 = |\mathbf{a}|^2 + |\mathbf{b}|^2 - 2|\mathbf{a}|\,|\mathbf{b}|\cos\alpha$

$\begin{aligned}|\mathbf{u}|^2 &= |\mathbf{b}|^2 + |\mathbf{c}|^2 - 2|\mathbf{b}|\,|\mathbf{c}|\cos\alpha \\ &= |\mathbf{a}|^2 + |\mathbf{b}|^2 - 2|\mathbf{a}|\,|\mathbf{b}|\cos\alpha \\ &= |\mathbf{v}|^2\end{aligned}$

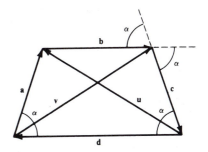

13. $\mathbf{u} = \mathbf{a} + \mathbf{b}, \mathbf{v} = \mathbf{a} + \mathbf{d}$

$\mathbf{d} = -n\mathbf{b}, r\mathbf{v} = \mathbf{d} + s\mathbf{u}$

$r(\mathbf{a} + \mathbf{d}) = -n\mathbf{b} + s(\mathbf{a} + \mathbf{b})$

$r(\mathbf{a} - n\mathbf{b}) = -n\mathbf{b} + s(\mathbf{a} + \mathbf{b}), (r - s)\mathbf{a} = (rn - n + s)\mathbf{b}$

$r - s = 0, rn - n + s = 0$

$rn + r - n = 0, r(1 + n) = n$

$r = \dfrac{n}{1 + n}, \; s = \dfrac{n}{1 + n}, \; 1 - r = 1 - s = \dfrac{1}{1 + n}$

$(1 - r)/r = (1 - s)/s = 1/n$

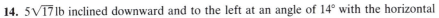

14. $5\sqrt{17}$ lb inclined downward and to the left at an angle of 14° with the horizontal

15. $10\sqrt{3}$ lb, 10 lb **16.** $\mathbf{v}_1 = \dfrac{48}{25}\mathbf{i} + \dfrac{64}{25}\mathbf{j}, \; \mathbf{v}_2 = \dfrac{252}{25}\mathbf{i} - \dfrac{189}{25}\mathbf{j}$

CHAPTER 3

Section 3.1 [pages 92–96]

1.

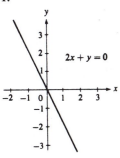

3.

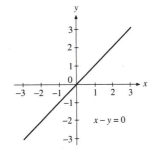

5.

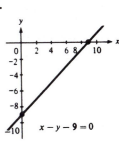

7.

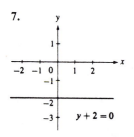

9.

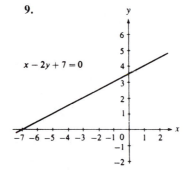

11.

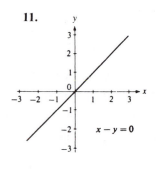

13.

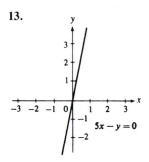

15.

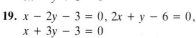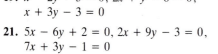

17. $2x + y - 6 = 0$, $x - 2y - 3 = 0$,
$3x - y + 1 = 0$

19. $x - 2y - 3 = 0$, $2x + y - 6 = 0$,
$x + 3y - 3 = 0$

21. $5x - 6y + 2 = 0$, $2x + 9y - 3 = 0$,
$7x + 3y - 1 = 0$

25. $3x - 2y + 5 = 0$

27. The given circles do not intersect.   **29.** $x - 3y + 3 = 0$

31. $7x - y - 20 = 0$, $7x - y - 20 = 0$. The triangle is isosceles.

33. The line determined by (x_2, y_2) and (x_3, y_3) is (by Problem 32)

$$\begin{vmatrix} x & y & 1 \\ x_2 & y_2 & 1 \\ x_3 & y_3 & 1 \end{vmatrix} = 0.$$

The point (x_1, y_1) is on that line if and only if it satisfies the above equation; i.e., if and only if

$$\begin{vmatrix} x_1 & y_1 & 1 \\ x_2 & y_2 & 1 \\ x_3 & y_3 & 1 \end{vmatrix} = 0.$$

23.

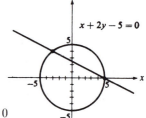

35. **(a)** Yes; $y = 3$.   **(b)** No; it is not a function.

37. The line becomes more nearly horizontal, always rising to the right.

39. The one with the positive slope rises as fast as the one with the negative slope falls.

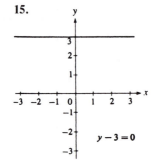

41. $3x + y = -10$ **43.** $3x + 4y = 20.42$

45. $7x + 11y = -16$ **47.** $2.1x + 2.3y = 10.4$

49. $D + 760P - 760 = 0$

51. 30. Demand decreases as the price increases; the supply increases as the price increases.

53. $D + 2000p - 100,000 = 0 \ (0 \le p \le 50)$

Section 3.2 [pages 102–105]

1.

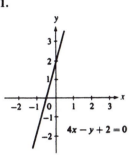

$4x - y + 2 = 0$

3.

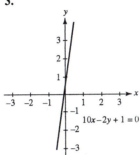

$10x - 2y + 1 = 0$

5.

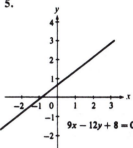

$9x - 12y + 8 = 0$

7.

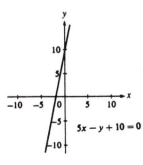

$5x - y + 10 = 0$

9.

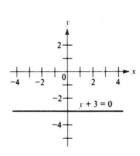

$y + 3 = 0$

11.

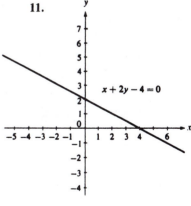

$x + 2y - 4 = 0$

13.

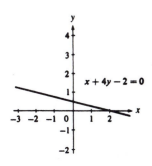

$x + 4y - 2 = 0$

15.

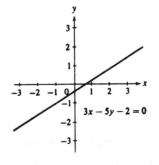

$3x - 5y - 2 = 0$

17.

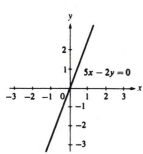

19.

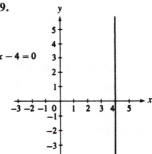

21. $2x - 5y + 11 = 0$

23. $x + y - 7 = 0$

25. $x + 3y - 6 = 0; x + y - 4 = 0$

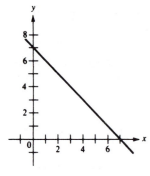

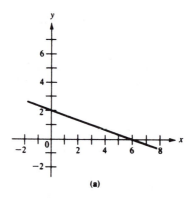

(a)

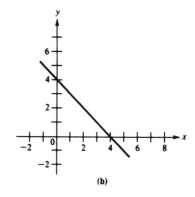

(b)

27. $4x + y - 3 = 0$ **29.** $(15/14, -5/14)$ **31.** $(-1, -7)$ **33.** $5/2$ **35.** 1

37.

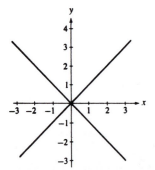

39.

41. A representative of **v** is $\overrightarrow{OP}$, where $O = (0, 0)$ and $P = (A, B)$. The slope of OP is B/A provided $A \neq 0$. The slope of $Ax + By + C = 0$ is $-A/B$ provided $B \neq 0$. Thus they are perpendicular if neither A nor B is 0. If $A = 0$, then OP is a vertical line and $Ax + By + C = 0$ a horizontal line; if $B = 0$, OP is horizontal and $Ax + By + C = 0$ vertical.

43. $\begin{vmatrix} x & y & 1 \\ x_1 & y_1 & 1 \\ x_2 & y_2 & 1 \end{vmatrix} = 0$ is linear in x and y. By Theorem 3.5 it represents a line. The points (x_1, y_1) and

(x_2, y_2) satisfy the equation. (If two rows of a determinant are equal, the value of the determinant is 0.) Thus the equation represents the line through (x_1, y_1) and (x_2, y_2).

45. $(8, 3), (3, -2), (-4, -1)$

47. The x coefficient is the slope of the line.

49.

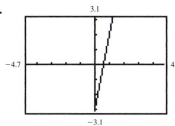

Intercepts: $(0, -3), (0.6, 0)$

$x/0.6 - y/3 = 1$

51.

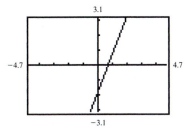

Intercepts: $(0, -1.78), (0.7008, 0)$

$x/0.7008 - y/1.78 = 1$

53. (a) The distance increases by 60 for each unit increase in time, and we are at the starting point at time 0. (b) The distance from the starting point decreases by 40 for each unit increase in time, and we are 1000 units from the starting point at time 0.

55. $V + 625t - 8000 = 0 \ (0 \le t \le 12)$

57. $3x + 4y - 144 = 0$. The slope is the increase in the number (decrease if it is negative) of deluxe models that can be made, corresponding to a unit increase in the number of standard models.

59. The slope is the cost of producing one more dinette set. The C intercept is the cost of operating, even when nothing is produced.

61. The p intercept is the lowest price at which none can be sold. The q intercept is the number that can be given away. The slope is the increase (decrease if the slope is negative) in demand corresponding to a unit increase in the price.

63. 4.2 **65.** 5.2 **67.** $k = 0.171, n = 2.217$

Section 3.3 [pages 112–115]

1. $\sqrt{2}$ **3.** $9/\sqrt{41}$ **5.** $2/5$ **7.** $18/\sqrt{29}$ **9.** $10/3$ **11.** $4/\sqrt{29}$ **13.** $7/2\sqrt{5}$

15. $8/\sqrt{5}$ **17.** $13/\sqrt{10}, 26/\sqrt{29}, 26/5$ **19.** 13 **21.** $7x - 7y + 2 = 0$

23. $39x + 13y - 24 = 0$ **25.** $2(\sqrt{2} - 1)x - 2y + (4 - 3\sqrt{2}) = 0$

27. $\pm 3/\sqrt{7}$ **29.** $\pm 3/2\sqrt{2}$ **31.** $x^2 - 2xy + y^2 + 2x + 2y - 1 = 0$ **33.** $(1, 1), (-2, 2), (3, -3), (6, 6)$

35. Either $A \ne 0$ or $B \ne 0$. Suppose $B \ne 0$. If P is on the same side as the origin O, then P and O are either both above or both below the line. Thus $Ax_1 + By_1 + C$ and $A \cdot 0 + B \cdot 0 + C = C$ have the same sign. If P and O are on opposite sides of the line, then one is above and the other below the line. Thus $Ax_1 + By_1 + C$ and $A \cdot 0 + B \cdot 0 + C = C$ have opposite signs.

If $B = 0$, then $A \ne 0$ and the same argument can be used, using right and left of the line instead of above and below (see Problem 34).

37. $m = -\dfrac{1}{\tan \alpha} = -\dfrac{\cos \alpha}{\sin \alpha}$

$y - p \sin \alpha = -\dfrac{\cos \alpha}{\sin \alpha}(x - p \cos \alpha)$

$y \sin \alpha - p \sin^2 \alpha = -x \cos \alpha + p \cos^2 \alpha$

$x \cos \alpha + y \sin \alpha - p = 0$

39. 60/13 miles

Section 3.4 [pages 122–124]

1. Lines through $(-1, 4)$; does not include $x = -1$

3. Lines with x intercept 2; does not include $x = 2$ **5.** All lines through the origin

7. Lines having both an x intercept and a y intercept; does not include any line through the origin

9. Lines containing the point of intersection of $2x + 3y + 1 = 0$ and $4x + 2y - 5 = 0$; does not include $4x + 2y - 5 = 0$

11. Lines with y intercept twice the x intercept; does not include any line through the origin

13. All lines with y intercept equal to their slope   **15.** $\{3x - 5y = k \mid k \text{ real}\}$

17. $\{y - 5 = m(x - 2) \mid m \text{ real}\} \cup \{x = 2\}$

19. $\{3x - 5y + 1 + k(2x + 3y - 7) = 0 \mid k \text{ real}\} \cup \{2x + 3y - 7 = 0\}$

21. $\{Ax + By = 0 \mid A, B \text{ real and not both } 0\}$ **23.** $\{Ax + By + C = 0 \mid |6A + C|/\sqrt{A^2 + B^2} = 5\}$

25. (a) $3x - 5y + 25 = 0$ (b) $5x + 3y - 49 = 0$ **27.** (a) $2x + y - 5 = 0$ (b) $x - 2y + 10 = 0$

29. $5x - y \pm 3\sqrt{26} = 0$ **31.** $15x - 8y - 43 = 0, 3x - 4y + 1 = 0$

33. $11x - 60y - 17 = 0, x - 7 = 0$ **35.** $16x - 8y - 27 = 0$

37. $x - 2y - 10 = 0, 3x + 2y - 6 = 0$

39. $x + 2y - 8 = 0, 9x + 2y - 24 = 0, (11 \pm 4\sqrt{7})x - 2y - (16 \pm 8\sqrt{7}) = 0$

41. $x + y - (4 + 2\sqrt{2}) = 0$ **43.** $\sqrt{3}x - y + (6 - 4\sqrt{3}) = 0, \sqrt{3}x + y - (6 + 4\sqrt{3}) = 0$

Review Problems [pages 124–125]

1. (a) $2x - 3y + 13 = 0$ (b) $2x - y - 6 = 0$ (c) $3x + 3y - 1 = 0$ (d) $x - 2 = 0$

2. (a) $3x - y - 10 = 0$ (b) $10x - 4y - 5 = 0$ (c) $y + 2 = 0$

3. (a) $m = 1/4, a = -1, b = 1/4$ (b) $m = -2/3, a = -5/2, b = -5/3$
   (c) $m = -5/2, a = 0, b = 0$ (d) No slope, $a = -1/3$, no y intercept

4. 6/13   **5.** $17/2\sqrt{10}$

6. (a) Set of lines through the point $(-3, 1)$; does not include the vertical line $x = -3$
   (b) Set of all lines with slope 3
   (c) Set of lines with y intercept -3; does not include the horizontal line $y = -3$

7. (a) $\{y + 1 = m(x - 5) \mid m \text{ real}\} \cup \{x = 5\}$ (b) $\{2x - 3y = k \mid k \text{ real}\}$
   (c) $\{Ax + By + C = 0 \mid |2A + 5B + C|/\sqrt{A^2 + B^2} = 3\}$

8. 1.416   **9.** $x + y - 1 = 0$

10. $3x - 2y + 7 = 0, x - 2y + 8 = 0, 3x + 2y - 10 = 0$ **11.** 4, $\sqrt{85}/2$, $\sqrt{181}/2$

12. $x - 1 = 0, 2x + 9y - 23 = 0, 10x + 9y - 31 = 0$ **13.** $(0, 4), (-4/3, 0)$

14. $x + 3y - 7 = 0$ **15.** $y - 1 = 0, 5x - 12y - 13 = 0$ **16.** $69x - 23y - 35 = 0$

17.

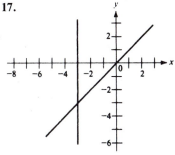

18. $m = 3.15 \pm 0.05, b = 4.4 \pm 0.5$

CHAPTER 4

Section 4.1 [pages 134–138]

1. $(x - 1)^2 + (y - 3)^2 = 25$
   $x^2 + y^2 - 2x - 6y - 15 = 0$

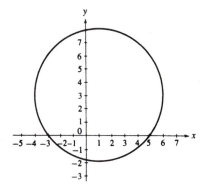

3. $(x - 5)^2 + (y + 2)^2 = 4,$
   $x^2 + y^2 - 10x + 4y + 25 = 0$

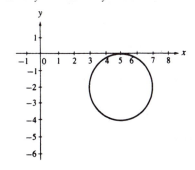

5. $(x - 1/2)^2 + (y + 3/2)^2 = 4$
   $2x^2 + 2y^2 - 2x + 6y - 3 = 0$

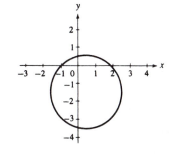

7. $(x - 4)^2 + (y + 2)^2 = 26$
   $x^2 + y^2 - 8x + 4y - 6 = 0$

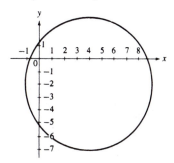

9. $x^2 + (y + 3/2)^2 = 25/4$
   $x^2 + y^2 + 3y - 4 = 0$

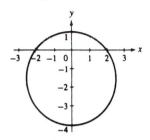

11. $(x - 3)^2 + (y - 3)^2 = 9$
   $x^2 + y^2 - 6x - 6y + 9 = 0$

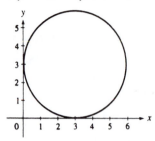

13. $(x - 4)^2 + (y - 1)^2 = 4$
   $x^2 + y^2 - 8x - 2y + 13 = 0$

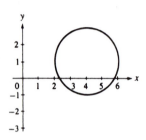

15. $(x - 4)^2 + (y + 4)^2 = 16$
   $x^2 + y^2 - 8x + 8y + 16 = 0$

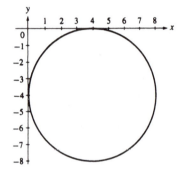

17. $(x - 1)^2 + (y - 2)^2 = 4$

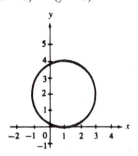

19. $(x + 3)^2 + y^2 = 25$

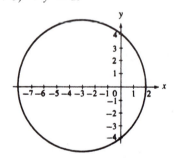

21. $(x - 1/2)^2 + (y - 3/2)^2 = 9/4$

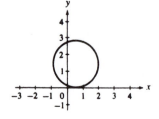

23. $(x - 4/5)^2 + (y - 2/5)^2 = 25$

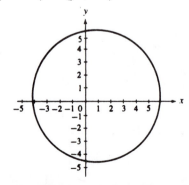

25. $(x - 1/3)^2 + (y + 1)^2 = -1/9$

27. $(x - 2/3)^2 + (y - 1/2)^2 = 0$

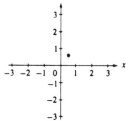

29. $(2, -1), (3, 3)$

31. $(1, 4), (3, 1)$

33. No solution; the circle and line do not intersect

35. $x + y + 1 = 0$

37. $x^2 + y^2 = a^2$

$$m_1 = \frac{y}{x + a}$$

$$m_2 = \frac{y}{x - a}$$

$$m_1 m_2 = \frac{y^2}{x^2 - a^2} = \frac{y^2}{-y^2} = -1$$

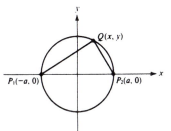

39. 0 if one circle is entirely inside the other; 1 if they are tangent internally; 2 if the circles intersect in two points; 3 if they are tangent externally; 4 if one circle is entirely outside the other.

41. (a) $D^2 + E^2 - 4AF > 0$ (b) $= 0$ (c) < 0; $h = -D/2A, k = -E/2A, r = \sqrt{D^2 + E^2 - 4AF}\,/2\,|A|$.

43.

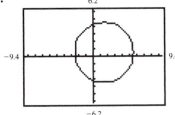

Center: $(1.5, 0.5)$
Radius $= 4$

45.

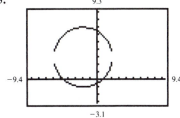

Center: $(-1.91, 3.04)$
Radius $= 3.99$

47.

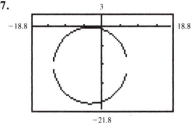

Center: $(-3.0, -10.22)$
Radius $= 11.15$

49.

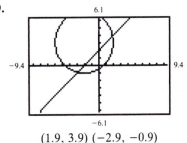

$(1.9, 3.9)\ (-2.9, -0.9)$

51.

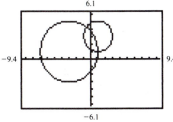

$(-0.6, 4.2), (1, 1)$

53. $x^2 + y^2 - 3y - 4 = 0$

55. $x^2 + y^2 - 2x - 3 = 0,\quad x^2 + y^2 + 2x - 3 = 0,$
$x^2 + y^2 - 2\sqrt{3}y - 1 = 0$

57. $4x^2 + 4y^2 - 20x - 75 = 0,$
$4x^2 + 4y^2 + 20x - 75 = 0$

Section 4.2 [pages 146–148]

1. $3x^2 + 3y^2 - 10x - 10y - 5 = 0$ 3. $5x^2 + 5y^2 + 13x - 17y - 40 = 0$

5. $x^2 + y^2 - 13x - y = 0$

7. $x^2 + y^2 + 4x - 10y + 4 = 0$, $x^2 + y^2 - 6x + 10y + 9 = 0$

9. $x^2 + y^2 - 4x + 8y - 5 = 0$ 11. $2x^2 + 2y^2 + 8x + 12y + 1 = 0, 2x^2 + 2y^2 - 12x - 48y + 81 = 0$

13. $x^2 + y^2 - 2x + 6y = 0$ 15. $x^2 + y^2 + 4x + 6y - 37 = 0$

17. $13x^2 + 13y^2 + 4x + 19y - 49 = 0$

19. $x^2 + y^2 + 6x + 6y + 9 = 0$, $x^2 + y^2 + 6x - 6y + 9 = 0$, $x^2 + y^2 - 6x + 6y + 9 = 0$,
    $x^2 + y^2 - 6x - 6y + 9 = 0$

21. $5x^2 + 5y^2 + 52x - 56y + 47 = 0, 5x^2 + 5y^2 - 32x + 56y - 37 = 0$

23. $x^2 + y^2 + 4x - 6y - 12 = 0$

25. There is no circle satisfying the given conditions.

27. $x^2 + y^2 - 6x + 2y + 3 = 0$

29. $13x^2 + 13y^2 + 20x - 22y - 35 = 0$, $x^2 + y^2 - 4x + 2y + 1 = 0$

31. Since the coordinates of P_1 satisfy the equations of both of the given circles, they must satisfy the equation of the family M no matter what value k might have. Thus the point P_1 belongs to every member of the family. Similarly, P_2 belongs to every member. Since the coefficients of x^2 and y^2 are the same, the members of M consist of circles containing P_1 and P_2, together with the line containing these two points (when $k = -A/A'$).

33. Since the three points are collinear, by Problem 33 of Section 3.1, the coefficient of $x^2 + y^2$ is 0. Now we need to show that either the coefficient of x or the coefficient of y is different from 0. Since the points are distinct, we cannot have both $x_1 = x_2 = x_3$ and $y_1 = y_2 = y_3$. Suppose y_1, y_2, and y_3 are not all the same. Suppose furthermore that $y_2 \neq y_3$.

$$\begin{vmatrix} x^2 + y^2 & y & 1 \\ x_2^2 + y_2^2 & y_2 & 1 \\ x_3^2 + y_3^2 & y_3 & 1 \end{vmatrix} = 0$$

represents a circle (since $y_2 \neq y_3$, the coefficient of $x^2 + y^2$, which is $y_2 - y_3$, is not 0) containing (x_2, y_2) and (x_3, y_3). If

$$\begin{vmatrix} x_1^2 + y_1^2 & y_1 & 1 \\ x_2^2 + y_2^2 & y_2 & 1 \\ x_3^2 + y_3^2 & y_3 & 1 \end{vmatrix} = 0,$$

then (x_1, y_1) is on this circle. But since the three points are given to be collinear, (x_1, y_1) cannot be on the circle. Thus the last determinant (which is the negative of the coefficient of x in the original determinant) is not 0.

35. $x - 3y - 7 \pm 13\sqrt{10} = 0$

Review Problems [page 148]

1. Circle with center $(5, -2)$ and radius 4   2. Point $(-3, 1)$ 3. Point $(1/3, -3/2)$

4. Circle with center $(1/2, -3/2)$ and radius $5/2$

5. No graph

6. $4x + 4y - 7 = 0$

7. $x^2 + y^2 - 2x - 4y = 0$

8. $x^2 + y^2 - 8x - 6y + 17 = 0$

9. $25x^2 + 25y^2 - 200x - 50y + 229 = 0$

10. $x^2 + y^2 - 20x - 8y + 100 = 0$, $x^2 + y^2 + 12x + 8y + 36 = 0$

11. $x^2 + y^2 - 4x + 2y - 12 = 0$

12. The circle has equation $x^2 + y^2 = r^2$.

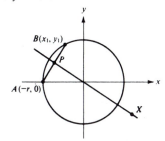

$$m_{AB} = \frac{y_1}{x_1 + r} \qquad m_{PX} = -\frac{x_1 + r}{y_1} \qquad P = \left(\frac{x_1 - r}{2}, \frac{y_1}{2} \right)$$

$$PX: \; y - \frac{y_1}{2} = -\frac{x_1 + r}{y_1} \left(x - \frac{x_1 - r}{2} \right)$$

This line contains the center $(0, 0)$ since substitution of $(0, 0)$ into this equation gives $x_1^2 + y_1^2 = r^2$, which is known to be true.

13. $4x - 3y - 25 = 0$

CHAPTER 5

Section 5.2 [pages 158–161]

1. Axis: x axis, $V(0, 0)$, $F(4, 0)$,
D: $x = -4$,   lr $= 16$

3. Axis: y axis, $V(0, 0)$,
$F(0, 1)$, D: $y = -1$,   lr $= 4$

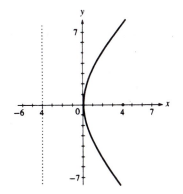

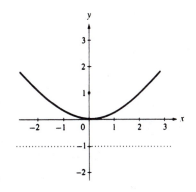

5. Axis: x axis, $V(0, 0)$, $F(5/2, 0)$,
D: $x = -5/2$,   lr $= 10$

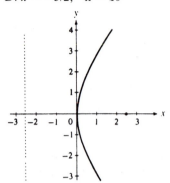

7. Axis: y axis, $V(0, 0)$, $F(0, 5/4)$,
D: $y = -5/4$,   lr $= 5$

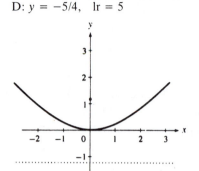

9. Axis: y axis, $V(0, 0)$, $F(0, -1/2)$,
D: $y = 1/2$,   lr $= 2$

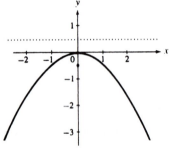

11. Axis: y axis, $V(0, 0)$, $F(0, 3/2)$,
D: $y = -3/2$,   lr $= 6$

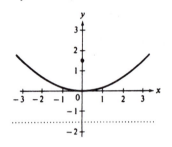

13. $y^2 = 25x$ **15.** $y^2 = 5x$, $y^2 = -5x$ **17.** $y^2 = -12x$ **19.** $x^2 = 4y/3$ **21.** $y^2 = 8x - 16$

23. $x^2 - 2xy + y^2 + 8x + 8y - 16 = 0$ **25.** $2x - y - 1 = 0$ **27.** $x + y - 4 = 0$

29. $\overline{CP} = \overline{PD}$

$$\sqrt{x^2 + (y - c)^2} = |y + c|$$

$$x^2 + y^2 - 2cy + c^2 = y^2 + 2cy + c^2$$

$$x^2 = 4cy$$

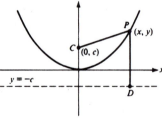

31. We assumed that $m^2x^2 + (4m^2 + 8m + 8)x + (4m^2 + 16m + 16) = 0$ is really a quadratic equation, which requires that $m \neq 0$.

33.

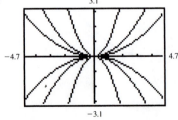

It determines the spread of the parabola; the larger the coefficient, the narrower the spread. The minus makes the parabola open in the negative direction.

35.

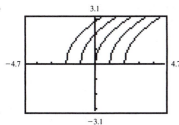

It moves the graph h units to the right (to the left if h is negative).

37. 0.75 inch   **39.** 9 inches from the vertex; $y^2 = 36x$

41. 6 inches from the axis of the parabola

43. **(a)** Decreased, because it will be closer to Saturn when retreating from it than when approaching it.
(b) Increased, because it will be closer to Saturn when approaching it that when retreating from it.

45. 100, 67.6, 42.4, 24.4, 13.6, 10, 13.6, 24.4, 42.4, 67.6, 100

47. 28.7 feet

Section 5.3 [pages 169–173]

1. $C(0, 0)$, $V(\pm 13, 0)$, $CV(0, \pm 5)$,
$F(\pm 12, 0)$,   lr $= 50/13$

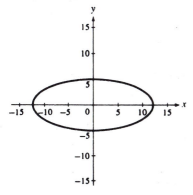

3. $C(0, 0)$, $V(\pm 5, 0)$, $CV(0, \pm 2)$
$F(\pm \sqrt{21}, 0)$,   lr $= 8/5$

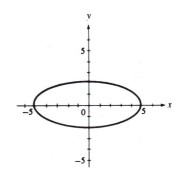

5. $C(0, 0)$, $V(0, \pm 7)$, $CV(\pm 5, 0)$,
$F(0, \pm 2\sqrt{6})$,   lr $= 50/7$

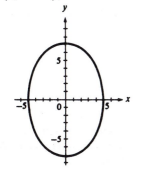

7. $C(0, 0)$, $V(0, \pm 3)$, $CV(\pm 2, 0)$,
$F(0, \pm \sqrt{5})$,   lr $= 8/3$

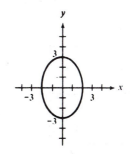

9. $C(0, 0)$, $V(0, \pm4)$, $CV(\pm3, 0)$,
  $F(0, \pm\sqrt{7})$,   lr $= 9/2$

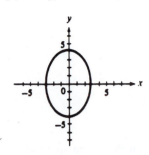

11. $x^2/144 + y^2/169 = 1$

13. $x^2/25 + y^2/10 = 1$

15. $x^2/36 + y^2/9 = 1$

17. $x^2/100 + y^2/64 = 1$

19. $x + 4y - 10 = 0$

21. $2x - 7y - 20 = 0$, $2x + y - 4 = 0$

23. For each line ABP, let h be the x intercept and let k be the y intercept. Now, $h^2 + k^2 = (a - b)^2$. Line ABP has the equation $x/h + y/k = 1$ for $h \neq 0, k \neq 0$.

$$x = \frac{h}{k}(k - y). \qquad (1)$$

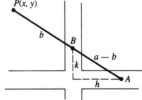

Also, P is a point on each member from the family of circles with center at $B(0, k)$ and radius b. Hence,

$$x^2 + (y - k)^2 = b^2. \qquad (2)$$

Note: P is on a line ABP and is on the corresponding circle. Using (1) and (2), we have

$$\frac{h^2}{k^2}(k - y)^2 + (y - k)^2 = b^2 \quad \text{or} \quad (y - k)^2\left(\frac{h^2}{k^2} + 1\right) = b^2$$

$$(y - k)^2\left(\frac{h^2 + k^2}{k^2}\right) = b^2 \quad (y - k)^2 = \frac{k^2b^2}{h^2 + k^2}.$$

Since $h^2 + k^2 = (a - b)^2$, we have

$$(y - k)^2 = \frac{k^2b^2}{(a - b)^2}, \qquad y - k = \frac{kb}{(a - b)}.$$

(Only one square root is used here. The sign of $y - k$ agrees with that of k.)

$$y = k + \frac{kb}{(a - b)} = \frac{ak - bk + bk}{a - b} = \frac{ak}{a - b}, \qquad \frac{y}{a} = \frac{k}{a - b}.$$

Using (1),

$$x = \frac{h}{k}(k - y) = \frac{h}{k}\left(\frac{kb}{b - a}\right) = \frac{hb}{b - a}, \qquad \frac{x}{b} = \frac{h}{b - a}.$$

Therefore,

$$\frac{x^2}{b^2} + \frac{y^2}{a^2} = \frac{h^2 + k^2}{(a - b)^2}.$$

But, from above, $h^2 + k^2 = (a - b)^2$. Thus,

$$\frac{x^2}{b^2} + \frac{y^2}{a^2} = \frac{(a - b)^2}{(a - b)^2} = 1,$$

which is an ellipse. Observe also that when $h = 0$ and $k = 0$, P is on this ellipse.

25. $\overline{PC} + \overline{PC'} = 2a$

$\sqrt{x^2 + (y - c)^2} + \sqrt{x^2 + (y + c)^2} = 2a$

$\sqrt{x^2 + (y - c)^2} = 2a - \sqrt{x^2 + (y + c)^2}$

$x^2 + y^2 - 2cy + c^2 = 4a^2 - 4a\sqrt{x^2 + (y + c)^2} + x^2 + y^2 + 2cy + c^2$

$4a\sqrt{x^2 + (y + c)^2} = 4a^2 + 4cy$

$\sqrt{x^2 + (y + c)^2} = a + \dfrac{c}{a}y$

$x^2 + y^2 + 2cy + c^2 = a^2 + 2cy + \dfrac{c^2}{a^2}y^2$

$x^2 + \dfrac{a^2 - c^2}{a^2}y^2 = a^2 - c^2$

$\dfrac{x^2}{a^2 - c^2} + \dfrac{y^2}{a^2} = 1$

Letting $b^2 = a^2 - c^2$, $\dfrac{y^2}{a^2} + \dfrac{x^2}{b^2} = 1$.

27. The square roots of the denominators are the intercepts.

29. The square roots of the denominators of the y^2 terms are the y intercepts.

31.

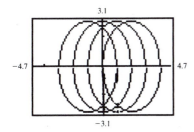

It moves the graph h units to the right (to the left if h is negative).

33. 0.0167, 186,006 miles, 185,980 miles

35. 69,800,000 kilometers, 46,000,000 kilometers

37. $x^2/2885^2 + y^2/701^2 = 1$. The sun is at $(-2798, 0)$.

39. Since the length of the loop of string is constant and the distance between the pins (CC') is constant, $PC + PC'$ is constant. This is the condition that determines an ellipse.
The pins are at the foci, and the loop has length $2a + 2c$.

41. $x^2/1762.5^2 + y^2/632.5^2 = 1$. The foci are at $(\pm1645, 0)$.

43. 11.2 feet above the ground

Section 5.4 [pages 181–184]

1. $C(0, 0)$, $V(\pm4, 0)$, $F(\pm5, 0)$,
A: $y = \pm3x/4$, lr $= 9/2$

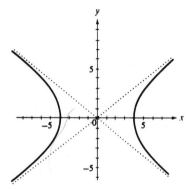

3. $C(0, 0)$, $V(0, \pm3)$, $F(0, \pm\sqrt{13})$,
A: $y = \pm3x/2$, lr $= 8/3$

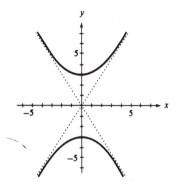

5. $C(0, 0)$, $V(\pm12, 0)$, $F(\pm13, 0)$,
A: $y = \pm5x/12$, lr $= 25/6$

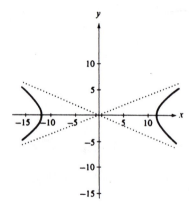

7. $C(0, 0)$, $V(0, \pm5)$, $F(0, \pm\sqrt{34})$,
A: $y = \pm5x/3$, lr $= 18/5$

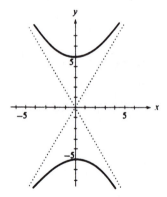

9. $C(0, 0)$, $V(\pm1, 0)$, $F(\pm\sqrt{5}, 0)$
A: $y = \pm2x$, lr $= 8$

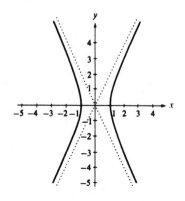

11. $C(0, 0)$, $V(\pm3, 0)$, $F(\pm3\sqrt{2}, 0)$,
A: $y = \pm x$, lr $= 6$

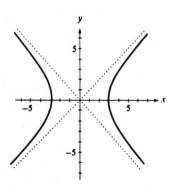

13. $C(0, 0)$, $V(0, \pm 5/2)$, $F(0, \pm \sqrt{34}/2)$
A: $y = \pm 5x/3$, lr $= 9/5$

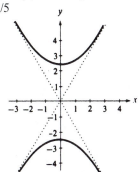

15. $x^2/4 - y^2/12 = 1$

17. $x^2/36 - y^2/16 = 1$

19. $16x^2 - 9y^2 = 144$

21. None

23. $y^2/9 - x^2/16 = 1$

25. $x^2/9 - y^2/16 = 1$

27. $13x - 5y - 144 = 0$

29. $5x - 4y - 9 = 0$

31. Using the hyperbola $x^2/a^2 - y^2/b^2 = 1$ (or $b^2x^2 - a^2y^2 = a^2b^2$), we find that its asymptotes are $bx + ay = 0$ and $bx - ay = 0$. If (x_0, y_0) is any point of the hyperbola, then the product of its distances from the asymptotes is

$$P = \frac{|bx_0 + ay_0|}{\sqrt{a^2 + b^2}} \cdot \frac{|bx_0 - ay_0|}{\sqrt{a^2 + b^2}} = \frac{|b^2x_0^2 - a^2y_0^2|}{a^2 + b^2}.$$

Since (x_0, y_0) is on the hyperbola, it satisfies the equation of the hyperbola, giving $b^2x_0^2 - a^2y_0^2 = a^2b^2$. Thus

$$P = \frac{|a^2b^2|}{a^2 + b^2} = \frac{a^2b^2}{a^2 + b^2},$$

which is seen to be independent of the point chosen.

33. The x intercepts are the square roots of the number in the denominator.

35. The lines are the asymptotes of the hyperbola.

37. The x intercepts are the square roots of the number in the denominator of the x^2 term.

39. The y intercepts are the square roots of the number in the denominator of the y^2 term.

41. All of the hyperbolas have foci at $(\pm 2\sqrt{3}, 0)$.

43. $x^2/4 - y^2/5 = 1$

45. $x^2/36^2 - y^2/70^2 = 1$, where x and y are in millions of miles. The sun is at $(79, 0)$.

47. $(-506, 543)$

Review Problems [page 185]

1. Axis: x axis, $V(0, 0)$, $F(-3/2, 0)$, D: $x = 3/2$, lr $= 6$

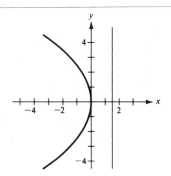

2. $C(0, 0)$, $V(\pm 2, 0)$, $F(\pm\sqrt{5}, 0)$, A: $y = \pm x/2$, lr $= 1$

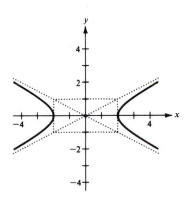

3. Axis: x axis
 $V(0, 0)$
 $F(4, 0)$
 D: $x = -4$
 lr $= 16$

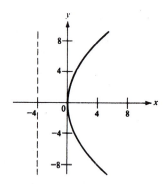

4. $C(0, 0)$, $V(\pm 3, 0)$, $CV(0, \pm 2)$, $F(\pm\sqrt{5}, 0)$, lr $= 8/3$

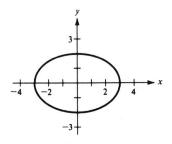

5. $C(0, 0)$
 $V(0, \pm 3)$
 $F(0, \pm 3\sqrt{2})$
 A: $y = \pm x$
 lr $= 6$

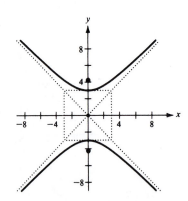

6. $C(0, 0)$, $V(0, \pm4)$, $CV(\pm2\sqrt{3}, 0)$, $F(0, \pm2)$, $lr = 6$

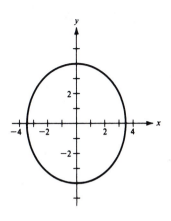

7. $y^2 = 20x$ **8.** $y^2 = 8x$

9. $16x^2 + 25y^2 = 1600$ **10.** $9x^2 + 4y^2 = 72$ **11.** $16x^2 - 9y^2 + 576 = 0$

12. $9x^2 - 8y^2 = 72$ **13.** $2x + 3y - 8 = 0$ **14.** The student's result is incorrect.

CHAPTER 6

Section 6.1 [pages 198–200]

1. $x'^2 = 8y'$

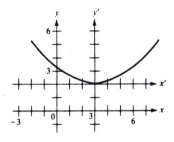

3. $(x + 3)^2/25 + y^2/16 = 1$,
   $x'^2/25 + y'^2/16 = 1$

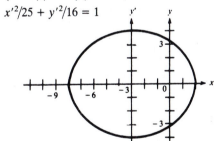

5. $(x + 2)^2/4 + (y + 1)^2/9 = 1$,
   $x'^2/4 + y'^2/9 = 1$

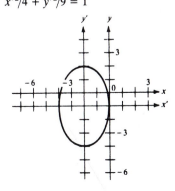

7. $(y - 1)^2 = 4(x - 2)$, $y'^2 = 4x'$

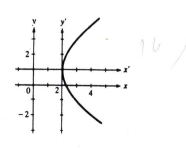

9. $(x + 3)^2/4 + (y - 1)^2/16 = 1$,
   $x'^2/4 + y'^2/16 = 1$

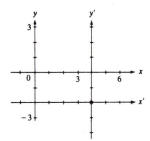

11. $(x + 5)^2/4 - (y - 4)^2/9 = 1$,
   $x'^2/4 - y'^2/9 = 1$

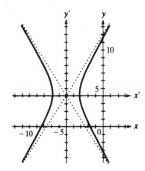

13. $9(x - 4)^2 + 4(y + 2)^2 = 0$,
   $9x'^2 + 4y'^2 = 0$

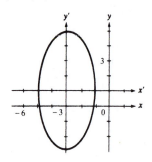

15. $(x - \frac{1}{2})^2 = y + \frac{3}{2}$, $x'^2 = y'$

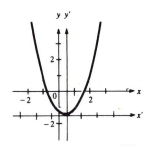

17. $(y - \frac{1}{2})^2/4 - (x + \frac{3}{2})^2/16 = 1$,
   $y'^2/4 - x'^2/16 = 1$

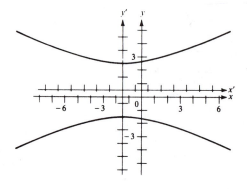

19. $25x'^2 + 4y'^2 = -25$

21. $y^2 - 8x - 10y + 33 = 0$

23. $3x^2 - y^2 - 12x + 2y - 1 = 0$

25. $9x^2 - 4y^2 - 24x - 8y - 184 = 0$

27. $x^2 + 36y^2 - 2x + 216y + 156 = 0$

29. $21x^2 - 4y^2 + 42x - 63 = 0$

31. The parabola must have an equation of the form $Ax^2 + Dx + Ey + F = 0$ and be satisfied by the three given points. The first row of the determinant assures us of the proper form. The three points must satisfy the given equation because replacing the x and y of the determinant by the coordinates of one of the points gives us two identical rows, which implies that the determinant is zero.

33. V: $(-1, 3)$

35. V: $(1, -2)$

37. Intersecting lines; $1.4x - y + 6.8 = 0, 1.4x + y - 1.2 = 0$

39. $y^2 = 120(x + 30)$, x, y in millions of miles.   **41.** $x^2 = -125(y - 70)/14$

43. $(x - 11.9)^2/57.9^2 + y^2/56.7^2 = 1$ or $(x + 11.9)^2/57.9^2 + y^2/56.7^2 = 1$, x, y in millions of kilometers

45. **(a)** Parabola: $y^2 = 40x$

  Hyperbola: $(x - 4)^2/25 - y^2/11 = 1$

  (b) Parabola: $y^2 = 40(x + 4)$

  Hyperbola: $x^2/25 - y^2/11 = 1$

Section 6.2 [pages 205–206]

1. $x'^2 - 2x'y' + 4y'^2 - 5 = 0$ **3.** $x'^2 + 4x'y' - y'^2 + 1 = 0$ **5.** $x'y' + 16 = 0$

7. $y' = x'^3 - x'$ **9.** $y' = x'^3$ **11.** $y' = x'^4 - 4x'^3 + 6x'^2$ **13.** $x'^2y' - 4 = 0$

15. $Ax^2 + Bxy + Cy^2 + Dx + Ey + F = 0$

  $x = x' + h, y = y' + k$

  $A(x' + h)^2 + B(x' + h)(y' + k) + C(y' + k)^2 + D(x' + h) + E(y' + k) + F = 0$

  This gives

  $Ax'^2 + Bx'y + Cy'^2 + (2Ah + Bk + D)x' + (Bh + 2Ck + E)y'$
  $+ (Ah^2 + Bhk + Ck^2 + Dh + Ek + F) = 0$

  The first three terms above prove that A, B, and C are invariant.

Section 6.3 [page 211]

1. $\sqrt{13}x' = 6$

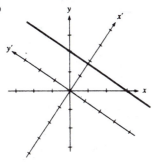

3. $x'^2 - y'^2 = 8$

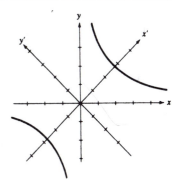

5. $\sqrt{2}y'^2 + x' = 0$

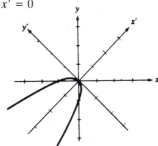

7. $x'^2 - 4y' = 0$

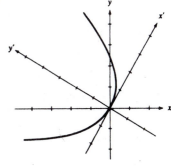

9. $17x'^2 - 9y'^2 = 8$

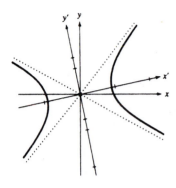

11. $17x'^2 + 7x'y' - 7y'^2 + 20 = 0$

13. $7x'^2 - 8x'y' + y'^2 - 10 = 0$

15. $3\sqrt{3}x'^2 - (6 - 8\sqrt{3})x'y'$
$\quad - (8 + 3\sqrt{3})y'^2 - 16 = 0$

17. $x = x'\cos\theta - y'\sin\theta, \ y = x'\sin\theta + y'\cos\theta$
$\quad x'^2\cos^2\theta - 2x'y'\sin\theta\cos\theta + y'^2\sin^2\theta + x'^2\sin^2\theta +$
$\quad 2x'y'\sin\theta\cos\theta + y'^2\cos^2\theta = 25$
$\quad x'^2(\cos^2\theta + \sin^2\theta) + y'^2(\sin^2\theta + \cos^2\theta) = 25$
$\quad x'^2 + y'^2 = 25$

Section 6.4 [pages 220–221]

1. $3x'^2 + y'^2 - 16y' = 0$

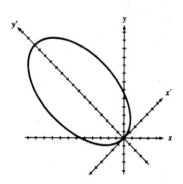

3. $4x'^2 - y'^2 - 16 = 0$

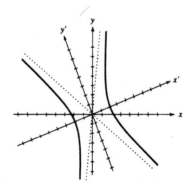

5. $5x'^2 + 20y'^2 = 20$

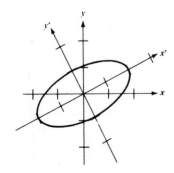

7. $4x'^2 + 9y'^2 - 36 = 0$

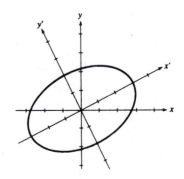

9. $x'^2 + 4x' - 5 = 0$

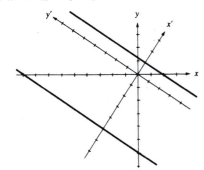

11. $y'^2 - 12x' = 0$

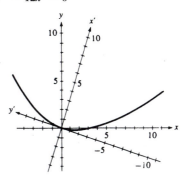

13. $x'^2 - y'^2 - 4 = 0$

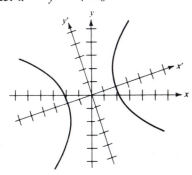

15. $x = x' \cos\theta - y' \sin\theta, \ y = x' \sin\theta + y' \cos\theta.$

$Ax'^2 \cos^2\theta - 2Ax'y' \sin\theta\cos\theta + Ay'^2 \sin^2\theta + Bx'^2 \sin\theta\cos\theta$
$+ B(\cos^2\theta - \sin^2\theta)x'y' - By'^2 \sin\theta\cos\theta + Cx'^2 \sin^2\theta$
$+ 2Cx'y' \sin\theta\cos\theta + Cy'^2 \cos^2\theta + Dx' \cos\theta - Dy' \sin\theta$
$+ Ex' \sin\theta + Ey' \cos\theta + F = 0$

$(A \cos^2\theta + B \sin\theta\cos\theta + C \sin^2\theta)x'^2 + [-2A \sin\theta\cos\theta$
$+ B(\cos^2\theta - \sin^2\theta) + 2C \sin\theta\cos\theta]x'y' + (A \sin^2\theta$
$- B \sin\theta\cos\theta + C \cos^2\theta)y'^2 + (D \cos\theta + E \sin\theta)x'$
$+ (-D \sin\theta + E \cos\theta)y' + F = 0$

$A' + C' = (A \cos^2\theta + B \sin\theta\cos\theta + C \sin^2\theta)$
$\qquad\qquad + (A \sin^2\theta - B \sin\theta\cos\theta + C \cos^2\theta)$
$\qquad\quad = A(\cos^2\theta + \sin^2\theta) + C(\sin^2\theta + \cos^2\theta)$
$\qquad\quad = A + C$

17. $\tan\theta_1 \cdot \tan\theta_2 = \dfrac{(C-A)^2 - [(C-A)^2 + B^2]}{B^2} = -1$

Since the product is negative, $\tan\theta_1$ and $\tan\theta_2$ must have opposite signs. If we look upon $\tan\theta_1$ and $\tan\theta_2$ as the slopes of lines l_1 and l_2, respectively, then l_1 and l_2 are perpendicular. Thus their inclinations, θ_1 and θ_2, differ by an odd multiple of 90°.

19.

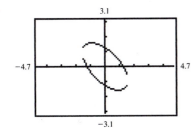

21.

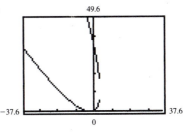

23.

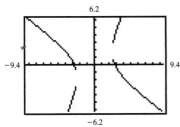

Review Problems [pages 221–222]

1. $C(1, -3)$, $V(1, -3 \pm 2)$,
$CV(1 \pm 1, -3)$, $F(1, -3 \pm \sqrt{3})$
lr = 1

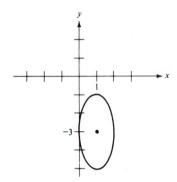

2. Axis: $y = -1$
$V(3, -1)$, $F(13/4, -1)$, D: $x = 11/4$, lr = 1

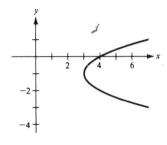

3. Axis: $y = -2$
$V(3, -2)$, $F(5, -2)$, D: $x = 1$,
lr = 8.

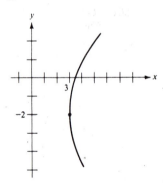

4. $C(-2, -4)$, $V(-2 \pm 4, -4)$,
$F(-2 \pm 5, -4)$,
A: $y + 4 = \pm(3/4)(x + 2)$, lr = 9/2,
$e = 5/4$.

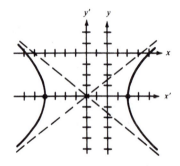

5. $C(-1, -2)$, $V(-1 \pm 5, -2)$,
$CV(-1, -2 \pm 3)$
$F(-1 \pm 4, -2)$, lr = 18/5.

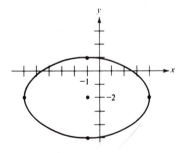

6. $C(-1, -1/2)$, $V(-1 \pm 4, -1/2)$,
$F(-1 \pm 5, -1/2)$,
A: $y = -1/2 \pm 3(x + 1)/4$, lr = 9/2,
$e = 5/4$.

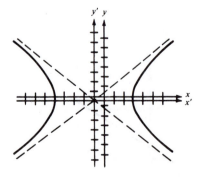

7. Axis: $x = 2$
$V(2, -1)$, $F(2, 0)$,
D: $y = -3$
lr = 4.

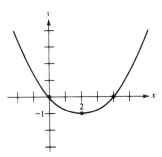

8. $3x'^2 + 4y'^2 = 0$

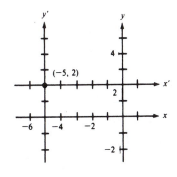

9. $2x + 3y + 5 = 0$,
$2x - 3y - 13 = 0$.

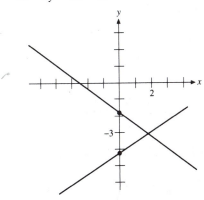

10. $x^2 + 16y^2 + 8x - 32y + 16 = 0$

11. $y^2 - 8x - 10y + 33 = 0$

12. $25x^2 - 16y^2 + 20x + 160y + 752 = 0$

13. $x^2 - 4x - 8y - 20 = 0$, $y^2 - x + 6y + 11 = 0$

14. $25x^2 + 16y^2 + 50x - 64y - 311 = 0$
$9x^2 + 25y^2 - 54x + 50y - 119 = 0$

15. $16x^2 - 9y^2 - 128x - 18y + 103 = 0$

16. $y' = x'^3 - 9x'$

17. $y' = x'^4 - 8x'^2$

18. $y'^2/7 - x'^2/6 = 1$

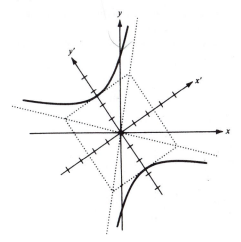

19. $x'^2/20 + y'^2/4 = 1$

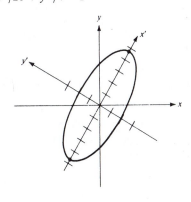

20. $y'^2 = -(x' - 2)$

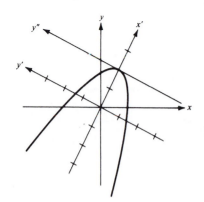

21.

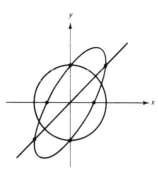

CHAPTER 7

Section 7.1 [pages 232–234]

1. y axis   **3.** None   **5.** x axis   **7.** Origin   **9.** Origin   **11.** $(-1, 0), (3, 0), (0, -3)$

13. $(6, 0), (-1, 0), (0, -6)$ **15.** $(0, 0), (1, 0), (-1, 0)$ **17.** $(-1, 0)$

19. $(-1/2, 0), (1, 0), (0, -1/2)$ **21.** $(-2, 0), (4, 0), (0, 8)$ **23.** $(-2, 0), (4, 0), (0, -16)$

25. None   **27.** $(-1/2, 0)$ **29.** $(0, 0)$

31. (a) Suppose we have symmetry about both axes. Then (x, y) on the curve implies that $(-x, y)$ is on the curve by symmetry about the y axis; this, in turn, implies that $(-x, -y)$ is on the curve by symmetry about the x axis. Thus we have symmetry about the origin.
   (b) Suppose we have symmetry about the x axis and the origin. Then (x, y) on the curve implies that $(x, -y)$ is on the curve by symmetry about the x axis; this, in turn, implies that $(-x, y)$ is on the curve by symmetry about the origin. Thus we have symmetry about the y axis.
   (c) Symmetry about the y axis and the origin implies symmetry about the x axis by a similar argument.

33. A graph can have (at least) two points of symmetry (a line, a sine curve, etc.), but it cannot have two and only two such points because two points of symmetry imply infinitely many such points.

35. Translate to make the axes the lines of symmetry and use the argument of Problem 31.

37. No, since (x, y) on the graph implies that $(x, -y)$ is on it. Since a function must be single-valued, this cannot be the graph of a function.

39. $(-3.5070, 0), (0.2219, 0), (1.2851, 0)$

41. $(-1.41, 0), (-1, 0), (0, 0), (1, 0), (1.41, 0)$

43. $(-1, 0), (7.98, 0)$

45. 5 seconds

47. 50 feet from the building

Section 7.2 [pages 243–247]

1.

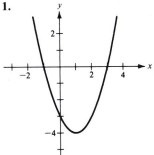

3.

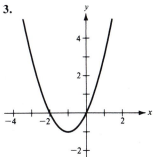

5.

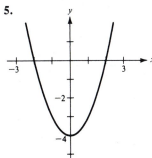

7.

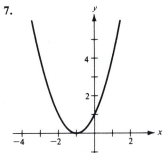

9.

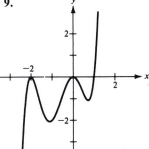

11.

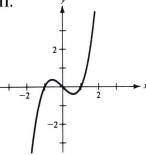

13.

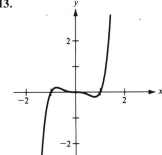

15.

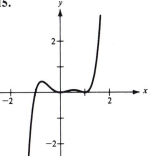

17.

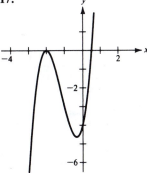

19.

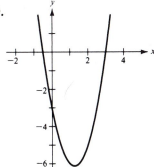

21.

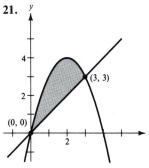

23.

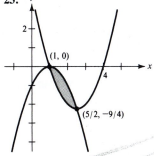

25.

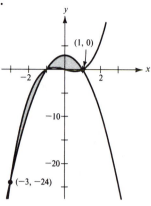

27.

Equation	Graph
a	b
b	e
c	f
d	a
e	c
f	d

29. (**a**) Improper behavior at the ends
(**b**) Not single-valued and not defined for all x
(**d**) Not single-valued and not defined for all x
(**f**) Improper behavior at the ends and not defined for all x

31.

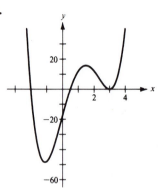

33.

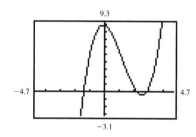

Intercepts: $(-1.414, 0), (2.232, 0), (2.832, 0), (0, 8.94)$

35.

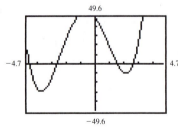

Intercepts $(-4.541, 0), (-2.646, 0), (1.541, 0), (2.646, 0), (0, 49)$

37. $(1.75, -0.105)$

39. $(268, 2683)$

41.

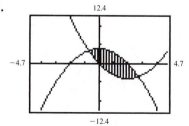

43.

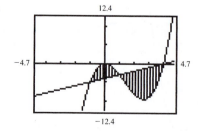

45. 125,000 square yards

Section 7.3 [pages 255–256]

1. $x = 0, x = 1, y = 0$ **3.** $x = -1, x = 2, y = 1$ **5.** $x = -2, y = 1$ **7.** $x = 3, y = 2$

9. $y = 1$ **11.** $x = 1$ **13.** $x = -2, y = x - 3$ **15.** $x = -1, y = 2$ **17.** $x = -1, y = 2x - 6$

19. $x = -3/2, x = -1, y = 0$ **21. (a)** $y = 0$ **(b)** $y = a_n/b_m$ **(c)** None

23.

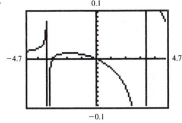

Asymptotes: $x = \pm 3.3$

25. 80,000; 9

Section 7.4 [pages 264–265]

The graphs here are not point-by-point plots. They are found using the methods of this chapter.

1.

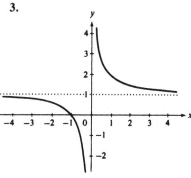

3.

5.

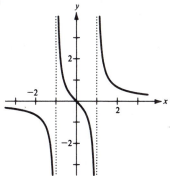

7.

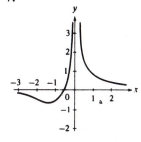

9.

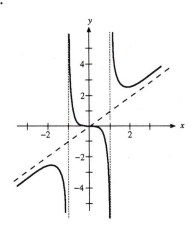

11.

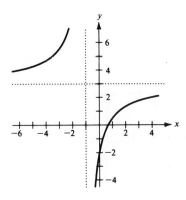

13.

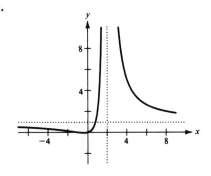

15.

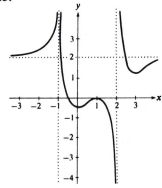

17.

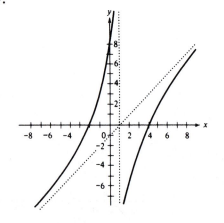

19.

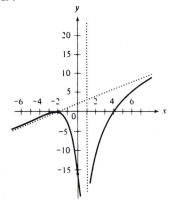

21.

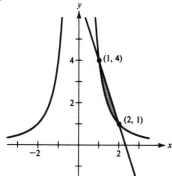

23.

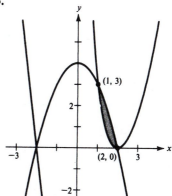

25.

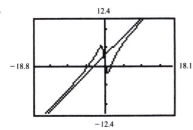

27. $(-2, 0)$, $(-3, 0)$, $(0, -2)$, $x = -4.3$, $x = -0.7$, $x = 1$, $y = 0$

29. $(-0.52, 0)$, $(0, 1)$, $x = 0.77$, $x = 1.78$, $y = 0$

31.

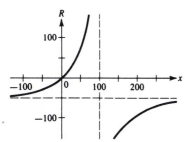

The significant portion is between $x = 0$ and $x = 100$.

Section 7.5 [pages 273–274].

The graphs here are not point-by-point plots. They are found using the methods of this chapter.

1.

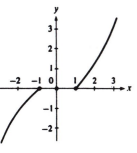

3.

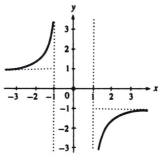

5.

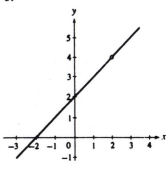

7.
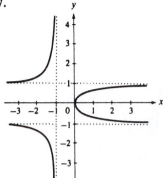

9.
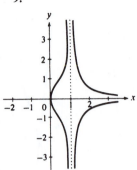

11.
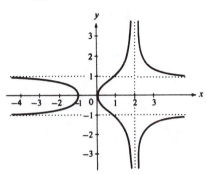

13.
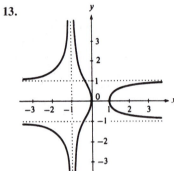

15.
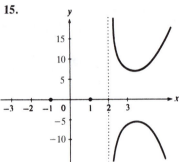

17.
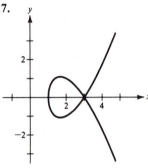

19.
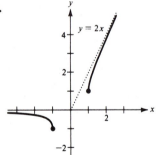

21.
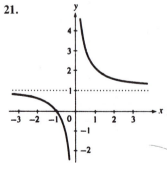

23.
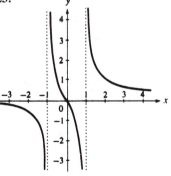

25.
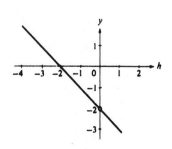

27.
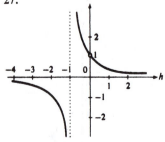

29. $y = x + \sqrt{x^2 + 1}$
$y - x = \sqrt{x^2 + 1}$
$y^2 - 2xy + x^2 = x^2 + 1$
$-2xy + y^2 - 1 = 0$
$B^2 - 4AC = (-2)^2 - 4 \cdot 0 \cdot 1 = 4$
Hyperbola since $B^2 - 4AC > 0$

31. $x \le -3 \quad x \ge 3$

33. $d = \sqrt{x^2(x + 8)^2 + 9}/x; \quad x = 1$

Section 7.6 [page 279]

1.

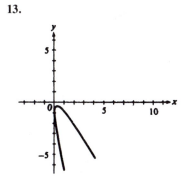

3.

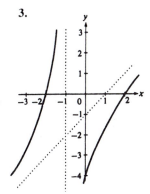

5.

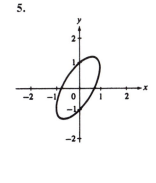

7.

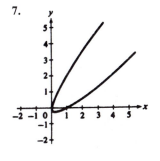

9.

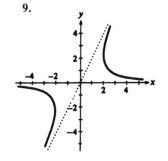

11.

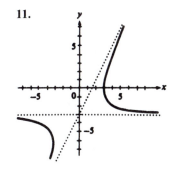

13.

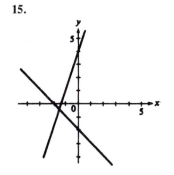

15.

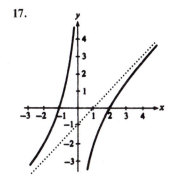

17.

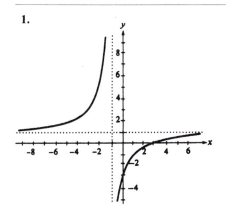

19.

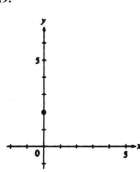

21. $\sqrt{x} + \sqrt{y} = \sqrt{a}$

$x + 2\sqrt{xy} + y = a$

$2\sqrt{xy} = a - x - y$

$4xy = a^2 + x^2 + y^2 - 2ax - 2ay + 2xy$

$x^2 - 2xy + y^2 - 2ax - 2ay + a^2 = 0$

$B^2 - 4AC = 4 - 4 \cdot 1 \cdot 1 = 0$

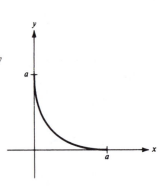

Review Problems [pages 279–280]

The graphs here are not point-by-point plots. They are found using the methods of this chapter.

1.

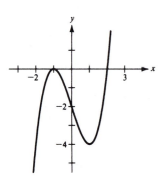

2.

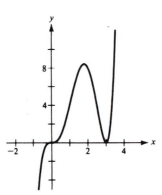

3.

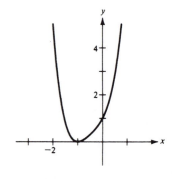

4.

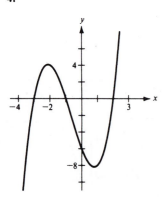

5.

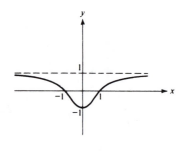

6.

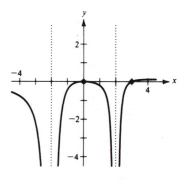

7.

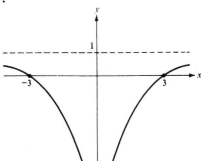

8.

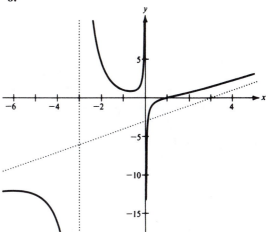

9.

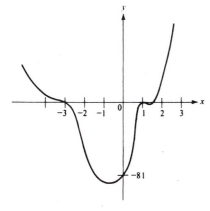

10.

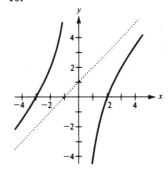

11.

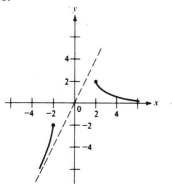

12.

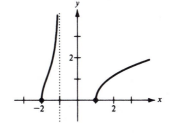

13.

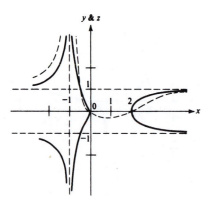

14.

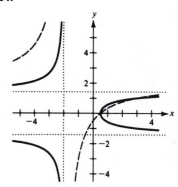

15.

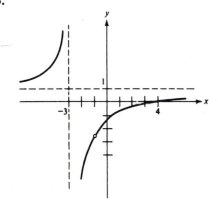

16.

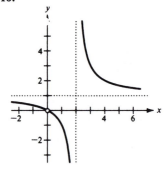

17.

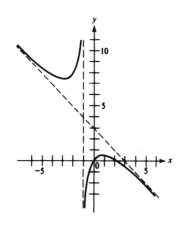

18.

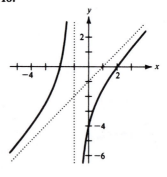

19.

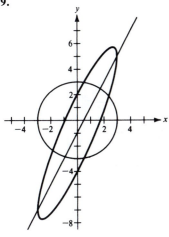

20.

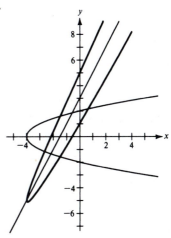

CHAPTER 8

Section 8.2 [pages 289–290]

1.

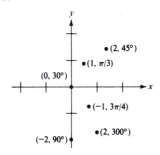

2. $(-4, 150°)$, $(-2, 60°)$, $(-1, 30°)$

3. $(4, 5\pi/3)$, $(3, 5\pi/3)$, $(0, 0)$

5. Circle

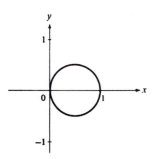

7. Cardioid

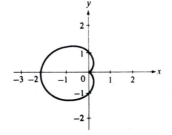

9. Cardioid

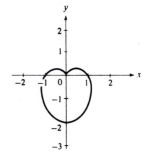

11. Four leafed rose

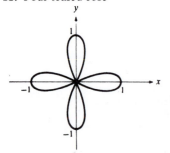

13. Three leafed rose

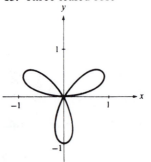

15. Five leafed rose

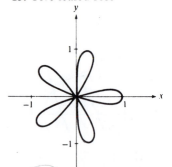

17. Limaçon

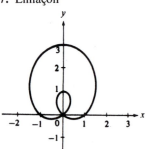

19. Limaçon

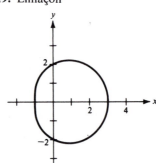

21.

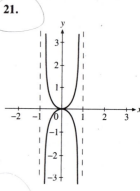

23.

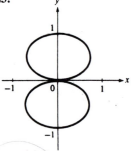

25.

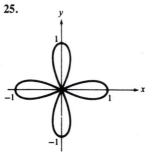

27.

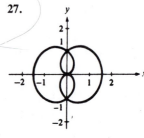

29. Spiral

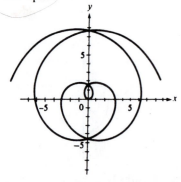

31.

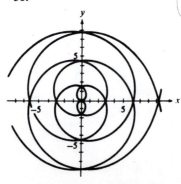

33. Hyperbola

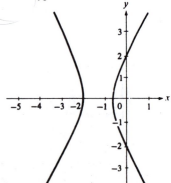

35. There are many answers. Two possible ones are:

$$\begin{cases} r \to -r \\ \theta \to -\theta, \end{cases} \quad \begin{cases} r \to r \\ \theta \to -\pi - \theta. \end{cases}$$

37.

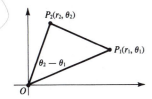

$$\overline{P_1P_2}^2 = \overline{OP_1}^2 + \overline{OP_2}^2 - 2\overline{OP_1}\,\overline{OP_2}\cos(\theta_1 - \theta_1)$$
$$d^2 = r_1^2 + r_2^2 - 2r_1r_2\cos(\theta_1 - \theta_2)$$

39. They all have the same shape; they differ only in their positions, each being rotated 90° clockwise from the previous one.

Section 8.3 [pages 294]

1. $(\sqrt{2}, \pi/4)$, $(\sqrt{2}, 7\pi/4)$ **3.** $(2, 0)$, $(2, \pi)$

5. $(1/2, \pi/3)$, $(1/2, 5\pi/3)$, $(0, \pi/2) = (0, 0)$ **7.** $(\sqrt{3}/2, \pi/3)$, $(-\sqrt{3}/2, 5\pi/3)$, $(0, 0)$

9. $(\sqrt{2}, \pi/4)$ **11.** $(3, 109.5°)$, $(3, 250.5°)$

13. $(2 + \sqrt{2}, \pi/4)$, $(2 - \sqrt{2}, 3\pi/4)$, $(2 - \sqrt{2}, 5\pi/4)$, $(2 + \sqrt{2}, 7\pi/4)$

15. $(1 - 1/\sqrt{2}, \pi/4)$, $(1 + 1/\sqrt{2}, 5\pi/4)$, $(0, \pi/2) = (0, 0)$ **17.** $(1, 0)$, $(-1, 0)$

19. $(0, 0)$, $(1, \pi/2)$ **21.** $(\sqrt{2}, \pi/4)$, $(-\sqrt{2}, 3\pi/4)$

Section 8.4 [pages 298–299]

1. $(-1, 0)$, $(\sqrt{3}/2, 3/2)$, $(1, 0)$, $(-1, 1)$

2. $(2, 7\pi/4)$, $(2, 2\pi/3)$, $(4, 0)$, $(\sqrt{2}, 5\pi/4)$, $(2, 3\pi/2)$

3. $r = 2\sec\theta$ **5.** $r = 1$ **7.** $r = \csc\theta\cot\theta$ **9.** $r = 4/(\cos\theta + 2\sin\theta)$

11. $\tan\theta = 3$ **13.** $x^2 + y^2 = a^2$ **15.** $\sqrt{3}x - y = 0$ **17.** $x^2 + y^2 - 4x = 0$

19. $r = (\cos\theta - \sin\theta)/(1 + 2\sin\theta\cos\theta)$ **21.** $r^2 - 2r(\sin\theta + \cos\theta) + 1 = 0$

23. $r^2 = \sec\theta\csc\theta$ **25.** $(x^2 + y^2)^3 = (x^2 - y^2)^2$

27. $(x^2 + y^2 - 2y)^2 = 9(x^2 + y^2)$ **29.** $(x^2 + y^2)(x^2 + y^2 - 1)^2 = y^2$

31. $y^2 = 2x + 1$ **33.** $x^2 + y^2 = 3x + 2y$

35. Circle with center $(h, 0)$ and radius h

37. Ellipse $x^2/a^2 + y^2/b^2 = 1$

39.

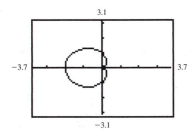

Section 8.5 [pages 303–305]

1. hyperbola, $(0, 0)$, $x = 2$, 2

3. ellipse, $(0, 0)$, $y = 2$, 2/3

5. parabola, $(0, 0)$, $y = 3$, 1

7. ellipse, $(0, 0)$, $y = 3$, 2/3

9. $r = 10/(3 + 2 \cos \theta)$

11. $r = 2/(1 + \sin \theta)$

13. $r = 25/(4 + 5 \cos \theta)$

15.

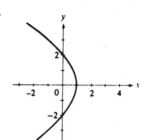

17.

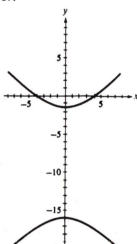

19.

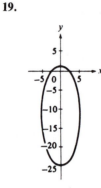

21.

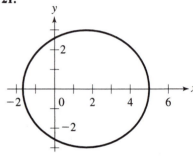

23.

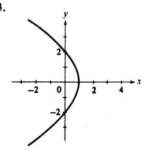

25. The shape of the conic is elliptic approaching a circle. The equation approaches $r = k$ $(k = ep)$, which is an equation of a circle.

27. $\cos(\alpha - \theta) = p/r$

$r \cos \theta \cos \alpha + r \sin \theta \sin \alpha = p$

$x \cos \alpha + y \sin \alpha - p = 0$

29. The line through (x_1, y_1) and parallel to the given line is $Ax + By - (Ax_1 + By_1) = 0$. Both lines can be put into the normal form by dividing by $\pm \sqrt{A^2 + B^2}$. Thus we have

$$\frac{A}{\pm \sqrt{A^2 + B^2}} x + \frac{B}{\pm \sqrt{A^2 + B^2}} y + \frac{C}{\pm \sqrt{A^2 + B^2}} = 0$$

and

$$\frac{A}{\pm\sqrt{A^2 + B^2}}x + \frac{B}{\pm\sqrt{A^2 + B^2}}y - \frac{Ax_1 + By_1}{\pm\sqrt{A^2 + B^2}} = 0.$$

By choosing the same sign in both cases, the polar coordinates of Q_1 and Q_2 are

$$Q_1: \left(\frac{C}{\pm\sqrt{A^2 + B^2}}, \alpha\right), \qquad Q_2: \left(-\frac{Ax_1 + By_1}{\pm\sqrt{A^2 + B^2}}, \alpha\right).$$

Thus the distance between them is

$$d = \left|\frac{C}{\pm\sqrt{A^2 + B^2}} - \left(-\frac{Ax_1 + By_1}{\pm\sqrt{A^2 + B^2}}\right)\right| = \frac{|Ax_1 + By_1 + C|}{\sqrt{A^2 + B^2}}.$$

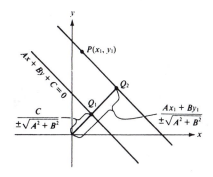

31. $r = 237{,}086.5/(1 - 0.06590\cos\theta)$

33. $r = 184{,}900/(43 - 2\cos\theta)$

Section 8.6 [pages 311–314]

1. $x = y^2 - 2y + 2$ **3.** $y = x^2 - 1$ **5.** $y = (x + 1)^2$

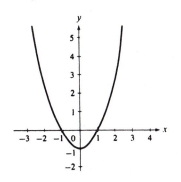

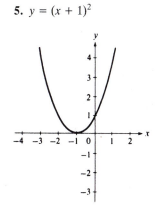

7. $x = 1 + 2r,\ y = 5 - 4r$ **9.** $x = 2 - 3r,\ y = 5 - 3r$ **11.** $x = 2 + 3r,\ y = 3$

13. $x^2 - 2xy + y^2 - 2x - 2y = 0$

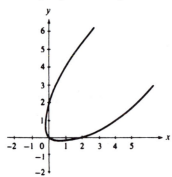

15. $y^3 = x^2$

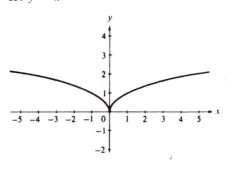

17. $(x - 2)^2 + (y + 1)^2 = 1$

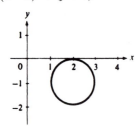

19. $(x - 3)^2/4 + (y - 2)^2/9 = 1$

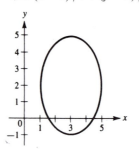

21. $y^2 = x^3$

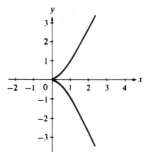

23. $x^2 + y^2 = 1$

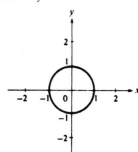

25.

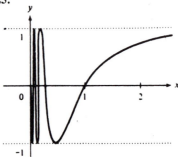

27.

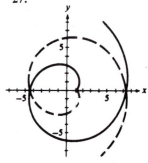

29.

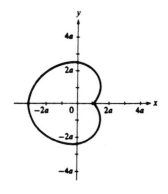

31.

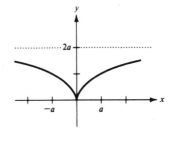

33. (a) $y = x^2$, $y \geq 0$

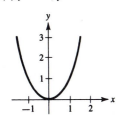

(b) $y = x^2$, $x \geq 0$, $y \geq 0$

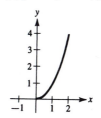

(c) $y = x^2$, $x > 0$, $y > 0$

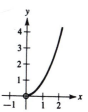

(d) $y = x^2$, $-1 \leq x \leq 1$, $0 \leq y \leq 1$

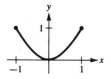

Review Problems [pages 315]

1.

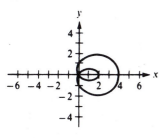

2.

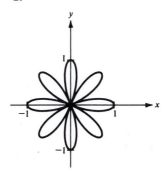

3.

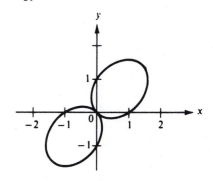

4.

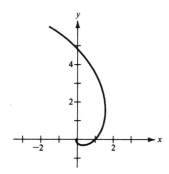

5.

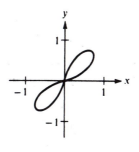

6.

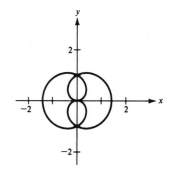

7. $(1, \pi/2)$, $(1, 3\pi/2)$ **8.** $(1 + 1/\sqrt{2}, 3\pi/4)$, $(1 - 1/\sqrt{2}, 7\pi/4)$, $(0, 3\pi/2) = (0, 0)$

9. $(-1, 3\pi/2)$, $(0, 0) = (0, 7\pi/6)$

10. $(1, 15°)$, $(1, 75°)$, $(1, 195°)$, $(1, 225°)$, $(-1, 285°) = (1, 105°)$, $(-1, 345°) = (1, 165°)$, $(-1, 105°) = (1, 285°)$, $(-1, 165°) = (1, 345°)$

11. $(0, \pi/2)$, $(1, 0)$

12. $(1 + \sqrt{2}, \pi/4) = (-1 - \sqrt{2}, 5\pi/4)$, $(1 - \sqrt{2}, 5\pi/4) = (-1 + \sqrt{2}, \pi/4)$, $(0, 3\pi/2) = (0, 0)$

13. **(a)** $(0, -1)$, $(-2, 2)$, $(-3\sqrt{3}, -3)$, $(\sqrt{3}, -3)$
    (b) $(2\sqrt{2}, 3\pi/4)$, $(-5, 0) = (5, \pi)$, $(2, -\pi/3)$, $(2\sqrt{29}, \text{Arctan } 2/5)$

14. $r = -3\sin\theta$ **15.** $r = 6/(3 + 2\sin\theta)$ **16.** $r = 6/(1 - \cos\theta)$

17. $y = x^2 + 2x - 5$ **18.** $(x - 1)^2/9 + (y - 2)^2/4 = 1$

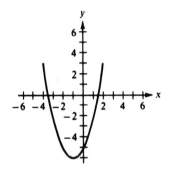
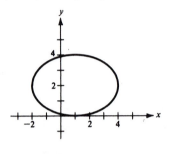

19. $x^2 - 2xy + y^2 - 2x - 2y + 1 = 0$ **20.** $y = \ln\ln x$

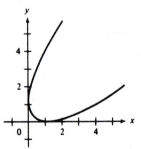
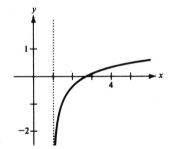

21. $x = 2r + 2$, $y = -8r + 3$ **22.** $x = -1 + 4r$, $y = 4 - 7r$

23. $(-13/38, 5/38)$ **24.** None   **25.** $(x^2 + y^2 - y)^2 = x^2 + y^2$

26. $16x^2 + 25y^2 + 96x - 256 = 0$ **27.** $F(0, 0)$, D: $y = -2$, $e = 1$

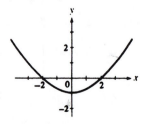

28. $F(0, 0)$, $e = 2$, D: $x = -1$, $C(-4/3, 0)$

29. $y^2 = 4x^2(1 - x^2)$

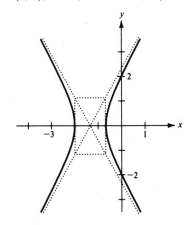

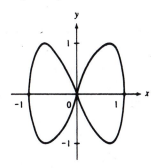

30.

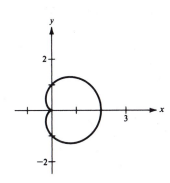

31. $(0, 3)$

CHAPTER 9

Section 9.1 [pages 321–322]

1. $5\sqrt{2}$ **3.** 5   **5.** $\sqrt{131}$ **7.** 5   **9.** 9   **11.** $(-2, 1, 2)$

13. $(1, 2, 2)$ **15.** $(-9, 7, -1)$ **17.** $(1, 1, 5)$ **19.** $(1, 3, 3/2)$

21. $(8, 4, -6)$ **23.** $(-2, 4, -3)$ **25.** $5, -3$ **27.** ± 5 **29.** $1, -3$

31. $-11, 5$

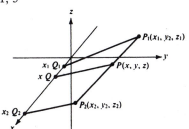

33. $r = \dfrac{\overline{P_1 P}}{\overline{P_1 P_2}} = \dfrac{\overline{Q_1 Q}}{\overline{Q_1 Q_2}} = \dfrac{x - x_1}{x_2 - x_1}$

$x - x_1 = r(x_2 - x_1)$

$x = x_1 + r(x_2 - x_1)$

By projecting onto the y axis and the z axis, we get

$y = y_1 + r(y_2 - y_1)$

$z = z_1 + r(z_2 - z_1)$.

Section 9.2 [page 327]

1. $-6\mathbf{i} - 2\mathbf{j} + 7\mathbf{k}$ 3. $-3\mathbf{i} - 4\mathbf{j} + 3\mathbf{k}$ 5. $\dfrac{4}{\sqrt{21}}\mathbf{i} + \dfrac{1}{\sqrt{21}}\mathbf{j} - \dfrac{2}{\sqrt{21}}\mathbf{k}$

7. $\dfrac{1}{\sqrt{35}}\mathbf{i} + \dfrac{5}{\sqrt{35}}\mathbf{j} - \dfrac{3}{\sqrt{35}}\mathbf{k}$ 9. $B = (4, 0, 8)$ 11. $A = (3, 1, 3)$

13. $A = (0, 6, -3/2)$, $B = (4, 4, -1/2)$ 15. $60°$ 17. $77°$

19. $3\mathbf{i} + 2\mathbf{j} + 6\mathbf{k}$, $-\mathbf{i} - 6\mathbf{j} + 4\mathbf{k}$, -1, not orthogonal

21. $3\mathbf{i} - 5\mathbf{j} + 5\mathbf{k}$, $\mathbf{i} + 3\mathbf{j} + 7\mathbf{k}$, 0, orthogonal

23. $\dfrac{4}{3}\mathbf{i} + \dfrac{4}{3}\mathbf{j} + \dfrac{4}{3}\mathbf{k}$ 25. $\dfrac{7}{6}\mathbf{i} - \dfrac{7}{3}\mathbf{j} + \dfrac{7}{6}\mathbf{k}$ 27. $\dfrac{1}{2}\mathbf{i} + \mathbf{j} + \dfrac{1}{2}\mathbf{k}$

29. Suppose we have a vector $\mathbf{v}$ in space. Let us consider the representative of $\mathbf{v}$ with its tail at the origin O. The head is at $P(a, b, c)$. Let us project P onto the coordinate axes, giving points $A(a, 0, 0)$, $B(0, b, 0)$, and $C(0, 0, c)$. Since $\overrightarrow{OA}$ represents a vector of length $|a|$ that is either in the direction of $\mathbf{i}$ or in the opposite direction, depending upon whether a is positive or negative, it represents $a\mathbf{i}$. Similarly, $\overrightarrow{OB}$ represents $b\mathbf{j}$ and $\overrightarrow{OC}$ represents $c\mathbf{k}$. It is clear that $\mathbf{v} = a\mathbf{i} + b\mathbf{j} + c\mathbf{k}$.

   Since the point P can be represented in rectangular coordinates by a triple (a, b, c) of numbers in one and only one way, the vector $\mathbf{v}$ has one and only one representation in component form.

31. $a_1\mathbf{i} + b_1\mathbf{j} + c_1\mathbf{k}$ is represented by $\overrightarrow{OP}$ from $(0, 0, 0)$ to (a_1, b_1, c_1) and $a_2\mathbf{i} + b_2\mathbf{j} + c_2\mathbf{k}$ by $\overrightarrow{OP_2}$ from $(0, 0, 0)$ to (a_2, b_2, c_2), or by $\overrightarrow{P_1P_3}$ from (a_1, b_1, c_1) to $(a_1 + a_2, b_1 + b_2, c_1 + c_2)$. Hence the sum is represented by $\overrightarrow{OP_3}$ or

$$(a_1\mathbf{i} + b_1\mathbf{j} + c_1\mathbf{k}) + (a_2\mathbf{i} + b_2\mathbf{j} + c_2\mathbf{k}) = (a_1 + a_2)\mathbf{i} + (b_1 + b_2)\mathbf{j} + (c_1 + c_2)\mathbf{k}.$$

Let $(a_1\mathbf{i} + b_1\mathbf{j} + c_1\mathbf{k}) - (a_2\mathbf{i} + b_2\mathbf{j} + c_2\mathbf{k}) = a\mathbf{i} + b\mathbf{j} + c\mathbf{k}$. Then

$a_1\mathbf{i} + b_1\mathbf{j} + c_1\mathbf{k} = (a_2\mathbf{i} + b_2\mathbf{j} + c_2\mathbf{k}) + (a\mathbf{i} + b\mathbf{j} + c\mathbf{k})$

$\qquad = (a_2 + a)\mathbf{i} + (b_2 + b)\mathbf{j} + (c_2 + c)\mathbf{k}$

$$\begin{array}{lll} a_1 = a_2 + a & b_1 = b_2 + b & c_1 = c_2 + c \\ a = a_1 - a_2 & b = b_1 - b_2 & c = c_1 - c_2 \end{array}$$

$a\mathbf{i} + b\mathbf{j} + c\mathbf{k}$ is represented by $\overrightarrow{OP}$ from $(0, 0, 0)$ to (a, b, c). Its length is

$$\sqrt{(a - 0)^2 + (b - 0)^2 + (c - 0)^2} = \sqrt{a^2 + b^2 + c^2}.$$

$\mathbf{v} = a\mathbf{i} + b\mathbf{j} + c\mathbf{k}$ is represented by $\overrightarrow{OP}$ from $(0, 0, 0)$ to (a, b, c). $\mathbf{w} = da\mathbf{i} + db\mathbf{j} + dc\mathbf{k}$ is represented by $\overrightarrow{OQ}$ from $(0, 0, 0)$ to (da, db, dc). Clearly these points lie on the same line; so $\mathbf{w}$ is in the same direction as or the opposite direction from $\mathbf{v}$, depending upon the sign of d. Furthermore

$$|\mathbf{w}| = \sqrt{d^2a^2 + d^2b^2 + d^2c^2} = \sqrt{d^2(a^2 + b^2 + c^2)} = |d|\,|\mathbf{v}|.$$

Hence $d(a\mathbf{i} + b\mathbf{j} + c\mathbf{k}) = da\mathbf{i} + db\mathbf{j} + dc\mathbf{k}$.

33. From Theorem 9.7(e) and Problem 32,

$$\begin{aligned} |\mathbf{u} + \mathbf{v}|^2 &= (\mathbf{u} + \mathbf{v}) \cdot (\mathbf{u} + \mathbf{v}) = |\mathbf{u}|^2 + 2\mathbf{u} \cdot \mathbf{v} + |\mathbf{v}|^2 \\ &= |\mathbf{u}|^2 + 2|\mathbf{u}| \cdot |\mathbf{v}| \cdot \cos\theta + |\mathbf{v}|^2 \le |\mathbf{u}|^2 + 2|\mathbf{u}| \cdot |\mathbf{v}| + |\mathbf{v}|^2 \\ &= (|\mathbf{u}| + |\mathbf{v}|)^2. \end{aligned}$$

Thus $|\mathbf{u} + \mathbf{v}| \le |\mathbf{u}| + |\mathbf{v}|$.

Section 9.3 [pages 333–334]

1. $84°, 64°, 27°$ **3.** $103°, 116°, 29°$ **5.** $125°, 125°, 125°$ **7.** $\{-4, 2, 4\}$

9. $\{2, 2, -2\}$ **11.** $\{5, 1, -2\}$ **13.** $(3, 8, 1), (4, 13, 3)$ **15.** $(3, 4, 5), (5, 5, 7)$

17. $(5, 3, -2), (9, 3, -3)$ **19.** x axis: $\{0°, 90°, 90°\}, \{1, 0, 0\}$

21. Parallel   **23.** Perpendicular

25. Coincident   **27.** None   **29.** Perpendicular   **31.** Coincident

Section 9.4 [pages 340–342]

1. $x = 5 + 3t, y = 1 - 2t, z = 3 + 4t; \dfrac{x - 5}{3} = \dfrac{y - 1}{-2} = \dfrac{z - 3}{4}$

3. $x = 5 + 4t, y = -2 + t, z = 1 - 2t; \dfrac{x - 5}{4} = \dfrac{y + 2}{1} = \dfrac{z - 1}{-2}$

5. $x = 1 + 2t, y = 1, z = 1 + t; \dfrac{x - 1}{2} = \dfrac{z - 1}{1}, y = 1$

7. $x = 4, y = 4, z = 1 + t; x = 4, y = 4$

9. $x = 4 + 2t, y = -3t, z = 5 + 4t; \dfrac{x - 4}{2} = \dfrac{y}{-3} = \dfrac{z - 5}{4}$

11. $x = 8 + 5t, y = 4 + 2t, z = 1 - t; \dfrac{x - 8}{5} = \dfrac{y - 4}{2} = \dfrac{z - 1}{-1}$

13. $x = 5, y = 1 + t, z = 3 + t; x = 5, y - 1 = z - 3$

15. $x = 1, y = -2 + t, z = 3; x = 1, z = 3$ **17.** Do not intersect

19. Do not intersect   **21.** The lines are identical   **23.** $(2, 1, -1)$ **25.** Perpendicular

27. None   **29.** Parallel   **31.** Perpendicular

33. x axis: $\{1, 0, 0\}; y = 0, z = 0$
y axis: $\{0, 1, 0\}; x = 0, z = 0$
z axis: $\{0, 0, 1\}; x = 0, y = 0$

Section 9.5 [pages 349–351]

1. $-5\mathbf{i} + 5\mathbf{j} + 5\mathbf{k}$ **3.** $2\mathbf{i} + \mathbf{j} + 7\mathbf{k}$ **5.** $-\mathbf{i} - \mathbf{j} + 3\mathbf{k}$ **7.** $\{1, -1, 0\}$

9. $\{1, 1, -1\}$ **11.** $\{8, 5, -9\}$ **13.** $x = 3 + t, y = 2 - 3t, z = 1 - 5t$

15. $x = 2 + t, y = 3 - 2t, z = 1 + 4t$ **17.** $x = 2 + 10t, y = t, z = 5 - 8t$

19. $24/\sqrt{30}$ **21.** $113/\sqrt{542}$ **23.** $\sqrt{26}$

25. The area of $\triangle ABC$ is $A = (1/2)AB \cdot BC \cdot \sin \angle BAC$. If $\mathbf{u}$ is represented by $\overrightarrow{AB}$, $\mathbf{v}$ by $\overrightarrow{AC}$, and $\theta = \angle BAC$, then $A = (1/2)|\mathbf{u}| \cdot |\mathbf{v}| \sin \theta = (1/2)|\mathbf{u} \times \mathbf{v}|$.

27. $\sqrt{893}/2$ **29.** $13/2$

31. $\mathbf{u} \cdot (\mathbf{v} \times \mathbf{w}) = (a_1\mathbf{i} + b_1\mathbf{j} + c_1\mathbf{k}) \cdot \begin{vmatrix} \mathbf{i} & \mathbf{j} & \mathbf{k} \\ a_2 & b_2 & c_2 \\ a_3 & b_3 & c_3 \end{vmatrix}$ **33.** 25   **35.** 53

$$= (a_1\mathbf{i} + b_1\mathbf{j} + c_1\mathbf{k}) \cdot \left(\begin{vmatrix} b_2 & c_2 \\ b_3 & c_3 \end{vmatrix}\mathbf{i} - \begin{vmatrix} a_2 & c_2 \\ a_3 & c_3 \end{vmatrix}\mathbf{j} + \begin{vmatrix} a_2 & b_2 \\ a_3 & b_3 \end{vmatrix}\mathbf{k} \right)$$

$$= a_1\begin{vmatrix} b_2 & c_2 \\ b_3 & c_3 \end{vmatrix} - b_1\begin{vmatrix} a_2 & c_2 \\ a_3 & c_3 \end{vmatrix} + c_1\begin{vmatrix} a_2 & b_2 \\ a_3 & b_3 \end{vmatrix}$$

$$= \begin{vmatrix} a_1 & b_1 & c_1 \\ a_2 & b_2 & c_2 \\ a_3 & b_3 & c_3 \end{vmatrix}$$

$$(\mathbf{u} \times \mathbf{v}) \cdot \mathbf{w} = \begin{vmatrix} \mathbf{i} & \mathbf{j} & \mathbf{k} \\ a_1 & b_1 & c_1 \\ a_2 & b_2 & c_2 \end{vmatrix} \cdot (a_3\mathbf{i} + b_3\mathbf{j} + c_3\mathbf{k})$$

$$= \left(\mathbf{i}\begin{vmatrix} b_1 & c_1 \\ b_2 & c_2 \end{vmatrix} - \mathbf{j}\begin{vmatrix} a_1 & c_1 \\ a_2 & c_2 \end{vmatrix} + \mathbf{k}\begin{vmatrix} a_1 & b_1 \\ a_2 & b_2 \end{vmatrix} \right) \cdot (a_3\mathbf{i} + b_3\mathbf{j} + c_3\mathbf{k})$$

$$= a_3\begin{vmatrix} b_1 & c_1 \\ b_2 & c_2 \end{vmatrix} - b_3\begin{vmatrix} a_1 & c_1 \\ a_2 & c_2 \end{vmatrix} + c_3\begin{vmatrix} a_1 & b_1 \\ a_2 & b_2 \end{vmatrix}$$

$$= \begin{vmatrix} a_3 & b_3 & c_3 \\ a_1 & b_1 & c_1 \\ a_2 & b_2 & c_2 \end{vmatrix} = \begin{vmatrix} a_1 & b_1 & c_1 \\ a_2 & b_2 & c_2 \\ a_3 & b_3 & c_3 \end{vmatrix}$$

$\mathbf{u} \cdot (\mathbf{v} \times \mathbf{w}) = (\mathbf{u} \times \mathbf{v}) \cdot \mathbf{w}$

Section 9.6 [pages 359–360]

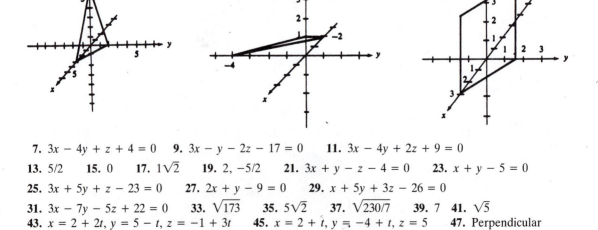

1.   **3.**   **5.**

7. $3x - 4y + z + 4 = 0$ **9.** $3x - y - 2z - 17 = 0$ **11.** $3x - 4y + 2z + 9 = 0$
13. 5/2   **15.** 0   **17.** $1\sqrt{2}$ **19.** 2, −5/2   **21.** $3x + y - z - 4 = 0$ **23.** $x + y - 5 = 0$
25. $3x + 5y + z - 23 = 0$ **27.** $2x + y - 9 = 0$ **29.** $x + 5y + 3z - 26 = 0$
31. $3x - 7y - 5z + 22 = 0$ **33.** $\sqrt{173}$ **35.** $5\sqrt{2}$ **37.** $\sqrt{230/7}$ **39.** 7   **41.** $\sqrt{5}$
43. $x = 2 + 2t, y = 5 - t, z = -1 + 3t$ **45.** $x = 2 + t, y = -4 + t, z = 5$ **47.** Perpendicular
49. None   **51.** Parallel   **53.** (1, 3, 7)   **55.** (4, 1, −1)   **57.** Yes

59. A vector perpendicular to the plane is $\mathbf{v} = A\mathbf{i} + B\mathbf{j} + C\mathbf{k}$. A, B, and C are not all 0. Suppose $A \neq 0$. Then a point in the plane is $(-D/A, 0, 0)$. The vector $\mathbf{u}$ from $(-D/A, 0, 0)$ to (x_1, y_1, z_1) is $\mathbf{u} = (x_1 + D/A)\mathbf{i} + y_1\mathbf{j} + z_1\mathbf{k}$.

$$d = \frac{|\mathbf{u} \cdot \mathbf{v}|}{|\mathbf{v}|} = \frac{|A(x_1 + D/A) + By_1 + Cz_1|}{\sqrt{A^2 + B^2 + C^2}} = \frac{|Ax_1 + By_1 + Cz_1 + D|}{\sqrt{A^2 + B^2 + C^2}}$$

Section 9.7 [pages 364–365]

1.

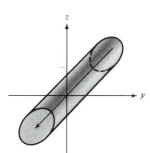

3.

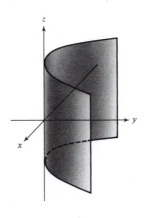

5.

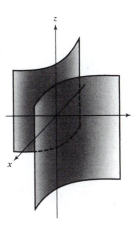

7.

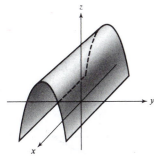

9.

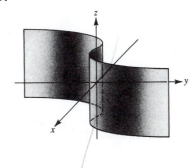

11. Sphere: $(1, 0, -2)$, 3   **13.** No locus   **15.** Sphere: $(-1/2, 3/2, -1)$, 2

17. Sphere: $(1/3, -1/3, -2/3)$, $2\sqrt{2}/3$ **19.** Point: $(1, 1/2, -2)$

21. $x^2 + y^2 + z^2 - 8x - 2y + 4z + 12 = 0$ **23.** $x^2 + y^2 + z^2 - 4x - 8y - 14z + 65 = 0$

25. $x^2 + y^2 + z^2 - 4x - 2y - 2z - 20 = 0$, $x^2 + y^2 + z^2 - 14y + 14z + 72 = 0$

27. $x^2 + y^2 + z^2 - 27x + 35y - 63z - 28 = 0$ **29.** $x^2 + y^2 + z^2 - 4x + 4y - 4z + 8 = 0$

31. From Theorem 9.28, a sphere of radius r and center (h, k, l) has equation

$(x - h)^2 + (z - k)^2 + (z - l)^2 = r^2$

$x^2 + y^2 + z^2 - 2hx - 2ky - 2lz + (h^2 + k^2 + l^2 - r^2) = 0$

This has the form

$$x^2 + y^2 + z^2 + G'x + H'y + I'z + J' = 0$$

Multiplying by nonzero constant A gives the form

$$Ax^2 + Ay^2 + Az^2 + Gx + Hy + Iz + J = 0$$

Section 9.8 [pages 376–377]

1. Ellipsoid

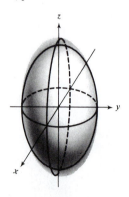

3. Circular paraboloid

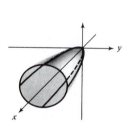

5. Circular cone

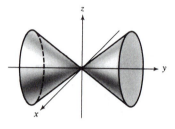

7. Circular paraboloid

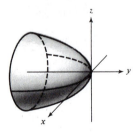

9. Hyperboloid of two sheets

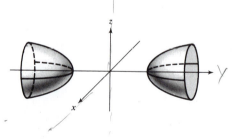

11. Hyperbolic paraboloid

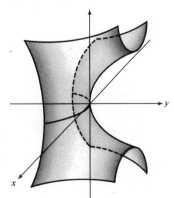

13. Hyperboloid of two sheets

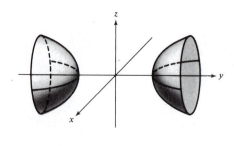

15. Circular cone

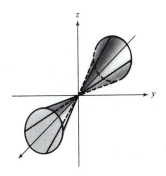

17. Hyperboloid of two sheets

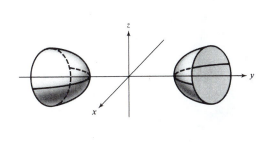

19. Hyperboloid of one sheet

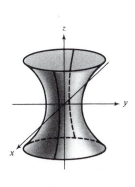

21. Hyperbolic paraboloid

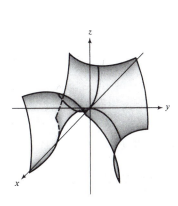

23. Circular cone

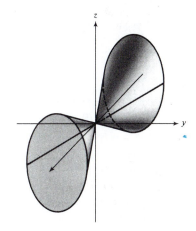

25. Hyperboloid of one sheet

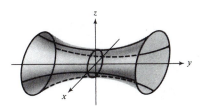

27. $z' = x'^2 + y'^2$

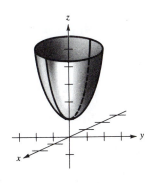

29. $x'^2 + 4y'^2 + 9z'^2 = 36$

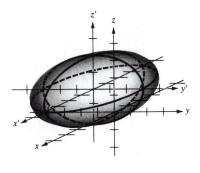

31.

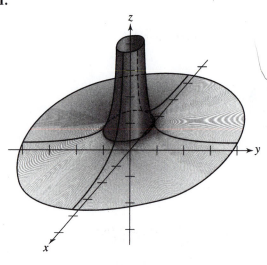

33.

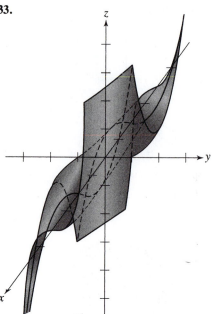

35. Circular cone with z axis as the axis of the cone

37. $8z = x^2 + y^2$, $(0, 0, 2)$

39. $x^2/25^2 + y^2/10^2 + z^2/6^2 = 1$

41. (a) $20z = x^2 + y^2$; $(z - 2)^2/4 - x^2/5 - y^2/5 = 1$
(b) $20(z + 2) = x^2 + y^2$; $z^2/4 - x^2/5 - y^2/5 = 1$

Section 9.9 [pages 382–383]

1. (a) $(\sqrt{2}, \sqrt{2}, 1)$ (b) $(-3/2, 3\sqrt{3}/2, -2)$ **2.** (a) $(\sqrt{2}, 45°, 3)$ (b) $(2, \pi/2, -2)$

3. (a) $(3/2\sqrt{2}, 3/2\sqrt{2}, 3\sqrt{3}/2)$ (b) $(0, 0, 1)$

4. (a) $(2\sqrt{2}, 45°, 90°)$ (b) $(3, \text{Arccos } (2/\sqrt{5}), \text{Arccos } (-2/3))$

5. (a) $(5, 30°, \text{Arccos } (4/5))$ (b) $(2\sqrt{2}, \pi/4, 3\pi/4)$ **6.** (a) $(2, 45°, 2\sqrt{3})$ (b) $(2, 2\pi/3, 0)$

7. $r = 2$, $\rho \sin \varphi = 2$ **9.** $r^2 = z$, $\rho = \csc \varphi \cot \varphi$

11. $r^2 \cos 2\theta - z^2 = 1$, $\rho^2(\sin^2 \varphi \cos 2\theta - \cos^2 \varphi) = 1$ **13.** $r^2 - z^2 = 1$, $\rho^2 = -\sec 2\varphi$

15. $z = 2xy$ **17.** $z = x^2 + y^2$ **19.** $xy = z(z \neq 0)$ **21.** $x^2 + y^2 + z^2 - x = 0$

23. $z = 0.05r^2$; $\rho = 20 \csc \varphi \cot \varphi$

25. Paraboloid: $20z = r^2$; $\rho = 20 \csc \varphi \cot \varphi$

Hyperboloid: $\dfrac{z^2}{4} - \dfrac{r^2}{5} = 1$; $\rho^2\left(\dfrac{\cos^2 \varphi}{4} - \dfrac{\sin^2 \varphi}{5}\right) = 1$

Review Problems [pages 383–385]

1. $(-1, 2, 2)$ **2.** $\frac{4}{29}\mathbf{i} + \frac{2}{29}\mathbf{j} + \frac{3}{29}\mathbf{k}$ **3.** $(60°, 120°, 45°)$ **4.** Perpendicular   **5.** Parallel

6. $x = 4 + t,\ y = t,\ z = 3;\ \dfrac{x - 4}{1} = \dfrac{y}{1},\ z - 3 = 0$

7. $x = 2 + 3t,\ y = -3 - 2t,\ z = 5 + t;\ \dfrac{x - 2}{3} = \dfrac{y + 3}{-2} = \dfrac{z - 5}{1}$

8. $(4, -2, 5)$ **9.** $(1, 5, -2)$ **10.** $6/\sqrt{14}$ **11.** $x - 2y - 4 = 0$

12. $2x - y - 3z + 4 = 0$ **13.** $1/3$ **14.** $5\sqrt{2}/6$ **15.** $2/\sqrt{3}$ **16.** $\sqrt{299/14}$ **17.** $68°$

18. Sphere: center $= (1, 2, -4),\ r = 4$ **19.** Point: $(1/2, 1, -3/2)$

20. Hyperboloid of one sheet

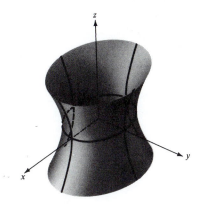

21. Hyperbolic cylinder

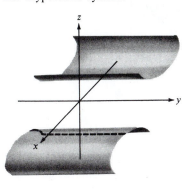

22. Elliptic cone

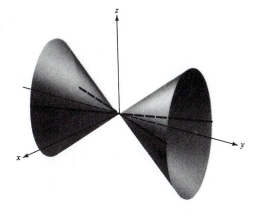

23. Hyperbolic paraboloid

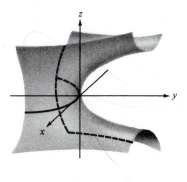

24. (a) $(2\sqrt{2}, \pi/4, 1),\ (3, \pi/4, \text{Arcsin } 2\sqrt{2}/3)$ (b) $(\sqrt{2}, 3\pi/4, -2),\ (\sqrt{6}, 3\pi/4, \text{Arcsin } 1/\sqrt{3})$

25. $z = x$ **26.** $z = 4r^2,\ \rho = \dfrac{1}{4}\csc\varphi\cot\varphi$ **27.** $x^2 + y^2 + z^2 - 2x + y = 0$

28. $-2, -10$ **29.** $A = (1/2, 9/2, -5),\ B = (7/2, 7/2, -1)$

30. $\mathbf{p} = -(5/18)\mathbf{i} - (10/9)\mathbf{j} + (5/18)\mathbf{k}$, $\mathbf{q} = (41/18)\mathbf{i} + (1/9)\mathbf{j} + (49/18)\mathbf{k}$
$\mathbf{p}$ is the projection of $\mathbf{u}$ upon $\mathbf{v}$

31. $x = 3 + 18t, y = 5 + 7t, z = -2 - 3t$ **32.** $x = 4, y = 2 + 2t, z = 3 + t$

33. $11x - 6y - 7z - 53 = 0$ **34.** $x = 3 + 11t, y = 1 + t, z = -1 - 7t$

35. Oblate spheroid

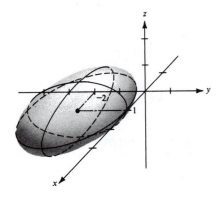

36. Let $\mathbf{u} = \mathbf{i}$, $\mathbf{v} = \mathbf{j}$, $\mathbf{w} = \mathbf{j} + \mathbf{k}$
$\mathbf{v} \times \mathbf{w} = \mathbf{i}$, $\mathbf{u} \times (\mathbf{v} \times \mathbf{w}) = \mathbf{0}$
$\mathbf{u} \times \mathbf{v} = \mathbf{k}$, $(\mathbf{u} \times \mathbf{v}) \times \mathbf{w} = -\mathbf{i}$,
$\mathbf{u} \times (\mathbf{v} \times \mathbf{w}) \neq (\mathbf{u} \times \mathbf{v}) \times \mathbf{w}$

Index